Dynamics, Bifurcation and Symmetry
New Trends and New Tools

NATO ASI Series

Advanced Science Institutes Series

A Series presenting the results of activities sponsored by the NATO Science Committee, which aims at the dissemination of advanced scientific and technological knowledge, with a view to strengthening links between scientific communities.

The Series is published by an international board of publishers in conjunction with the NATO Scientific Affairs Division

A	**Life Sciences**	Plenum Publishing Corporation
B	**Physics**	London and New York
C	**Mathematical and Physical Sciences**	Kluwer Academic Publishers
D	**Behavioural and Social Sciences**	Dordrecht, Boston and London
E	**Applied Sciences**	
F	**Computer and Systems Sciences**	Springer-Verlag
G	**Ecological Sciences**	Berlin, Heidelberg, New York, London,
H	**Cell Biology**	Paris and Tokyo
I	**Global Environmental Change**	

NATO-PCO-DATA BASE

The electronic index to the NATO ASI Series provides full bibliographical references (with keywords and/or abstracts) to more than 30000 contributions from international scientists published in all sections of the NATO ASI Series.
Access to the NATO-PCO-DATA BASE is possible in two ways:

– via online FILE 128 (NATO-PCO-DATA BASE) hosted by ESRIN,
Via Galileo Galilei, I-00044 Frascati, Italy.

– via CD-ROM "NATO-PCO-DATA BASE" with user-friendly retrieval software in English, French and German (© WTV GmbH and DATAWARE Technologies Inc. 1989).

The CD-ROM can be ordered through any member of the Board of Publishers or through NATO-PCO, Overijse, Belgium.

Dynamics, Bifurcation and Symmetry

New Trends and New Tools

edited by

Pascal Chossat

Institut Non Linéaire de Nice,
CNRS – Université de Nice,
Sophia Antipolis, France

Springer Science+Business Media, B.V.

Proceedings of the NATO Advanced Research Workshop on
Dynamics, Bifurcation and Symmetry. New Trends and New Tools
E.B.T.G. Conference
Cargèse, France
September 3–9, 1993

A C.I.P. Catalogue record for this book is available from the Library of Congress.

ISBN 978-0-7923-2958-9   ISBN 978-94-011-0956-7 (eBook)

DOI 10.1007/978-94-011-0956-7

Printed on acid-free paper

Table of contents

vi

PREFACE

This book collects contributions to the conference "Dynamics, Bifurcation and Symmetry, new trends and new tools", which was held at the Institut d'Etudes Scientifiques de Cargèse (France), September 3-9, 1993. The first aim of this conference was to gather and summarize the work of the European Bifurcation Theory Group after two years of existence (the EBTG links european laboratories in five countries via an EC grant). Thanks to a NATO ARW grant, the conference developed into an international meeting on bifurcation theory and dynamical systems, with the participation of leading specialists not only from Europe but also from overseas countries (Canada, USA, South America). It was a great satisfaction to notice the active, and quite enthusiastic participation of many young scientists. This is reflected in the present book for which many contributors are PhD students or post-doc researchers. Although several "big" themes (bifurcation with symmetry, low dimensional dynamics, dynamics in EDP's, applications, ...) are present in these proceedings, we have divided the book into corresponding parts. In fact these themes overlap in most contributions, which seems to reflect a general tendancy in nonlinear science.

I am very pleased to thank for their support the NATO International Exchange Scientific Program as well as the EEC Science Program, which made possible the success of this conference. It is also my pleasure to thank the staff of the IESC (Cargese) and of the Institut Non Linéaire de Nice, for their valuable help in organizing the conference, and J.M. Gambaudo and J. Montaldi for their advice and help during the preparation of these Proceedings.

LIST OF PARTICIPANTS

Andrei Afendikov. Keldysh Institute of Applied Mathematics, Russian Academy of Sciences, 4, Miusskaya sq., 125047 Moscow (Russia)

Dieter Armbruster. Department of Mathematics, Arizona State University, Tempe, Arizona 85287-1804 (U.S.A.)

J.C. Artes. Departamento de Matematiques, Facultat de Ciencias, Universita Autonoma de Barcelona, Campus Universitario, 08193 Bellaterra, Barcelona (Spain)

P.J. Aston. Department of Mathematical and Computing Sciences, University of Surrey Guildford, Surrey GU2 5XH (U.K.)

Claude Baesens. Mathematics Institute, University of Warwick, Coventry CV4 7AL (U.K.)

Rodrigo Bamon. Depto de Matematicas, Facultad de Ciencias, Universidad de Chile, Casilla 653, Santiago (Chile)

Slimane Benmiled. I.N.L.N., CNRS - Université de Nice, 1361 Route des Lucioles, Sophia-Antipolis, 06560 Valbonne (France)

Bjorn Birnir. The University of Iceland Science Institute, 3 Dunhaga, IS-107 Reykjvik (Iceland)

Braasma. Departement Wiskunde, Limburgs Universitair Centrum, Universitaire Campus, 3590 Diepenbeek (Belgium)

Ignacio Bosch-Vivancos. I.N.L.N., CNRS - Université de Nice 1361 Route des Lucioles, Sophia-Antipolis, 06560 Valbonne (France)

Jorge Buescu. Mathematics Institute, University of Warwick, Coventry CV4 7AL (U.K.)

Sofia Castro. Departamento de Matematica, Faculdade de economia do porto, rua Dr. Roberto Frias, 4200 Porto (Portugal)

Pascal Chossat. I.N.L.N. CNRS - .Université de Nice, 1361 Route des Lucioles, Sophia-Antipolis, 06560 Valbonne (France)

Pierre Collet. Physique Thorique Ecole Polytechnique, 91128 Palaiseau cedex (France)

Olivier Courcelle. I.N.L.N., CNRS - Université de Nice, 1361 Route des Lucioles, Sophia-Antipolis, 06560 Valbonne (France)

Gerhard Dangelmayr. Institut für Informationsverarbeitung, Universität Tuebingen, Koestlinstrasse 6 7400 Tuebingen (Germany)

Michael Dellnitz. Institut für Angewandte Mathematik, Universität Hamburg, Bundesstrasse 55 2000 Hamburg 13 (Germany)

Frederic Dias. I.N.L.N., CNRS - Université de Nice, 1361 Route des Lucioles, Sophia-Antipolis, 06560 Valbonne (France)

Freddy Dumortier. Departement Wiskunde, Limburgs Universitair Centrum, Universitaire Campus, 3590 Diepenbeek (Belgium)

Bernold Fiedler. Institt fr Mathematik I, Freie Universitt Berlin, Arnimallee 2-6, 1000 Berlin 33 (Germany)

Mike Field. University of Houston Department of Mathematics, Houston, TX 7204-3476 (USA)

Jacques Furter. Mathematics Institute, University of Warwick, Coventry, CV4 7AL (U.K.)

Jean-Marc Gambaudo. I.N.L.N., CNRS - Université de Nice, 1361 Route des Lucioles, Sophia-Antipolis, 06560 Valbonne (France)

Karin Gatermann. Konrad-Zuse-Zentrum Heilbronner Str. 10 , 1000 Berlin 31 (Germany)

Martin Golubitsky. University of Houston, Department of Mathematics, Houston, TX 77204-3476 (USA)

Gabriela Gomes. Mathematics Institute, University of Warwick, Coventry, CV4 7AL (U.K.)

John Guckenheimer. Dept. of Mathematics Cornell University Ithaca, N.Y. 14853-7901 (USA)

Frederic Guyard. I.N.L.N. CNRS - .Université de Nice, 1361 Route des Lucioles, Sophia-Antipolis, 06560 Valbonne (France)

Ale-Jan Homburg. Dept. of Mathematics, University of Groningen (The Netherlands)

Mariana Haragus. I.N.L.N., CNRS - Université de Nice, 1361 Route des Lucioles, Sophia-Antipolis, 06560 Valbonne (France)

Gérard Iooss. I.N.L.N., CNRS - Université de Nice, 1361 Route des Lucioles, Sophia-Antipolis, 06560 Valbonne (France)

Greg King. Mathematics Institute, University of Warwick, Coventry, CV4 7AL (U.K.)

Juergen Knobloch. Institut für Mathematik, TH Ilmenau PSF 327, 6300 Ilmenau (Germany)

Muriel Koenig. I.N.L.N., CNRS - Université de Nice, 1361 Route des Lucioles, Sophia-Antipolis, 06560 Valbonne (France)

Boris Kolev. I.N.L.N., CNRS - Université de Nice, 1361 Route des Lucioles, Sophia-Antipolis, 06560 Valbonne (France)

Jeroen Lamb. Institute for Theoretical Physics University of Amsterdam, Valcke-nierstraat 65, 1018 XE Amsterdam (The Netherlands)

W.F. Langford. Dept. of Mathematics University of Guelf , Guelf, Ontario NIG 2W1 (Canada)

Reiner Lauterbach. Institut für Angewandte Analysis und Stochastik, Hansvogteiplatz 5-7, 1086 Berlin (Germany)

Christian Leis. Institt fr Mathematik I, Freie Universitt Berlin, Arnimallee 2-6, 1000 Berlin 33 (Germany)

Jerôme Los. I.N.L.N., CNRS - Université de Nice, 1361 Route des Lucioles, Sophia-Antipolis, 06560 Valbonne (France)

Miriam Manoel. University of Warwick, Mathematics Institute, Coventry CV4 7AL (U.K.)

Jan-Cees van der Meer. Faculteit Wiskunde en Informatica T.U., Eindhoven, Postbus 513, 5600 MB Eindhoven (The Netherlands)

Ian Melbourne. University of Houston, Department of Mathematics, Houston, TX 77204-3476 (USA)

David Menasce. I.N.L.N., CNRS - Université de Nice, 1361 Route des Lucioles, Sophia-Antipolis, 06560 Valbonne (France)

Vincent Naudot. Universit de Bourgogne, Laboratoire de Topologie, Dijon (France)

James A. Montaldi. I.N.L.N. CNRS - .Université de Nice, 1361 Route des Lucioles, Sophia-Antipolis, 06560 Valbonne (France)

Basil Nicolaenko. Department of Mathematics Arizona State University, Tempe Arizona 85287-1804 (U.S.A.)

Juliana Oprea. I.N.L.N., CNRS - Université de Nice, 1361 Route des Lucioles, Sophia-Antipolis, 06560 Valbonne (France)

Marie-Christine Perouème. I.N.L.N., CNRS - Université de Nice, 1361 Route des Lucioles, Sophia-Antipolis, 06560 Valbonne (France)

Enrique Ponce. Dept. Applied Mathematics, E.S. Ingenieros Industriales, Avda. Reina Mercedes s/n 41012-SEVILLA (Spain)

Emmanuel Protte. I.N.L.N., CNRS - Université de Nice, 1361 Route des Lucioles, Sophia-Antipolis, 06560 Valbonne (France)

David Rodriguez. Institut für Informationsverarbeitung Universität Tuebingen, Koestlin-strasse 6 7400 Tuebingen (Germany)

Jan Sanders. Faculteit Wiskunde en Informatica Vrije Universiteit Amsterdam, De Boelelaan 1081a, 1081 HV Amsterdam (The Netherlands)

Steve Schecter. Department of Mathematics, North Carolina State University, Box 8205 Raleigh, NC 27695-8208 (U.S.A.)

Arnd Scheel. Institt fr Mathematik I, Freie Universitt Berlin, Arnimallee 2-6, 1000 Berlin 33 (Germany)

Juergen Scheurle. Institut für Angewandte Mathematik, Universität Hamburg, Bundesstrasse 55, 2000 Hamburg 13 (Germany)

Dana Schlomiuk. Dept. de Mathematiques et de Statistiques, Universite de Montreal, C.P. 6128, succ. A Montreal, Quebec H2C 3J7 (Canada)

Jorge Sotomayor. Inst. de Matematica Pura e Aplicada, Jardim Botanico, Estrada Dona Castorina, CEP 22460 Rio de Janeiro (Brasil)

E. Sousa Dias. Mathematics Institute, University of Warwick, Coventry, CV4 7AL (U.K.)

Ian Stewart. Mathematics Institute, University of Warwick , Coventry CV4 7AL (U.K.)

Marco A. Teixeira. IMECC Universidade Estadual de Campinas, C.P. 6065, 13081-970 Campinas SP (Brasil)

Charles Tresser. IBM, Watson Center, Po Box 218 , Yorktown Heights N.Y. 10598 (USA)

Hans True. Technical University of Denmark, Lab. of Applied Mathematics, Bldg. 303, 2800 Lyngby (Denmark)

Stephan van Gils. Dept. of Applied Mathematics Universiteit Twente, Postbus 217, 7500 AE Enschede (The Netherlands)

Andre Vanderbauwhede. Department of Pure Mathematics and Computer Algebra, University of Gent, Krijgslaan 281 9000 Gent (Belgium)

Michael Wegelin. Institut für Informationsverarbeitung Universität Tuebingen, Koestlir strasse 6, 7400 Tuebingen (Germany)

Dave Wood. Institute of Mathematics, University of Warwick, Coventry CV4 7AL (U.K.)

LIST OF CONTRIBUTORS

Andrei Afendikov. Keldysh Institute of Applied Mathematics, Russian Academy of Sciences, 4, Miusskaya sq., 125047 Moscow (Russia)

Dieter Armbruster. Department of Mathematics, Arizona State University, Tempe, Arizona 85287-1804 (U.S.A.)

Peter Ashwin. Mathematics Institute, University of Warwick, Coventry, CV4 7AL (U.K.)

P.J. Aston. Department of Mathematical and Computing Sciences, University of Surrey Guildford, Surrey GU2 5XH (U.K.)

Hanna Brands. Institute for Theoretical Physics University of Amsterdam, Valckenierstraat 65, 1018 XE Amsterdam (The Netherlands)

Jorge Buescu. Mathematics Institute, University of Warwick, Coventry CV4 7AL (U.K.)

Sofia Castro. Departamento de Matematica, Faculdade de economia do porto, rua Dr. Roberto Frias, 4200 Porto (Portugal)

Pascal Chossat. I.N.L.N., CNRS - Université de Nice, 1361 Route des Lucioles, Sophia-Antipolis, 06560 Valbonne (France)

Pierre Collet. Physique Thorique Ecole Polytechnique, 91128 Palaiseau cedex (France)

Gerhard Dangelmayr. Institut für Informationsverarbeitung, Universität Tuebingen, Koestlinstrasse 6 7400 Tuebingen (Germany)

Michael Dellnitz. Institut für Angewandte Mathematik, Universität Hamburg, Bundesstrasse 55 2000 Hamburg 13 (Germany)

Benoit Dionne. Department of Mathematics, University of Ottawa, Ottawa, Ontario K1N 6N5 (Canada)

Mike Field. University of Houston Department of Mathematics, Houston, TX 7204-3476 (USA)

Karin Gatermann. Konrad-Zuse-Zentrum Heilbronner Str. 10 , 1000 Berlin 31 (Germany)

C. Geiger. Institut für Informationsverarbeitung, Universität Tuebingen, Koestlinstrasse 6 7400 Tuebingen (Germany)

Martin Golubitsky. University of Houston, Department of Mathematics, Houston, TX 77204-3476 (USA)

Gabriela Gomes. Mathematics Institute, University of Warwick, Coventry, CV4 7AL (U.K.)

John Guckenheimer. Dept. of Mathematics Cornell University Ithaca, N.Y. 14853-7901 (USA)

Frederic Guyard. I.N.L.N., CNRS - Université de Nice, 1361 Route des Lucioles, Sophia-Antipolis, 06560 Valbonne (France)

A. Jacquemard. Laboratoire de Topologie, Université de Bourgogne , BP 128, 21004 Dijon (France)

Juergen Knobloch. Institut für Mathematik, TH Ilmenau PSF 327, 6300 Ilmenau (Germany)

Jeroen Lamb. Institute for Theoretical Physics University of Amsterdam, Valckenierstraat 65, 1018 XE Amsterdam (The Netherlands)

W.F. Langford. Dept. of Mathematics University of Guelf , Guelf, Ontario NIG 2W1 (Canada)

Reiner Lauterbach. Institut für Angewandte Analysis und Stochastik, Hansvogteiplatz 5-7 1086 Berlin (Germany)

Victor LeBlanc. Dept. of Applied Mathematics, University of Waterloo, Waterloo, Ontario N2L 3G1 (Canada)

Jan-Cees van der Meer. Faculteit Wiskunde en Informatica T.U., Eindhoven, Postbus 513, 5600 MB Eindhoven (The Netherlands)

Ian Melbourne. University of Houston, Department of Mathematics, Houston, TX 77204-3476 (USA)

Alexander Mielke. Institut für Angewandte mathematik, Universitat Hannover, Welfengarten 1, 30167 Hannover (Germany)

James A. Montaldi. I.N.L.N., CNRS - Université de Nice, 1361 Route des Lucioles, Sophia-Antipolis, 06560 Valbonne (France)

Basil Nicolaenko. Department of Mathematics Arizona State University, Tempe, Arizona 85287-1804 (U.S.A.)

David Rodriguez. Institut für Informationsverarbeitung Universität Tuebingen, Koestlin strasse 6 7400 Tuebingen (Germany)

Jan Sanders. Faculteit Wiskunde en Informatica Vrije Universiteit Amsterdam, De Boelelaan 1081a, 1081 HV Amsterdam (The Netherlands)

Juergen Scheurle. Institut für Angewandte Mathematik, Universität Hamburg, Bundesstrasse 55, 2000 Hamburg 13 (Germany)

N. Smaoui. Department of Mathematics Arizona State University Tempe, Arizona 85287-1804 (U.S.A.)

E. Sousa Dias. Mathematics Institute, University of Warwick, Coventry, CV4 7AL (U.K.)

Ian Stewart. Mathematics Institute, University of Warwick , Coventry CV4 7AL (U.K.)

Marco A. Teixeira. IMECC Universidade Estadual de Campinas, C.P. 6065, 13081-970 Campinas SP (Brasil)

Hans True. Technical University of Denmark, Lab. of Applied Mathematics, Bldg. 303, 2800 Lyngby (Denmark)

Stephan van Gils. Dept. of Applied Mathematics Universiteit Twente, Postbus 217, 7500 AE Enschede (The Netherlands)

Andre Vanderbauwhede. Department of Pure Mathematics and Computer Algebra, University of Gent, Krijgslaan 281 9000 Gent (Belgium)

Michael Wegelin. Institut für Informationsverarbeitung Universität Tuebingen, Koestlin-strasse 6, 7400 Tuebingen (Germany)

A SPATIAL CENTER MANIFOLD APPROACH TO A HYDRODYNAMICAL PROBLEM WITH O(2) SYMMETRY[*]

ANDREI AFENDIKOV[†] and ALEXANDER MIELKE[‡]

Abstract: We consider bifurcations from the 3D Poiseuille flow between parallel plates. In the "classical" statement the problem possesses $SO(2) \times O(2)$ symmetry group and from the beginning seems to be completely analogous the Couette-Taylor problem, but in fact it is quite different as in the most physically interesting range of parameters the most dangerous are pure 2D disturbances. In contrast to the classical studies, we make no assumptions on the behavior in the spanwise direction, except the uniform closeness of the bifurcating solution to the basic flow. However, we impose time periodicity as well as spatial periodicity with period $2\pi/\alpha$ in streamwise direction. This allows to apply the "spatial dynamics" approach taking the spanwise variable as an evolutionary one. For a certain range of parameters α, we are able to reduce the bifurcation problem to a spatial center manifold on which the flow is described by a steady Ginzburg-Landau equation. All relevant coefficients can be taken from the analysis of the purely 2D Poiseuille problem. For small β the study of the reduced problem demonstrates that both, spirals and ribbons, bifurcate subcritical, in contrast to the Couette-Taylor problem. These are solutions which are additionally $2\pi/\beta$ periodic in the spanwise direction.

1 The three-dimensional Poiseuille problem

In the theory of hydrodynamic stability there are several problems where the linear theory gives the Reynolds number of the instability threshold which is much higher than experimentally observed. A famous example is the Poiseuille problem of pressure driven flows between the parallel plates. The widespread hypothesis is that this phenomena cor-

---

[*]Research partially supported by Alexander-von-Humboldt-Stiftung and by the European Science Programme SC1-CT91-0670

[†]Keldysh Institute of Applied Mathematics, Miusskaya sq. 4, Moscow, RUSSIA

[‡]Institut für Angewandte Mathematik, Universität Hannover, Welfengarten 1, D-30167 Hannover, GERMANY

1

P. Chossat (ed.), Dynamics, Bifurcation and Symmetry, 1–10.
© 1994 *Kluwer Academic Publishers.*

responds to the existence of subcritical solution branches (cf. [EK91]) as well as on the nonnormality of the linearization at the Poiseuille flow (cf. [TTRD93]). On the one hand classical linear theory claims that the most dangerous in the Poiseuille problem are pure 2D perturbations [Lin55]. That is why the 2D bifurcation problem was carefully treated is several papers, see [CJ73, AV91] and references therein. On the other hand experimentally observed flows have a visible spatial structure [NII75] and hence it is natural to investigate the bifurcation problem in the full 3D setting.

In the Poiseuille problem the flow domain is $Q = I\!R \times (-1, 1) \times I\!R$ with the spatial variables (x, y, z). At $y = \pm 1$ the flow field is zero. *In x-direction (downstream) we assume periodicity with period $2\pi/\alpha$.*

The trivial basic flow is the Poiseuille flow V which reads in nondimensional form

$$V(y) = (U(y), 0, 0)^T \quad \text{with } U(y) = 1 - y^2.$$

Then, for the perturbations of the Poiseuille flow we have the problem

$$\left\{ \begin{array}{rcl} \dfrac{\partial v}{\partial t} + (V \cdot \nabla)v + (v \cdot \nabla)V + (v \cdot \nabla)v + \operatorname{grad} p - \dfrac{1}{R}\Delta v & = & 0, \\ \operatorname{div} v & = & 0. \end{array} \right. \tag{1.1}$$

The boundary conditions are

$$v(t, x, y, z) = v(t, x + 2\pi/\alpha, y, z), \qquad v|_{y=\pm 1} = 0. \tag{1.2}$$

The general local problem of hydrodynamic stability is to describe the behavior of perturbations of the basic flow near the instability threshold. The classical approach to the problem is to impose spatial periodicity along the unbounded z coordinate with the period $2\pi/\beta$. This approach is natural in the case when the minimal critical Reynolds number $R_0(\beta)$ corresponds to $\beta_0 \neq 0$ as it is for instance in the Couette-Taylor problem. If $\beta_0 = 0$ the idea of fixing rather small β in order to obtain relevant description of flows in extended domains, does not seem to be very fruitful. At least the limiting behavior of solutions with $\beta \to 0$ is somewhat delicate as this limit is singular.

In order to analyze this limiting behavior and to start discussion of perturbations which are not periodic in the spanwise direction we use an approach which allows arbitrary z-dependence of the perturbations. Yet, we restrict the temporal behavior to be either stationary or periodic. Then, time plays the role of a compact cross-sectional variable while the unbounded axial variable z takes over the role as the evolutionary variable, leading to the so-called *spatial dynamics* formulation which was introduced in the hydrodynamic stability theory firstly in [Ki82] and extended to viscous fluid flows in [IMD89, IM91].

In this paper we give all the physical and computational details involved in the analysis of the bifurcations from the Poiseuille flow. For the mathematical details we refer the reader to [AM93].

As U is x- and z-independent, the problem has the following symmetries: translation invariance along x and z direction and reflection symmetry S,

$$S \begin{pmatrix} v_1(t,x,y,z) \\ v_2(t,x,y,z) \\ v_3(t,x,y,z) \end{pmatrix} = \begin{pmatrix} v_1(t,x,y,-z) \\ v_2(t,x,y,-z) \\ -v_3(t,x,y,-z) \end{pmatrix}.$$

This means that all mappings in the left-hand side of (1.1) commute with

$$\rho_b \; : \; v(t,x,y,z) \to v(t,x+b,y,z),$$

translations τ_a in z direction $\tau_a : z \to z + a$ and reflection S. Obviously

$$S^2 = 1, \quad S\tau_a = \tau_{-a}S, \quad \rho_b S = S\rho_b, \quad \rho_b\tau_a = \tau_a\rho_b. \tag{1.3}$$

If the flow is periodic in spanwise direction, we take $b \in S^1$ and the symmetry group of the problem is $SO(2) \times O(2)$. Thus, the problem is analogous to the classical Couette-Taylor problem (nonaxisymmetric case), which was investigated independently in [ABY82], [CI85] and [GS86]. Two different kinds of time periodic solutions which appear in the bifurcation analysis are called "spirals" (travelling waves) and "ribbons" (standing waves) (cf. in [CI85]). In the 3D Poiseuille problem the same sort of solutions were found in [Dh83] and [Br89]. In the latter work the spirals are called oblique travelling waves (OTW) and the ribbons are called standing travelling waves (STW). Numerical calculations of the global branches of spirals and ribbons together with the discussion of the relevance of the results to the experiment are given in [EK91].

The two-dimensional Poiseuille problem (no z-dependence) is treated in [AV91]. It is demonstrated there that on the neutral curve in the (R,α)-plane there is a point $T_3 = (R_3,\alpha_3)$ such that the arising Hopf bifurcation of $2\pi/\alpha$-periodic solutions is subcritical for $\alpha > \alpha_3$ and supercritical for $\alpha < \alpha_3$.

It is claimed sometimes that by the well-known Squire's theorem [Sq33] the most dangerous modes for instability in Poiseuille flow are two-dimensional. However, this result has to be used with care. Firstly the limit for $\beta \to 0$ is singular and just taking in corresponding problems $\beta = 0$ we get formal results which should be carefully interpreted. Secondly if we prescribe α in advance, this statement is even no longer true. We will show below that there is the point $T_4 = (R_4,\alpha_4)$ on the 2D neutral curve, such that $\alpha < \alpha_4$ implies that the most dangerous modes are three-dimensional with z-periodicity $2\pi/\beta$ where $\beta = \beta(\alpha) > 0$. Hence, the neural curve of the three-dimensional problem coincides with the 2D neutral curve only for $\alpha \geq \alpha_4$ but lies to the left for $\alpha < \alpha_4$, see Fig. 0. According to the Fig. 0., *the study of large wavelength perturbations is physically only motivated for $\alpha \geq \alpha_4$*. For smaller α the relevant perturbations have shorter wavelength and should be treated as in [IM91].

Under the assumption $v(t,x,y,z) = \chi(y)\exp(\lambda t + i\alpha x + i\beta z)$ the linearization of (1.1)

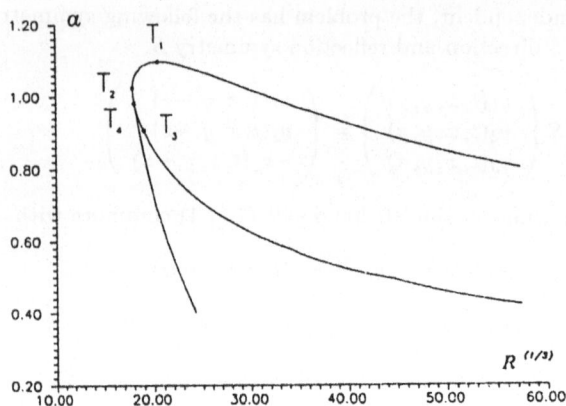

Fig. 0: Neutral curve for the 2D and 3D Poiseuille flow
$T_3 = (R_3, \alpha_3) \approx (6842.2, 0.90667), \ T_4 = (R_4, \alpha_4) \approx (5874.7, 0.98787)$

and (1.2) yields the spectral problem of stability in the form

$$\left. \begin{aligned}
\lambda\chi_1 + \imath\alpha U\chi_1 + \chi_2 U' + \imath\alpha p &= R^{-1} l_{\alpha,\beta}\chi_1, \\
\lambda\chi_2 + \imath\alpha U\chi_2 + p' &= R^{-1} l_{\alpha,\beta}\chi_2, \\
\lambda\chi_3 + \imath\alpha U\chi_3 + \imath\beta p &= R^{-1} l_{\alpha,\beta}\chi_3, \\
\imath\alpha\chi_1 + \imath\beta\chi_3 + \chi_2' &= 0,
\end{aligned} \right\} \quad \text{for } y \in (-1,1), \qquad (1.4)$$

$$\chi = 0, \quad \text{for } y = \pm 1,$$

where $\chi = (\chi_1, \chi_2, \chi_3)^T$ and $l_{\alpha,\beta} = \frac{d^2}{dy^2} - (\alpha^2 + \beta^2)$.

The spectral problem (1.4) might be reduced by the Squire transformation [Sq33] to the Orr-Sommerfeld problem with $k = (\alpha^2 + \beta^2)^{1/2}$ and $l_k = \frac{d^2}{dy^2} - k^2$

$$\begin{cases}
\imath k \tilde{R}((U - \sigma/\imath k) l_k \varphi - U''\varphi) = l_k^2 \varphi, & \sigma = \lambda\dfrac{k}{\alpha} \\
\varphi|_{y=\pm 1} = \dfrac{d\varphi}{dy}\Big|_{y=\pm 1} = 0,
\end{cases} \qquad (1.5)$$

if we define the 2D Reynolds number $\tilde{R} = \alpha R/k$ and the 2D eigenvalue parameter $\sigma = k\lambda/\alpha$. Thus, the neutral curve of stability of the 3D problem (1.4) is determined by the relation

$$\text{Re } \sigma_0(k, \tilde{R}) = 0, \qquad (1.6)$$

where σ_0 is the eigenvalue of the Orr-Sommerfeld problem with minimal real part.

Taking α as fixed the relation $\lambda(\beta, R) = \frac{\alpha}{k}\sigma(k, \frac{\alpha}{k}R)$, with $k^2 = \alpha^2 + \beta^2$, leads to the expansion

$$\lambda(R, \beta) = \imath w_0 + e_{10}(\alpha, R_0)(R - R_0) + e_{02}(\alpha, R_0)\beta^2 + \text{h.o.t.} \qquad (1.7)$$

Thus, the neutral stability curve $R = R(\beta)$, defined by $\operatorname{Re}\lambda(R,\beta)0$, is even in β and has an extremum at $\beta = 0$. It is a local minimum if $\operatorname{Re} e_{02} < 0$ and a local maximum if $\operatorname{Re} e_{02} > 0$. The numerical calculations in [AM93] show $\operatorname{Re} e_{02}(\alpha, R(0))(\alpha - \alpha_4) \leq 0$ with equality only for $\alpha = \alpha_4$.

2 Spatial dynamical system formulation

Let us look now for solutions of (1.1) and (1.2), which are $2\pi/\omega$ periodic in time. The unbounded axial coordinate z will play the role of the evolutionary variable. By setting $\theta = (v, W)^T$, where

$$W = R^{-1}\frac{\partial v}{\partial z} - Pe_3 , \quad e_3 = (0,0,1)^T, \tag{2.1}$$

the equation (1.1) takes the form

$$\frac{d\theta}{dz} = \mathcal{L}_{\mu,\partial_t}\theta + \mathcal{B}(\theta,\theta). \tag{2.2}$$

Note that the construction of W is such that there are no z-derivatives in the right-hand side. The linear operator $\mathcal{L}_{\mu,\partial_t}\theta$ reads

$$\mathcal{L}_{\mu,\partial_t}\theta = \begin{pmatrix} W^* \\ \partial_t v - A_0 v - \nabla_\Omega(W_3 + \operatorname{div}_\Omega v) \end{pmatrix}, \quad \text{where}$$

$$W^* = (RW_1, RW_2, -\operatorname{div}_\Omega v)^T, \quad \nabla_\Omega = (\tfrac{\partial}{\partial x}, \tfrac{\partial}{\partial y}, 0)^T, \quad \operatorname{div}_\Omega v = \tfrac{\partial v_1}{\partial x} + \tfrac{\partial v_2}{\partial y}.$$

With the notation $\Delta_\Omega = \frac{\partial^2}{\partial x^2} + \frac{\partial^2}{\partial y^2}$, the operator A_0 is defined by $A_0 v = \frac{1}{R}\Delta_\Omega v - (V \cdot \nabla_\Omega)v - (v \cdot \nabla_\Omega)V$. For the bilinear operator $\mathcal{B}$ we have

$$\mathcal{B}(\theta,\theta) = \begin{pmatrix} (0,0,0)^T \\ v_1 v_{1,x} + v_2 v_{1,y} + R v_3 W_1 \\ v_1 v_{2,x} + v_2 v_{2,y} + R v_3 W_2 \\ v_1 v_{3,x} + v_2 v_{3,y} - v_3 \operatorname{div}_\Omega v \end{pmatrix}.$$

Using the general approach in [IM91] it is possible to show that all small bounded solutions of (2.2) lie on a finite dimensional spatial center manifold. Note that these solutions are time periodic as well as periodic in the downstream variable x. In the present case it turns out that all bifurcating solutions will be, in fact, travelling waves, i.e. they depend only on $x - ct$. This is a consequence of the fact that the eigenfunctions constructed below have this property and that the center manifold reduction retains this symmetry property.

We exploit this property by restricting the whole analysis to travelling waves. This simply amounts in replacing $\frac{\partial}{\partial t}u$ by $-c\frac{\partial}{\partial x}u$. In this way the linear operator $\mathcal{L}_{\mu,c} : D(\mathcal{L}_{\mu,c}) \subset X \to X = [H_0^1(\Omega)]^3 \times [L_2(\Omega)]^3$ is defined with

$$D(\mathcal{L}_{\mu,c}) = \{\theta \in [H_0^2(\Omega)]^3 \times [H^1(\Omega)]^3 : w_1, w_2, v_{1,x} + v_{2,y} \in H_0^1(\Omega)\}. \tag{2.3}$$

where $H_0^k(\Omega) = \{u \in H^k(\Omega) : u(x, y) = u(x + 2\pi/\alpha, y), u(x, \pm 1) = 0\}$.
¿From the eigenvalue problem

$$(\mathcal{L}_{\mu,c} - \imath\beta)(v, W)^T = 0 \tag{2.4}$$

we find, after elimination of W, the problem

$$\left.\begin{aligned}
-\tfrac{1}{R}(\Delta_\Omega - \beta^2)v + (U - c)v_x + U'v_2e_1 + \nabla_\Omega p + e_3\imath\beta p &= 0, \\
v_{1,x} + v_{2,y} + \imath\beta v_3 &= 0,
\end{aligned}\right\} \quad \text{for } (x, y) \in \Omega, \tag{2.5}$$
$$v(x + 2\pi/\alpha, y) = v(x, y), \quad v = 0 \quad \text{for } y = \pm 1.$$

We may expand v into a Fourier series $v = \sum_{-\infty}^{\infty} u_n(y)e^{\imath\alpha nx}$. As U is independent of x the system decouples. All the cases $n \neq 0$ can be treated in the same way, as α can be replaced by $\tilde{\alpha} = n\alpha$. Hence, it is enough to deal with $n = 1$. There, we recover the spectral problem (1.4), which in turn is equivalent to (1.5). That is why for $R = R_0$ the linear operator $\mathcal{L}_{0,c_0}$ has no eigenvalues $\imath\beta$ for $\beta \neq 0$, but an additional eigenvector Ψ_0 corresponding to the eigenvalue $\beta = 0$,

$$\mathcal{L}_{0,c_0} \Phi_{2,3} = 0 \quad \text{where } \Phi_2 = \text{Re } \Psi_0, \qquad \Phi_3 = \text{Im } \Psi_0$$

with $\Psi_0(x, y) = e^{\imath\alpha x}(\imath\varphi'(y), \alpha\varphi(y), 0, 0, 0, R_0 p(y))^T$. The function $\varphi(y)$ is the eigenfunction of the Orr-Sommerfeld operator and the "pressure" can be determined from (1.4). The numerical calculations in [AM93] gave $e_{02}(\alpha, R_0) \neq 0$ along the neutral curve of the Orr-Sommerfeld problem. Using Theorem 2.1 from the same paper we know that the Jordan block has length 2. To calculate the generalized eigenvectors $\Phi_{4,5}$ satisfying $\mathcal{L}_{0,c_0} \Phi_{4,5} = \Phi_{2,3}$, let us denote by $K(t, y)$ the Green's function of the problem

$$\psi'' - \alpha^2\psi = f, \quad \text{for } y \in (-1, 1); \qquad \psi|_{y=\pm 1} = 0.$$

Then, $\Phi_4 = \text{Re } \Psi_1$ and $\Phi_5 = \text{Im } \Psi_1$ where

$$\Psi_1(x, y) = e^{\imath\alpha x}(0, 0, R_0 \int_{-1}^{1} K(t, y)p(t)\, dt, \ \imath\varphi'(y), \ \alpha\varphi(y), 0)^T.$$

For $n = 0$ and $\beta \neq 0$, it is easy to demonstrate that $u_1 = p = u_2 = u_3 = 0$. But if $\beta = 0$, then the vector $\Phi_0 = (0, 0, 0, 0, 0, 1)^T$ corresponding to the variations of the pressure is in $\ker \mathcal{L}_{\mu,c}$ and, gives rise to the Jordan block of length 2, as there is a generalized eigenvector $\Phi_1 = (0, 0, \tfrac{1}{2}U(y), 0, 0, 0)^T$ satisfying the equation $\mathcal{L}_{\mu,0}\Phi_1 = \Phi_0$. This vector measures the mean flux in z direction.

3 Center manifold reduction

The travelling waves under consideration satisfy the partial differential equation

$$\frac{d}{dz}\theta = \mathcal{L}_{\mu,c}\theta + \mathcal{B}_\mu(\theta, \theta), \tag{3.1}$$

The basic space and the linear operator $\mathcal{L}_{\mu,c}$ are defined as in (2.3).

It was demonstrated in the previous section that $\mathcal{L}_{0,c_0}$ has (for fixed α) the eigenvalue $k = 0$ with the generalized eigenspace

$$X_0 = \mathrm{Span}\{\Phi_0, \Phi_1, \mathrm{Re}\Psi_0, \mathrm{Im}\Psi_0, \mathrm{Re}\Psi_1, \mathrm{Im}\Psi_1\} \subset X.$$

Using the small parameter $\varepsilon = c - c_0$ the unfolding of the fourfold eigenvalue 0 (neglecting the trivial zero eigenvalue due to the pressure indeterminacy) is as follows. By the equivalence of the classical and the spatial dynamics descriptions we see that $\mathcal{L}_{\mu,c_0+\varepsilon}$ has four unique small eigenvalues $k \in \mathbb{C}$ which are solutions of

$$\imath\alpha(c_0 + \varepsilon) = \frac{1}{\sqrt{1 + k^2/\alpha^2}}\sigma_0(\sqrt{\alpha^2 + k^2}, \sqrt{1 + k^2/\alpha^2}(R_0 + \mu)),$$

where σ_0 was the neutral curve from (1.6). Since $e_{02} \neq 0$ we are able to solve for $k^2 \in \mathbb{C}$ to obtain

$$- e_{02}k^2 = a(\mu, \varepsilon) = E\mu - \imath\alpha\varepsilon + \mathcal{O}(\mu^2 + \varepsilon^2). \tag{3.2}$$

The function $a(\mu, \varepsilon)$ will appear later in the reduced equation.

The unique spectral projection $P_0 : X \to X$ (with $P_0 X = X_0$ and $\mathcal{L}_{\mu,c_0} P_0 = P_0 \mathcal{L}_{\mu,c_0}$) defines the splitting $\theta = \theta_0 + \theta_1$ with $\theta_0 = P_0\theta \in X_0$ and $\theta_1 \in X_1 = (I - P_0)X$. The appropriate version of the center manifold theorem [Mi88] allows to reduce the study of all solutions of (3.1) which are *uniformly small* in $z \in \mathbb{R}$. These solutions lie in the *spatial center manifold* $\mathcal{M}_{\mu,\varepsilon}$ which is a smooth graph over the center space X_0. Hence, the solutions have the form $\theta(z) = \theta_0(z) + h(\mu, \varepsilon, \theta_0(z))$ and satisfy the "reduced problem"

$$\frac{d}{dz}\theta_0 = \mathcal{L}_{\mu,c_0+\varepsilon}\theta_0 + P_0\mathcal{B}_\mu(\theta_0 + h(\mu, \varepsilon, \theta_0), \theta_0 + h(\mu, \varepsilon, \theta_0)).$$

Introducing coordinates in X_0 via $\theta_0 = \alpha_1\Phi_0 + \alpha_2\Phi_1 + A\Psi_0 + \overline{A}\overline{\Psi}_0 + B\Psi_1 + \overline{B}\overline{\Psi}_1$ the reduced problem can be written as the ODE

$$\left.\begin{array}{rcl} \frac{d}{dz}\alpha_1 & = & \alpha_2 + f_1(\mu, \varepsilon, \alpha_2, A, \overline{A}, B, \overline{B}), \\ \frac{d}{dz}\alpha_2 & = & f_2(\mu, \varepsilon, \alpha_2, A, \overline{A}, B, \overline{B}), \\ \frac{d}{dz}A & = & B + f_3(\mu, \varepsilon, \alpha_2, A, \overline{A}, B, \overline{B}), \\ \frac{d}{dz}B & = & f_4(\mu, \varepsilon, \alpha_2, A, \overline{A}, B, \overline{B}), \end{array}\right\} \tag{3.3}$$

where $\alpha_1, \alpha_2, f_1, f_2 \in \mathbb{R}$, and $A, B, f_3, f_4 \in \mathbb{C}$. Moreover,

$$f_j(\mu, \varepsilon, \alpha_2, A, \overline{A}, B, \overline{B}) = \mathcal{O}(|(\alpha_2, A, B)|^2 + |(\alpha_2, A, B)| \cdot |(\mu, \varepsilon)|).$$

Note that α_1 does not appear since the full problem (and hence the reduced one) is invariant under the addition of a constant to the pressure.

Because of the symmetries of the problem (2.2) the reduced system (3.3) is reversible with respect to the involution $S = \mathrm{diag}\,(1, -1, 1, -1)$. This means

$$f_j(\mu, \varepsilon, -\alpha_2, A, \overline{A}, -B, -\overline{B}) = (-1)^j f_j(\mu, \varepsilon, \alpha_2, A, \overline{A}, B, \overline{B}), \quad \text{for } j = 1, 2, 3, 4.$$

The translations along the x direction acts, due to space periodicity in x, as the $SO(2)$-action

$$\tau_a : (\alpha_1, \alpha_2, A, B) \rightarrow (\alpha_1, \alpha_2, e^{ia}A, e^{ia}B) \tag{3.4}$$

As we have mentioned earlier the vector Φ_1 corresponds to the mean flux. The component $\alpha_2(z)$ is proportional to the flux through the cross-section $\Omega \times \{z\}$. Because of incompressibility we find $\alpha_2 =$const., whence $f_2 \equiv 0$. Thus, α_2 can be treated as a parameter, which we restrict to the value $\alpha_2 = 0$.

We eliminate of B from (3.3) to obtain a second order equation for $A(z)$. Using the reversibility and the symmetry (3.4) an expansion leads to

$$\begin{aligned}
0 = {} & a(\mu, \varepsilon)A + e_{02}\frac{d^2 A}{dz^2} + b(\mu, \varepsilon)|A|^2 A + \\
& + b_1(\mu, \varepsilon)\left|\frac{dA}{dz}\right|^2 A + b_2(\mu, \varepsilon)\left(\frac{dA}{dz}\right)^2 \overline{A} + \mathcal{O}(|(A, \tfrac{dA}{dz})|^5).
\end{aligned} \tag{3.5}$$

Here we have used the fact that we know the four eigenvalues $\pm(a(\mu, \varepsilon)/e_{02})^{1/2}$ and $\pm(\overline{a(\mu, \varepsilon)/e_{02}})^{1/2}$ of the linear part exactly, namely from (3.2).

4   Analysis of spanwise periodic solutions

As an application of the spatial center manifold reduction we study the bifurcation of solutions which are periodic in z direction with a large period $2\pi/\beta$. Here $\beta > 0$ is small, since the reduced system is only defined for such solutions (dA/dz has to be small). The limit $\beta \rightarrow 0$ is somewhat degenerate and delicate but using the spatial center manifold constructed above, we are in a position to handle precisely this case.

The symmetry (3.4) implies the existence of solutions $d(\varepsilon(\mu), \mu)e^{\pm i\beta z}$. These are *spiral solutions*; flow patterns depending only on $\alpha x + \beta z - ct$ and y.

In order to find other $2\pi/\beta$ periodic small solutions of (3.5) we apply the Liapunov-Schmidt reduction in the space of C^2 functions which are $2\pi/\beta$ periodic. The linearization of (3.5) at the solution $A = 0$ has the spectrum $-m^2\beta^2 e_{02} + a(\mu\varepsilon)$, $m \in \mathbb{Z}$, with eigenvectors $e^{im\beta z}$. Because of relation (3.2) there is a smooth function $(\mu, \varepsilon) = n(\beta^2) = \mathcal{O}(\beta^2)$, such that $a(n(\beta^2)) = \beta^2 e_{02}$. Hence, at $(\mu_c, \varepsilon_c) = n(\beta^2)$ the linearization has the eigenvalue 0 with the complex two-dimensional kernel spanned by $e^{\pm i\beta z}$. Thus, the Liapunov-Schmidt reduction gives $A(z) = d_1 e^{i\beta z} + d_2 e^{-i\beta z} +$ h.o.t. The bifurcation equations take the form

$$d_1\left(a - \beta^2 e_{02} + b(|d_1|^2 + 2|d_2|^2) + \mathcal{O}(|d|^4 + \beta^2|d|^2)\right) = 0,$$
$$d_2\left(a - \beta^2 e_{02} + b(2|d_1|^2 + |d_2|^2) + \mathcal{O}(|d|^4 + \beta^2|d|^2)\right) = 0.$$

Note that the system is equivariant under the reversibility action $(d_1, d_2) \mapsto (\overline{d}_2, \overline{d}_1)$.

The spiral solutions are recovered by letting

$$d_2 = 0 \ \text{ and } \ |d_1|^2 = \frac{\text{Re}\,(\beta^2 e_{02} - a(\mu, \varepsilon))}{\text{Re}\,b(\mu, \varepsilon)} + \mathcal{O}(\mu^2 + \beta^4)$$

or vice versa. Another solution appears in the form

$$\bar{d}_1 = d_2 \ \text{ and } \ |d_{1,2}|^2 = \frac{\text{Re}\,(\beta^2 e_{02} - a(\mu, \varepsilon))}{3\text{Re}\,b(\mu, \varepsilon)} + \mathcal{O}(\mu^2 + \beta^4).$$

It is even and $2\pi/\beta$-periodic in z and is a standing time-periodic flow pattern. These solutions are called *ribbons*.

Therefore, for β small enough the direction of bifurcation is the same for both types of solutions, spirals and ribbons. This contradicts the numerical results in [Br89].

It is worth to mention that the coefficient $b(0,0)$ can be found from the analysis of the plane Poiseuille flow as stationary solutions of the (3.4) are exactly the same solutions as studied in [AV91]. As a sequel from the results of [AV91] we have the following statement.

Theorem. *In the 3D Poiseuille problem with $\alpha \geq \alpha_4$ both, "spirals" and "ribbons", are bifurcating in the subcritical direction. Hence, both types of solutions are unstable in the usual sense.*

For the last statement see the classical literature for bifurcations in $SO(2) \times O(2)$ equivariant systems, e.g. [ABY82], [AB86], [CI85] and [GS86].

References

[ABY82] A.L. Afendikov, K.I. Babenko, S.P. Yuriev: On the bifurcation of Couette flow between counter-rotating cylinders in the case of the double eigenvalue. *Soviet. Phys. Dokl.* **27** (9), 706-709, 1982.

[AB86] A.L. Afendikov, K.I. Babenko: Bifurcation in the presence of the symmetry group and loss of stability of some plane viscous fluid flows. *Soviet Math. Dokl.* **33** (3), 742-747, 1986.

[AV91] A.L. Afendikov, V.P. Varin: An analysis of periodic flows in the vicinity of the plane Poiseuille flow. *Europ. J. Mech. B/Fluids* **10** (6), 577-603, 1991.

[AM93] A. Afendikov, A. Mielke: Bifurcations of Poiseuille flow between parallel plates: three-dimensional solutions with large spanwise wavelength. *Arch. Rational Mech. Anal.* 1993. Submitted.

[Br89] T.J. Bridges: The Hopf bifurcation with symmetry for the Navier-Stokes equations in $(L_p(\Omega))^n$, with applications to plane Poiseuille flow. *Arch. Rational Mech. Anal.* **106**, 335-376, 1989.

[CJ73] T.S. Chen, D.D. Joseph: Subcritical bifurcation of the plane Poiseuille flow. *J. Fluid Mech.* **58**, 337-351, 1973.

[CI85] P. Chossat, G. Iooss: Primary and secondary bifurcations in the Couette-Taylor problem. *Japan J. Appl. Math.* **2** (1), 37-68, 1985.

[Dh83] M. Dhanak: On certain aspects of the 3D instability of parallel flows. *Proc. Royal Soc. London Ser. A*, **385**, 53-84, 1983.

[EK91] U. Ehrenstein, W. Koch: Three-dimensional wavelike equilibrium states in plane Poiseuille flow. *J. Fluid Mechanics*, **288**, 111-148, 1991.

[GS86] M. Golubitsky, I. Stewart: Symmetry and stability in Couette-Taylor flow. *SIAM J. Math. Anal.* **17**, 249-288, 1986.

[GSS88] M. Golubitsky, I. Stewart, D. Schaeffer: *Singularities and groups in bifurcation theory*. Vol. *II*, Appl. Math. Sci. **69**, Springer. New York, 1988.

[IMD89] G. Iooss, A. Mielke, Y. Demay: Theory of the steady Ginzburg-Landau equation in hydrodynamic stability problems. *Europ. J. Mech. B/Fluids* **8** (3), 229-268, 1989.

[IM91] G. Iooss, A. Mielke: Bifurcating time periodic solutions of Navier-Stokes equations in infinite cylinders. *J. Nonlinear Science* **1**, 106-146, 1991.

[Ki82] K. Kirchgässner: Wave solutions of reversible systems and applications. *J. Diff. Eqns.* **45**, 113-127, 1982.

[Lin55] C.C. Lin: *The theory of hydrodynamic stability*. Cambridge University Press, 1955

[Mi88] A. Mielke: Reduction of quasilinear elliptic equations in cylindrical domains with applications. *Math. Meth. Appl. Sciences*, **10**, 51-66, 1988.

[NII75] M. Nishioka, S. Iida, Y. Ichikawa: An experimental investigation of the stability of plane Poiseuille flow. J. Fluid Mech. **72**, 731-751, 1975

[Sq33] H.B. Squire: On the stability of the three-dimensional disturbances of viscous flow between parallel walls. *Proc. Roy. Soc. London Ser. A*, **142**, 621-628, 1933.

[TTRD93] L.N. Trefethen, A.E. Trefethen, S.C. Reddy, and T.A. Driscoll: Hydrodynamic stability without eigenvalues. *Science* **261**, 578-584, 1993.

ANALYZING BIFURCATIONS IN THE KOLMOGOROV FLOW EQUATIONS

D. ARMBRUSTER, B. NICOLAENKO, N. SMAOUI
Department of Mathematics,
Arizona State University
and
P. CHOSSAT
Institut Non-Linéaire de Nice
CNRS - Université de Nice Sophia-Antipolis

Abstract. Simulations of forced 2-D Navier-Stokes equations are analyzed. The forcing is spatially periodic and temporally steady. Two regimes are analyzed: a bursting regime and a regime that exhibits discrete traveling waves. A Karhunen Loeve analysis is used to identify the structures in phase space that generate the PDE behavior. Their relationship to the invariant subspaces generated by the symmetry group is discussed.

1 Introduction

We consider the Kolmogorov flow equations, i.e. the 2-D Navier-Stokes equations with a force which is assumed stationary and spatially biperiodic. We focus in this note to the simpler case of a force of the form $(\frac{8}{Re}\cos 2y, 0)$ in x, y coordinates (R_e is the Reynolds number). Then the basic Kolmogorov flow has a streamfunction formulation of $\phi = \sin 2y$. It is invariant under translations in the x-direction and π-periodic in the y-direction. In the usual stream function and vorticity representation of the flow, the equations for a perturbation of the Kolmogorov flow reduce to

$$\frac{\partial \triangle \phi}{\partial t} = \triangle^2\phi - 2R_e\frac{\partial}{\partial x}[\triangle\phi + 4\phi]\cos 2y - R_e[\frac{\partial}{\partial x}(\triangle\phi\frac{\partial\phi}{\partial y}) - \frac{\partial}{\partial y}(\triangle\phi\frac{\partial\phi}{\partial x})] \quad (1)$$

where ϕ is the perturbation of the stream function from the trivial Kolmogorov flow. Boundary conditions are assumed periodic, both in x and y directions. Numerical

11

P. Chossat (ed.), Dynamics, Bifurcation and Symmetry, 11–33.

studies have shown that this equation exhibits a remarkable sequence of symmetry breaking bifurcations, until a chaotic state is reached, which has the form of an intermittent switching between two vortex-like coherent structures (periodically repeated over the plane). This problem is a prototype for the study of transitions to coherent structures in turbulent flow. Our aim is to understand the behavior of the PDE in terms of concepts from dynamical systems theory. In order to do so we are relying heavily on two tools: We are going to exploit as much as possible the symmetries in the problem and we will use the Karhunen-Loeve (KL) decomposition ([12], [10]) as a probe into the behavior of the PDE (see e.g. [9], [1]). We would like to emphasize the almost experimental nature of the way this study was undertaken: Numerical results focused through the K-L analysis suggest a certain phase space setup. With that kind of setup in mind we then determine the group representations on the phase space and its invariant subspaces. Armed with that knowledge we would do more K-L tests that would either confirm or disprove the original ideas. This process converges to the almost complete understanding of the behavior of the 2-d Navier Stokes equation for small Reynolds numbers that we will discuss below.

Let us first describe the symmetries of the equation. These symmetries follow from the euclidean invariance of the Navier-Stokes equations, restricted by the form of the force. It is a simple exercise to check that all the symmetries are generated by the following transformations:

$$
\begin{array}{llll}
T_\xi & : & x & \rightarrow & x + \xi \\
r & : & (x, y) & \rightarrow & (-x, -y + \frac{\pi}{2}) \\
s & : & (y, \phi) & \rightarrow & (-y, -\phi) \\
t & : & y & \rightarrow & y + \pi
\end{array}
$$

Notice that r is a rotation by π in the x, y-plane, centered at $(0, \frac{\pi}{4})$. r does not commute with any of the other group elements but $t = (rs)^2 = (sr)^2$. The symmetry s is peculiar in that it involves ϕ itself. This fact will play a crucial role in our analysis. The complete symmetry group of the Kolmogorov equations with periodic boundary conditions is the group $G = G_1 \cup sG_1$, where $G_1 = O(2) \times Z_2$, $O(2)$ being generated by $\{T_\xi, \xi \in S^1\}$ and r, and Z_2 being the 2-element group generated by t. Notice that, since $(rs)^4 = id$ G can also be written as the semidirect product $D_4 \dotplus SO(2)$.

In this note we want to focus our attention on two different regimes both of which exhibit behavior in the PDE simulations that appears puzzling: At $R_e = 25.70$ the flow appears to be laminar as can be seen by the time series for the vorticity in Figure 1. Increasing the Reynolds number to 25.77 we notice bursts which have been described earlier in related simulations [11], citerand. These bursts are very short

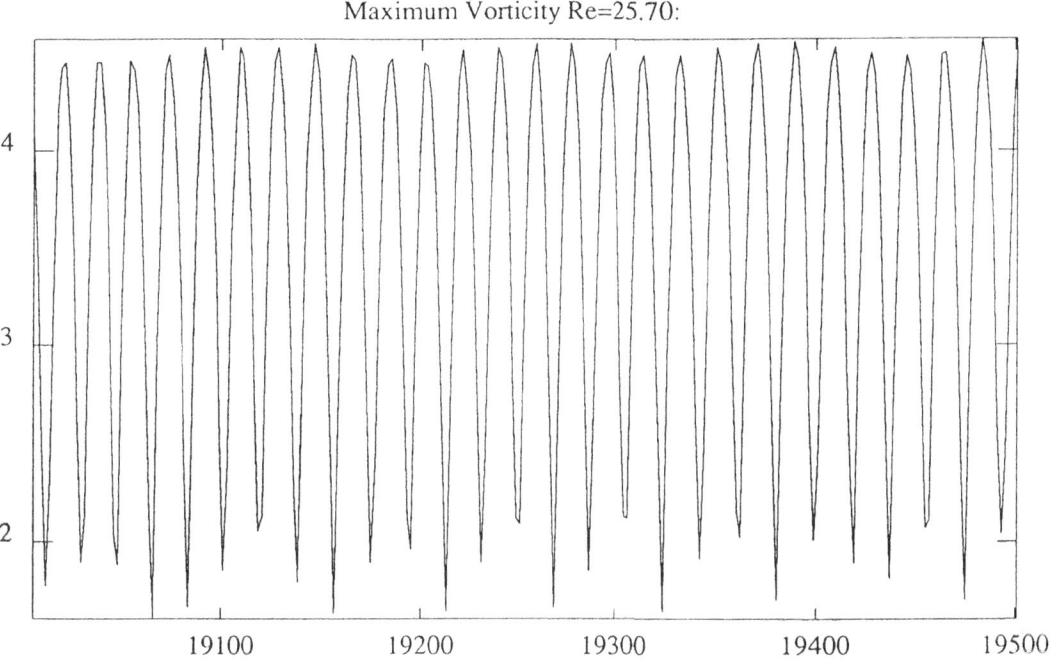

Figure 1: Maximum vorticity over the spatial domain for $Re = 25.70$

14

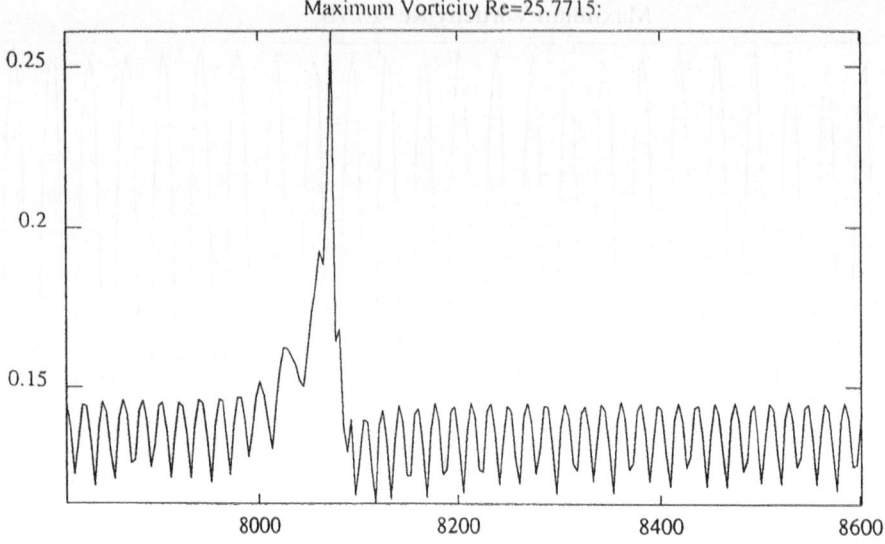

Figure 2: Maximum vorticity over the spatial domain for $Re = 25.7715$

lived and occur intermittently. After the burst the flow settles back to a laminar state, (see Figure 2). The whole sequence and the fact that an O(2)-symmetry plays a major role in the symmetry group suggests that these transitions are generated by structurally stable heteroclinic cycles [1]. We will show below that this cannot be true. The other regime that we are interested occurs around $R_e = 17$. Here the numerical simulations suggest a bifurcation from a standing wave to a traveling wave in the y-direction. However, the symmetry group does not have a translation element in that direction. We will resolve that issue in section 5.

In section 2 we present the linear stability analysis of the trivial Kolmogorov flow and suggest a phase space for a dynamical system description of the flow up to $R_e \approx 30$. In section 3 we discuss the resulting group representations on that phase space and its fibration by invariant subspaces. Section 4 presents the K-L analysis of the laminar flow at 25.0 and the bursting behavior at 25.7. Section 5 discusses a structurally unstable heteroclinic bifurcation at $R_e = 17.0$. We conclude with a preview of further work.

2 The spectral analysis of the trivial solution

Simple inspection of the Fourier amplitudes of the first nontrivial numerical solution suggests that the structure is generated by long wavelength or small wavevectors. We therefore expand the streamfunction into Fourier modes $e^{i(k_x x + k_y y)}$ with
$k_x^2 + k_y^2 \leq 10$ and produce a Galerkin projection. The trivial solution of the resulting
sytem of ODEs can easily be analyzed. We find that it is stable for $R_e < \frac{5}{\sqrt{6}}$ and it
shows a steady state bifurcation at $R_e = \frac{5}{\sqrt{6}}$. This value agrees well with our numerical findings. Since the trivial solution is invariant under our symmetry group it is in
particular also invariant under the O(2)-symmetry. Hence the steady state bifurcation is created through a double zero eigenvalue and the associated eigenfunction is
complex of the form $e^{ix}(1 - i/2\cos 2y)$. There is a second bifurcation from the already
unstable trivial solution at $R_e = \frac{60}{\sqrt{83}}$. This time the eigenvalues are purely imaginary
and hence, due to the O(2)-symmetry we have a 4-dimensional eigenspace generated
by $\{e^{ix}\cos y, e^{-ix}\sin(y)\}$. As a result we have a O(2)-symmetric Hopf bifurcation.
The Galerkin projection does not show any other bifurcation from the zero solution.
This agrees with a "Spectral Barrier Theorem" by Foias and Nicolaenko [5]. This
theorem is established within the context of generalized Kolmogorov flows:

$$\frac{\partial \triangle \phi}{\partial t} = \triangle^2 \phi - R_e J(\phi, \phi) + \Phi_F \tag{2}$$

where $J(\phi, \phi)$ stands for the nonlinear term in the right hand side of equation (1).
The forcing stream function $\Phi_F(x, y)$ is chosen in such a way that, with periodic
boundary conditions, it is an eigenfunction of the Laplacian

$$\triangle \Phi_F = -\lambda_F \Phi_F. \tag{3}$$

The corresponding base Kolmogorov flow is

$$\phi_k = -\frac{1}{\lambda_F^2} \Phi_F \tag{4}$$

since $J(\Phi_F, \Phi_F) = J(\phi_k, \phi_k) = 0$. Consider now the linearization of (2) around the
base flow (4) together with the associated eigenvalue problem:

$$\lambda \triangle \psi = \triangle^2 \psi - R_e J(\phi_k, \psi) - R_e J(\psi, \phi_k) \tag{5}$$

(e.g. this is the eigenvalue problem associated with the linearization of (1) at $\phi = 0$). We have:

Spectral Barrier Theorem
The spectrum of (5) is a pure point spectrum with complex conjugate eigenvalues. For $R_e \geq 0$, $\lambda = -\lambda_F$ remains a fixed eigenvalue. As R_e increases *no eigenvalue can cross the line* $Re\lambda = Re - \lambda_F$ in the complex spectral plane.

Consequently, as R_e increases the only eigenvalues of the Laplacian that can generate bifurcations are those smaller in modulus than λ_F. In our specific Kolmogorov flow, $\lambda_F = 4$ and the only eigenvalues which will ultimately bifurcate are $\lambda = -1, -2$ (corresponding to wavevectors $k = (0,1), (1,0)$ and $(\pm 1, \pm 1)$. Furthermore one has

Theorem: [5] There exists R_e^* such that for all $R_e \geq R_e^*$, the dimension of the unstable manifold of ϕ_k stays constant and is equal to the sum of dimensions of all eigenspaces of the Laplacian corresponding to eigenvalues with $|\lambda| < -\lambda_F$.

These results for the linear stability of the trivial solution may suggest that we should look for the explanation of our bursting behavior in the framework of hetero-clinic cycles generated by the interaction of a Hopf and a steady state bifurcation with O(2)-symmetry - a topic that is well understood [8]. Unfortunately the Navier Stokes equation do not behave as expected: The next bifurcation from the nontrivial steady state occurs at about $R_e = 7.5$ to a structure that travels in the x-direction. Numeri-cally we find that as one gets closer and closer to bifurcation the traveling wave speed seems to go to zero. This indicates that we have a bifurcation from a group orbit of steady states [7] i.e. a structure that preserves the SO(2)-symmetry but breaks the reflection symmetry of the original O(2)-symmetry. This can clearly not happen with the eigenfunctions that locally describe the Hopf-steady state interaction. We therefore make an Ansatz to include two modes that break the r-symmetry into our phase space. The K-L analysis below suggests that the modes $\cos y$ and $\sin y$ are good representatives.

3 Group representation and invariant subspaces

For the purpose of analyzing the PDE dynamics we don't need to find a explicit ODE model and hence we do not need to explicitly determine the eigenfunctions whose amplitudes span the phase space. Whatever these amplitudes are it is prudent to choose them in a way that they generate a simple representation of the symmetry group. Our considerations in the previous chapter suggests to define an eight dimensional representation space by

- (x_1, x_2) transforming as $(\cos y, \sin y)$ (stable modes independent of x)

- z_0 transforming as $e^{ix}[1 - i \cos 2y]$ and $\bar{z}_0$ (steady-state unstable modes)

- z_1 transforming as $e^{ix} \cos y$, $z_2 = r z_1$, $\bar{z}_1$ and $\bar{z}_2$ (oscillatory unstable modes)

This makes an 8-dimensional phase space, in which the group G acts as follows:

$$T_\xi(x_1, x_2, z_0, z_1, z_2) = (x_1, x_2, e^{i\xi} z_0, e^{i\xi} z_1, e^{-i\xi} z_2) \tag{6}$$

$$r(x_1, x_2, z_0, z_1, z_2) = (x_2, x_1, \bar{z}_0, z_2, z_1) \tag{7}$$

$$s(x_1, x_2, z_0, z_1, z_2) = (-x_1, x_2, -z_0, -z_1, z_2) \tag{8}$$

$$t(x_1, x_2, z_0, z_1, z_2) = (-x_1, -x_2, z_0, -z_1, -z_2) \tag{9}$$

It remains in this section to describe the G-orbit structure induced in R^8 by the transformations (2)-(5). This is a crucial information because of the well-known fact that the flow of G-equivariant differential equations respect this structure. Let us define the isotropy group of $x \in R^8$ to be $G_x = \{g \in G/gx = x\}$, and $Fix(G_x) = \{y \in R^8/G_x y = y\}$. For any isotropy subgroup H of G, the space $Fix(H)$ is invariant by the flow of any G-equivariant ODE. Given $g \in G$ and a solution $x(t)$ in $Fix(H)$, $gx(t)$ is another solution lying in $Fix(gHg^{-1})$. Hence it is only important to classify conjugacy classes of isotropy subgroups. These classes are called isotropy types and are partially ordered by set inclusion. This allows us to form the "lattice" of isotropy types, which gives the essential informations that we need in the next section. Figure 3 shows this lattice, where each isotropy type is defined by one representative. The trivial isotropy type is omitted.

The following invariant subspaces are associated with the isotropy subgroups of G appearing in Figure 3 (subspaces are indicated by their non-zero coordinates):

1. Dimension 1:

    - $Fix(SO(2) \times Z_2(s)) = \{x_2\}$
    - $Fix(O(2)) = \{x_1 = x_2\}$
    - $Fix(\Sigma_1) = \{Re z_0\}$, Σ_1 generated by r and sT_π
    - $Fix(\Sigma_2) = \{z_1\}$, Σ_2 generated by st and sT_π
    - $Fix(\Sigma_3) = \{z_1 = z_2\}$, Σ_3 generated by r and tT_π

2. Dimension 2:

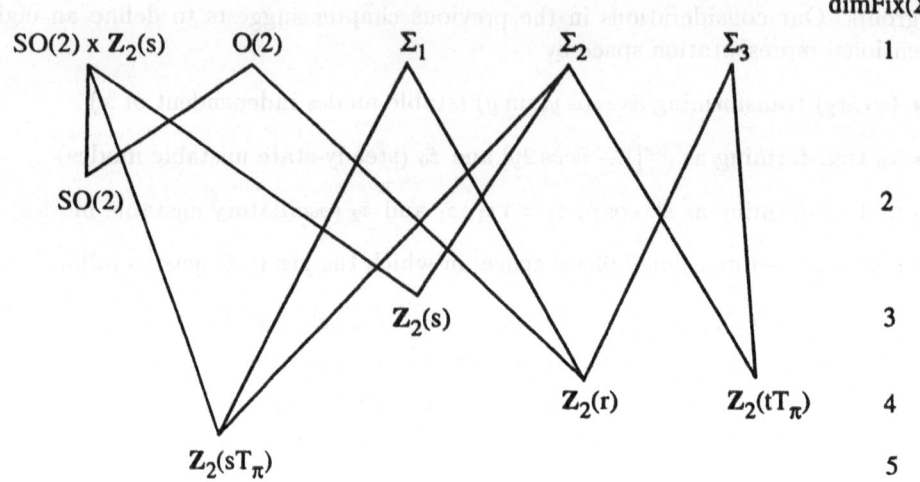

Figure 3: the lattice of isotropy types for the G-action in R^8

- $Fix(SO(2)) = \{x_1, x_2\}$

3. Dimension 3: $Fix(Z_2(s)) = \{x_2, z_2\}$

4. Dimension 4:

   - $Fix(Z_2(r)) = \{x_1 = x_2, Rez_0, z_1 = z_2\}$
   - $Fix(Z_2(tT_\pi)) = \{z_1, z_2\}$

5. Dimension 5: $Fix(sT_\pi) = \{x_2, z_0, z_1\}$.

In this list, $Z_2(k)$ indicates the 2-element group generated by $k \in G$. The calculation of the lattice and of the invariant subspaces is a lengthy, but straightforward resolution of the equation $gX = X$, $g \in G$, $X \in R^8$.

Note a couple of equivalent subspaces: Acting with r on a solution in $Fix(sT_\pi)$ leads to a solution in $Fix(rsT_\pi r) = Fix(tsT_\pi)$, spanned by the modes $\{x_1, \bar{z}_0, z_2\}$. Acting with r on a solution in $Fix(Z_2(s))$ leads to a solution in $Fix(rsr) = Fix(ts)$, spanned by the modes $\{x_1, z_1\}$.

Note also the typical solutions in some of those subspaces that will show up in our simulations:

- $Fix(\Sigma_1) = \{Rez_0\}$: The typical solution is a fixed point generated by the steady state bifurcation from zero.

- $Fix(\Sigma_2) = \{z_1\}$ The typical solution is a unstable traveling wave generated via the Hopf bifurcation from zero.

- $Fix(\Sigma_3) = \{z_1 = z_2\}$ The typical solution is a unstable standing wave generated via the Hopf bifurcation from zero.

- $Fix(Z_2(r)) = \{x_1 = x_2, Rez_0, z_1 = z_2\}$: Solutions here are generated via mode interaction of branches with higher isotropy.

- $Fix(sT_\pi) = \{x_2, z_0, z_1\}$: Different types of solutions exist here. A important one is a traveling wave, generated via a bifurcation from the circle of fixed points in $Fix(\Sigma_4)$ that breaks the relection symmetry r. A nonlocal "normal form" analysis presented elsewhere shows that this is the solution type that we find numerically near $R_e \approx 8$.

4 Karhunen-Loeve analysis of the bursting behavior

4.1 The laminar cases

The Karhunen Loeve analysis is a data analysis that extracts the dominant spatial structures out of a complicated spatio-temporal dataset. Those structures are called the coherent sturctures and are calculated as the eigenfunctions of a correlation matrix. For details of this method and our approach to it see [2], [9],[10] [12]. At $R_e = 25.70$ the PDE simulation indicates a quasiperiodic solution which consists of a traveling structure plus additional time dependent behavior. Converting the complex Fourier amplitude of the mode e^{ix} into polar form we see that the phase changes linearly. We extract the slope of the phase change which determines the wavespeed of the traveling (see Figure 4). It is well known that a K-L analysis of a traveling wave just gives the Fourier modes of that traveling wave. Since we are, however, interested in the spatial structure that travels and its symmetries we transform our simulation data into a moving coordinate frame, moving with the pre-determined wavespeed. With those "untraveled data" we perform a KL analysis and find three eigenfunctions that capture 99.8% of the energy of the datafield (see Figure 5). Specifically, the first eigenfunction captures 95.8% of the energy. Inspection shows that it lies in the invariant subspace $Fix(sT_\pi)$ and therefore represents the traveling backbone of the structure. The next two eigenfunctions capture 3.4% and 0.6% of the energy, respectively. Those two eigenfunctions span a plane in phase space orthogonal to $Fix(sT_\pi)$. These findings confirm that the laminar flow can be described as a modulated traveling wave which in phase space is represented by a torus. Since most

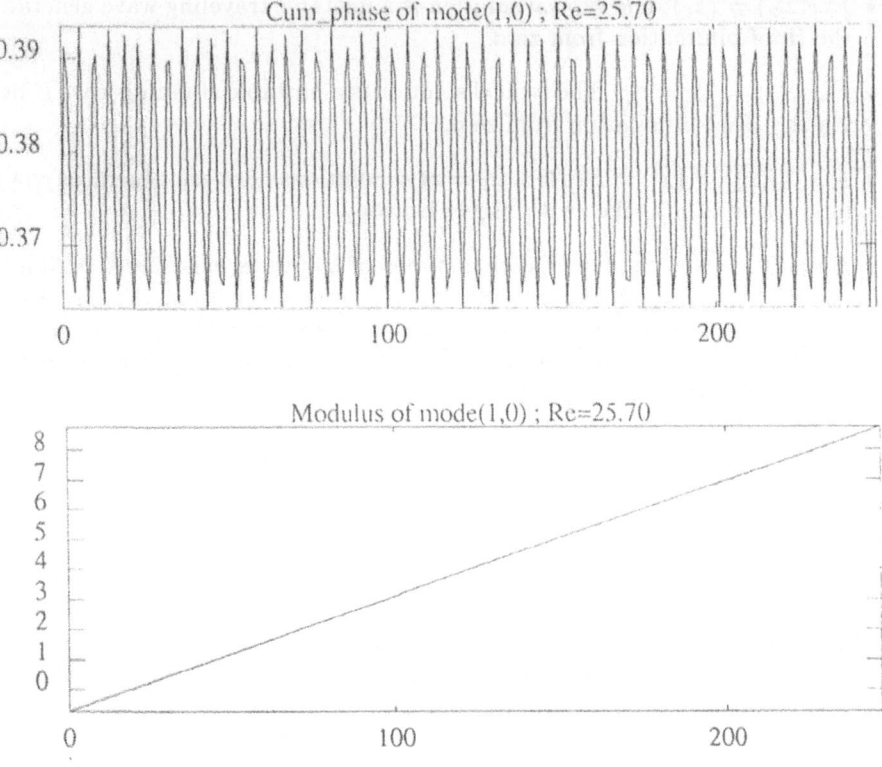

Figure 4: Amplitude and phase for the mode e^{ix} for $R_e = 25.70$

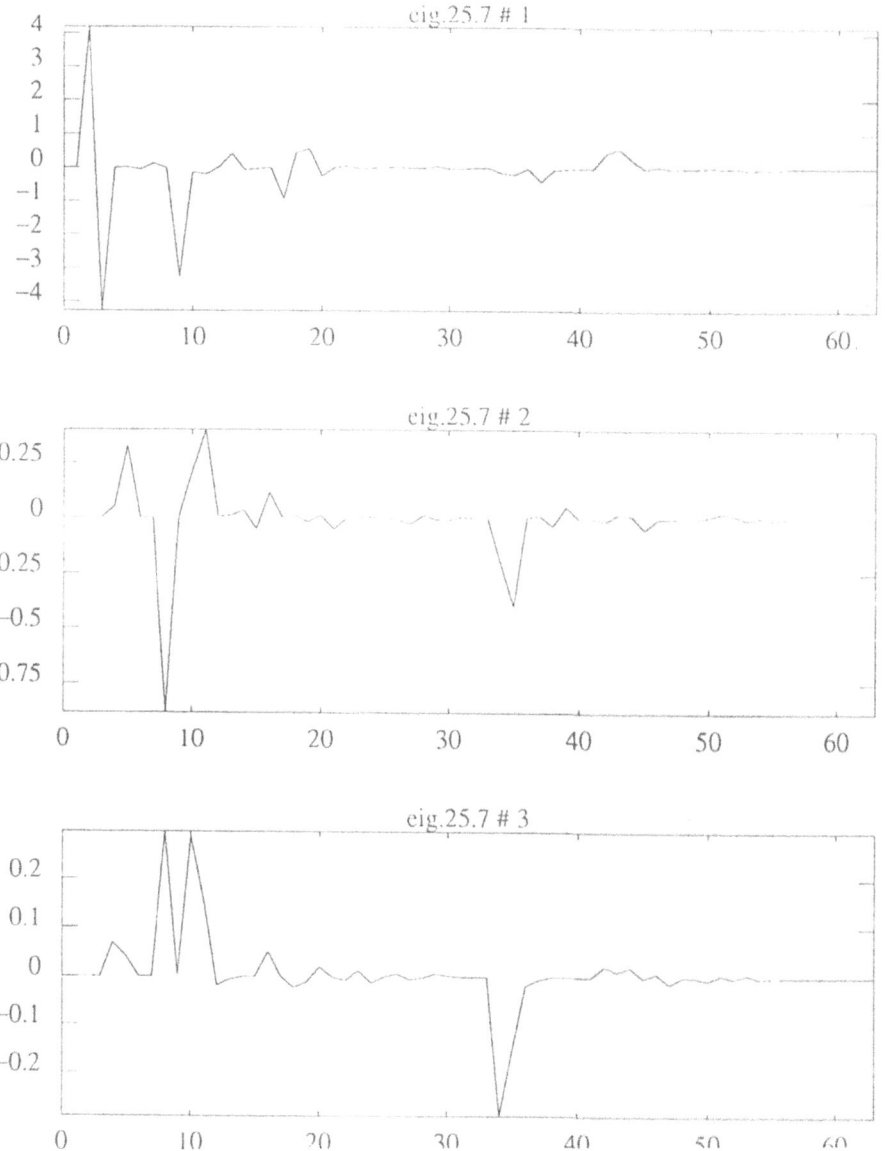

Figure 5: The first three KL-eigenfunctions of the data analysis at $Re = 25.70$. The data are given as the Fourier amplitudes of a spectral code. Hence the eigenfunctions also consist of a vector of Fourier amplitudes. The labeling on the x-axis is done as follows: $(k_x = 0..3, k_y = 0, k_x = 0..3, k_y = 1$ etc until $k_y = 3$ and then repeat for $k_y = -1.. - 3$.

of the energy is in the first eigenmode the relationship between the two radii of the torus is about 1:20. Clearly our data came from one particular simulation and therefore pertain to one single modulated traveling wave. However, since the system is symmetric we can act with the whole symmetry group on the torus and see whether we get different tori. In particular we can act with t and r on the solution: Focusing on the first eigenfunction which lies in $Fix(sT_\pi)$ we have nonzero components in (x_2, z_0, z_1). Acting with the symmetry st on it transforms such a solution into one with the coordinates $(-x_2, -z_0, z_1)$. This is still a solution in $Fix(sT_\pi)$ but it is clearly a different one from the first. Also, as discussed above, acting with r on $Fix(sT_\pi)$ leads to $Fix(tsT_\pi)$ which has nonzero coordinates $(x_1, \bar{z}_0, z_2)$. Notice that the translation group is acting in the x-direction which implies that all the traveling is also in that direction. Now the symmetry operation r changes x to -x, hence it reverses the traveling direction.

As a conclusion of the analysis of the laminar case we can state: The laminar flow is a quasiperiodic flow. It may occur in four different ways:

- Traveling in the positive x-direction with a spatial structure in y dominated by $\sin y$ (the x_2 mode is positive).

- Traveling in the positive x-direction with a spatial structure in y dominated by $-\sin y$ (the x_2 mode is negative).

- Traveling in the negative x-direction with a spatial structure in y dominated by $\cos y$ (the x_1 mode is positive).

- Traveling in the positive x-direction with a spatial structure in y dominated by $-\cos y$ (the x_1 mode is negative)

4.2   Bursting

A slightly higher Reynolds number $R_e = 25.7715$ leads to the bursting behavior described above. Since the Reynolds number differs so little it is reasonable to assume that there exists a close connnection between the stable laminar flow at $Re = 25.70$ and the laminar phases in between bursts at $R_e = 25.7715$. If we look close enough at the Fourier amplitudes for the modes e^{ix} and e^{iy} (Figure 6) we notice that the laminar phases in between bursts very much look like one of the 4 possible modulated traveling waves, described in the last section. To test this idea we perform K-L analyses on the the time series in between bursts. We cut the data in such a way that they look perfectly laminar, i.e. we cannot visually detect the coming of a burst. Again we go into a co-moving coordinate frame. If we do that for each of the different

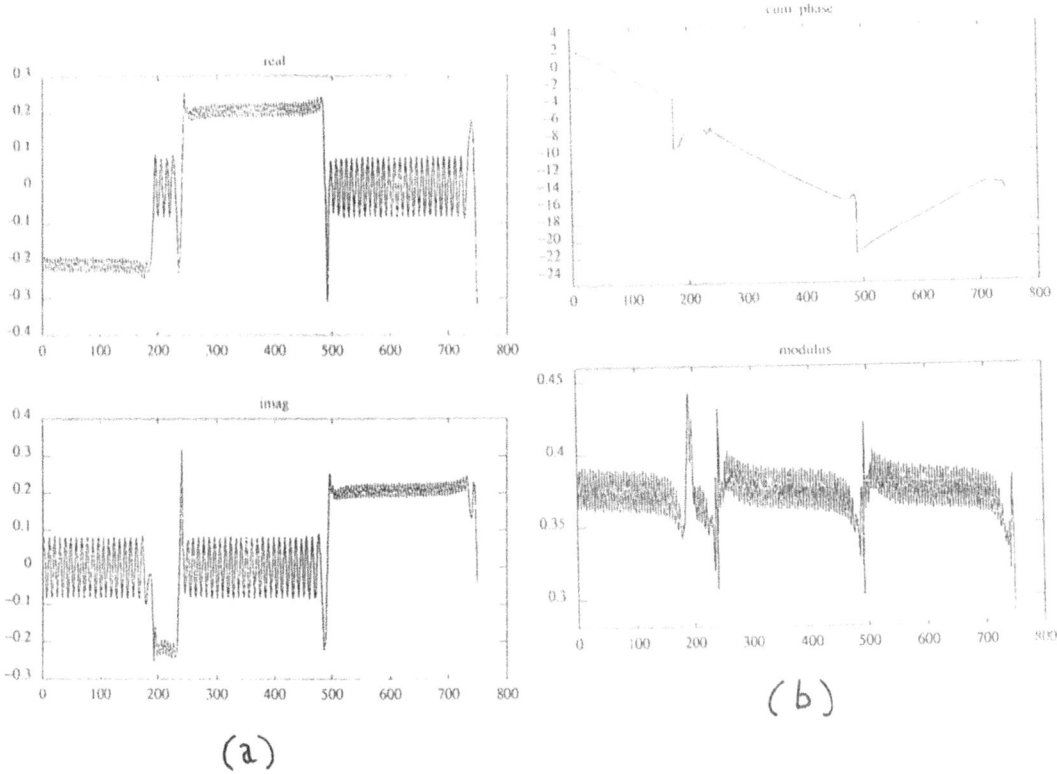

(a)

(b)

Figure 6: a) Real and imaginary part of the Fourier amplitude e^{iy}, b) amplitude and phase for the Fourier amplitude e^{ix} for $R_e = 27.7715$

laminar phases we find that the first 5 eigenfunctions for each dataset map onto each other under the symmetry operation rT_ξ and tT_η, respectively, where the shifts ξ and η are determined by phase shifts of the modulated traveling waves due to different initial conditions. This confirms that the unstable limit sets in the bursting regime are related to each other via the symmetries of the system. In order to more clearly determine the limit sets and their unstable manifolds we try to separate them: Taking the dataset 1 in Figure 6 we determine its most dominant eigenfunction. This eigenfunction is almost identical to the first eigenfunction of the laminar phase up to a phase shift. By matching the amplitudes and phase of the e^{ix} mode we can determine this phase shift. We then shift the first 5 eigenmodes of the laminar case by the so determined phase shift and project the data onto these modes. Subtracting this projection from the datafield we are left with those components of the laminar states between the bursts that differ from the modulated traveling waves states. A reasonable assumption about these components is that they describe the unstable manifold that leads towards bursting. A further K-L analysis of these data reveals the following facts:

- 99.5 of the energy in the unstable manifold is contained in 3 eigenfunctions.

- The first eigenfunction lies in $Fix(sT_\pi)$.

- The second and third eigenfunction do not lie in any invariant subspace.

- The first three eigenfunctions of the unstable manifolds of different laminar phases map onto each other under the same group operations that map the corresponding modulated traveling waves.

- Projecting the data for the unstable manifold onto the three K-L eigenfunctions we can display the time series of the amplitudes of these eigenfunctions. Figure 7 shows a linear and oscillatory exponential growth, respectively, further confirming the association of these modes with an unstable manifold.

- The eigenfunctions of the unstable manifolds seem to be the same independent of the laminar phase that is reached after the burst.

4.3 Structurally stable heteroclinic cycle?

The above observations allow us to speculate about the nature of the transition between the two flow states: Clearly the bursting behavior is not associated with structurally stable heteroclinic cycles. In order for those to occur the quiescent phases in between bursts need to sit in invariant subspaces and the connection between

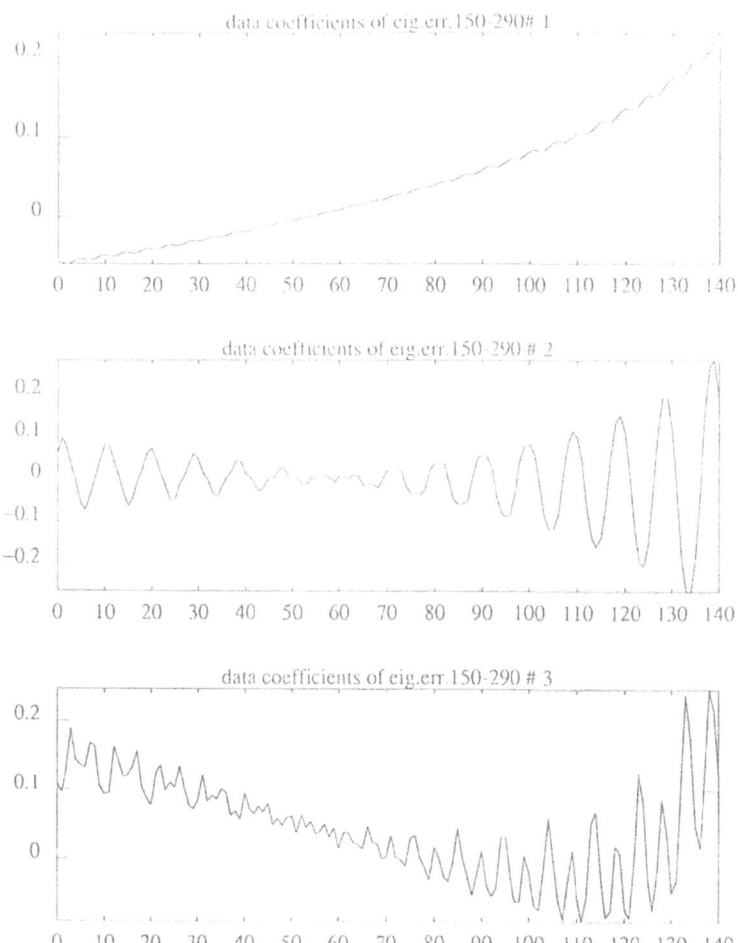

Figure 7: Amplitudes of the first three eigenfunctions of the unstable manifold of the laminar phase as a function of time

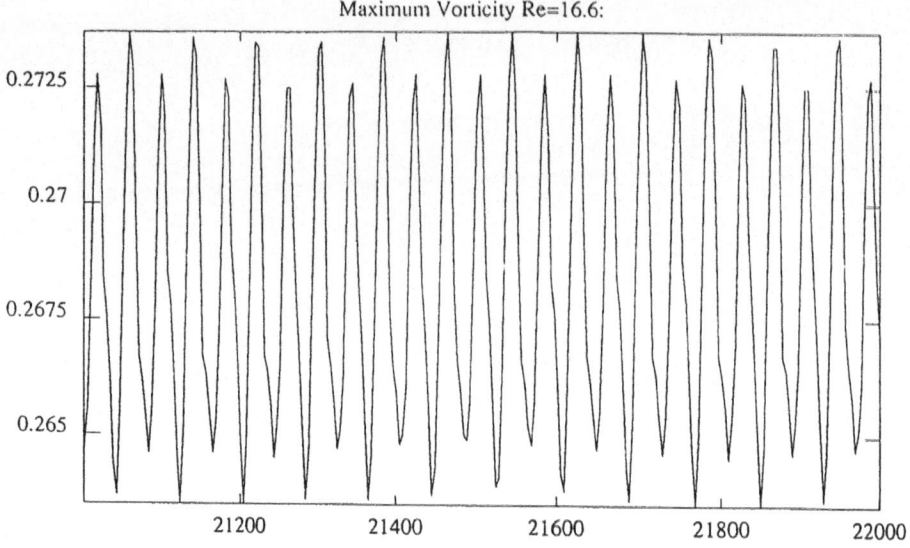

Figure 8: Maximum vorticity over the spatial domain for $Re = 16.60$

them must again lie in invariant subspaces. Neither is the case here: While the traveling backbone sits in $Fix(sT_\pi)$ or $Fix(tsT_\pi)$ the modulation lies in a space orthogonal to this subspace. Secondly there is no invariant subspace that contains the unstable manifold of the previous laminar phase and the next laminar phase. All those observations make it highly unlikely that a structurally stable heteroclinic cycle exist. The data however are consistent with a *symmetry increasing bifurcation* through a structurally unstable heteroclinic cycle that exists around $R_e \approx 25.73$. The next section will describe a similar but simpler transition.

5 A structurally unstable heteroclinic cycle

At $Re = 16.60$ the flow is periodic as can be seen in Figure 8. Doing a phase plot of the data in the (x_1, x_2) coordinates (recall that these are the amplitudes of $\cos y$ and $\sin y$) we find a limit cycle centered around the diagonal $x_1 = x_2$ (Figure 9). In fact

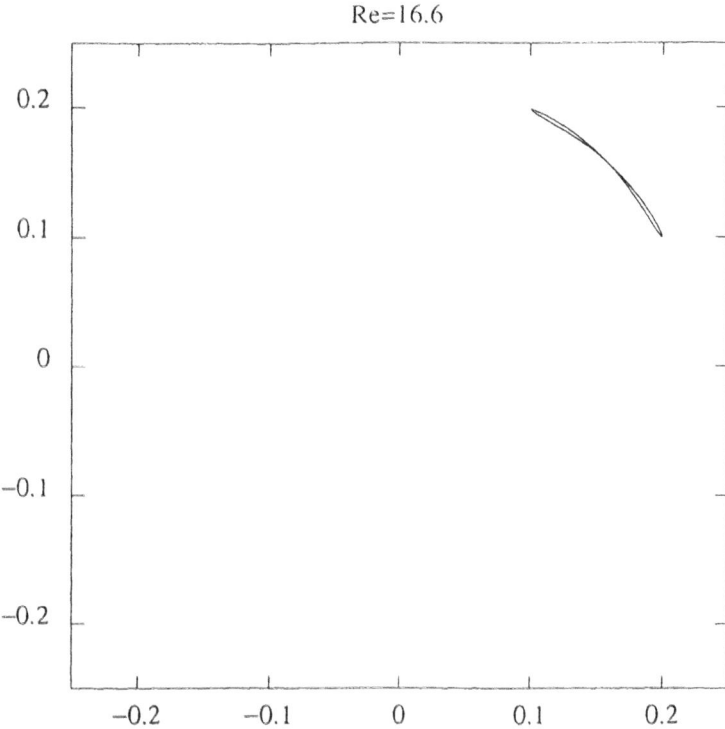

Figure 9: Phase plot of the simulation at $Re = 16.60$ in the (x_1, x_2) coordinates

a K-L analysis reveals that the first eigenfunction is in $Fix(rT_\xi)$ with $\xi = 4.7$ which changes with different initial conditions. This fixed point subspace is a conjugate to $Fix(r)$ and we can continue the discussion without loss of generality for $\xi = 0$. The next two eigenfunctions are orthogonal to this subspace. It turns out that the limit cycle is invariant under r as a set. I.e. if we reflect a snapshot with r and phase shift the time by half the period of the oscillation then we get the same snapshot back. Again we can act with the whole symmetry group on this limit cycle. Since it is a standing wave we can shift it arbitrarily and get a whole torus of limit cycles. Also r, s and st act like a D_4 symmetry on the (x_1, x_2) plane. We therefore get four different limit cycles centered on the diagonals in Figure 9. Increasing the Reynolds number to 16.97 we see that the period of the limit cycle becomes longer and longer and the time series of the amplitude of the e^{iy} mode approaches a long plateau. (Fig 10 a). At that time the solution is almost in $Fix(sT_\pi)$ or $Fix(tsT_\pi)$. Projecting a longer time series into the (x_1, x_2) plane we see that the two limit cycles in the first and fourth quadrant have merged. Waiting even longer we find that the system randomly switches between all four of the possible limit cycles (Fig 10 b). Note that in that projection the fixed point subspaces $Fix(sT_\pi)$ and $Fix(tsT_\pi)$ are given by the x_1 and x_2 axes respectively. Either numerical error or a small region of a heteroclinic tangle near the transition generates the random switching. Increasing the Reynolds number to $Re = 17.5$ the solution settles to a Z_4 symmetric solution: The projection into the (x_1, x_2) plane shows a limit cycle that is invariant under rotations by 90 degrees (Figure 11). The limit cycle is traversed either in clockwise or counterclockwise direction depending on initial conditions. A amplitude-phase description of complex amplitude of e^{iy} will therefore clearly give a phase that is monotonically increasing (or decreasing) with time. Therefore, although we do not have a continuous group symmetry in the y-direction the solution represents a discrete traveling wave like they have been observed in coupled oscillator lattices [6]. The breaking of the heteroclinic orbit that connects the four traveling waves in the point subspaces $Fix(sT_\pi)$ and $Fix(tsT_\pi)$ therefore creates a symmetry increasing bifurcation from an attractor that has the symmetry of $Z_2(r) \times SO(2)$ as a set, to an attractor that has $Z_4 \times SO(2)$ as a set, although there are no invariant subspaces with these groups as fixed point subgroups.

6 Conclusion

We have shown in two examples how the interplay of the symmetry of the PDE, the representation and the fixed point subspaces that it creates and the data analysis via the Karhunen-Loeve decomposition allows us to understand the structure of the

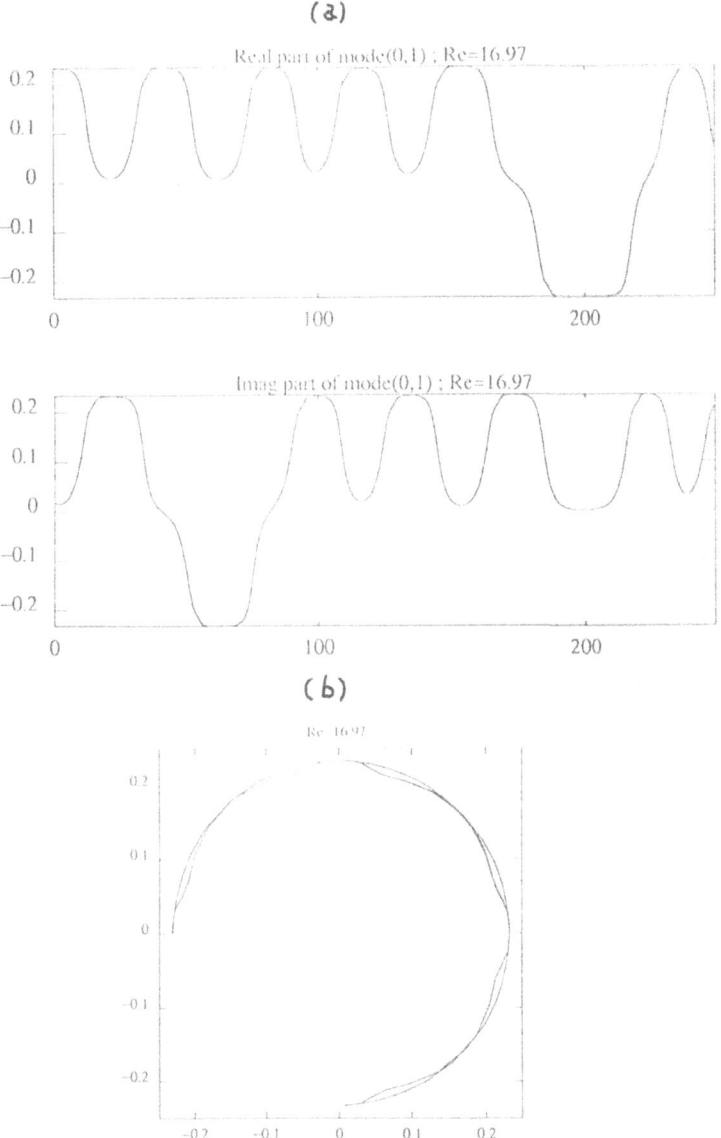

Figure 10: a) Real and imaginary part of the Fourier amplitude e^{iy}; b) phase plot in the (x_1, x_2) coordinates for $R_e = 16.97$

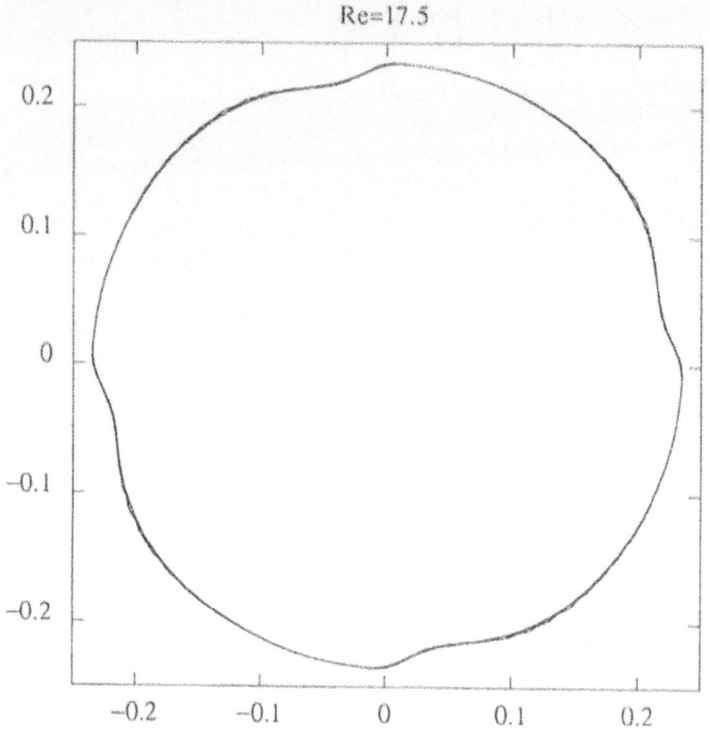

Figure 11: Projection of a limit cycle in the (x_1, x_2) plane for $R_e = 17.5$

limit sets of the PDE simulations and make progress in identifying the nonlocal bifurcations involved. We have learned two important facts:

In contrast to the Kuramoto-Sivashinsky equation where the bifurcation structure can be understood for a range of parameters by the interaction of the unstable manifold of the trivial solution [3], the Navier Stokes equation in two dimension is determined strongly by the interaction of the modes that are in the stable manifold of the zero solution.

Bursting behavior does not seem to be related to structurally stable heteroclinic orbits.

In a more detailed paper [13] we will show that all the limit sets occuring in the simulations for $R_e < 30$ can be understood in the context of the representation of the symmetry group discussed in Section 3.

Our study opens more questions than it answers so far. Work currently is in progress on Galerkin projections on the K-L eigenmodes to generate low dimensional models that show the behavior of the PDE and that are easier to visualize. Such models have to be equivariant with respect to the symmetry group and are essential to check the conclusions of the symmetry increasing bifurcation at $R_e \approx 17$ and to come up with a global bifurcation that generates the bursting behavior. Another area of research that we are currently pursueing is to determine unambiguously the symmetry types of the various attractors in the Kolmogorov flow simulations. We proceed here along the theory developed in [4]. Finally, this study suggests to develop normal forms for the type of symmetry increasing bifurcations that can be seen here. Such normal form are essential for a complete understanding of the bifurcation and its behavior under perturbations.

6.1   Acknowledgement

Helpful conversations with I. Melbourne and M. Golubitsky are gratefully acknowledged. N.S. and B.N where partially supported through the Airforce Office of Scientific Research, D.A. was partially supported through NSF grants DMS 9017174 and DMS 9101964. D.A., P.C. and B.N. would like to thank the Fields Institute, Waterloo, Canada for its hospitality during a workshop in spring 1993 where part of this work has been done. P.C. was partially supported through a grant from the C.N.R.S. (C.N.R.S.-N.S.F. cooperation) for his visits to the Math. Dept. of ASU at Tempe.

References

[1] D.Armbruster, R.Heiland, E.Kostelich, b.Nicolaenko: Phase-space analysis of bursting behavior in Kolmogorov flow, Physica D **58** 392 (1992)

[2] D.Armbruster, R.Heiland, E.Kostelich: KLTOOL a tool to analyze spatiotemporal complexity, preprint 1993

[3] D.Armbruster, J.Guckenheimer, P.J.Holmes: Kuramoto-Sivashinsky dynamics on the center-unstable manifold, SIAM J. Appl. Math. **49**, 676 (1989)

[4] M.Dellnitz, M.Golubitsky, M.Nicol: Symmetry of attractors and the Karhunen-Loeve decomposition, preprint 1993

[5] C.Foias, B.Nicolaenko, in preparation 1994

[6] M.Golubitsky, I.Stewart: Hopf bifurcations with dihedral group symmetry: coupled nonlinear oscillators. In: M.Golubitsky and J. Guckenheimer, eds: Multiparameter bifurcation theory, Contemp. Math. **56** 131 (1986)

[7] M. Krupa: Bifurcations of relative equilibria, SIAM J. Math. Anal. **21** (6) 1453 (1990)

[8] I.Melbourne, P.Chossat, M.Golubitsky Heteroclinic cycles involving periodic solutions in mode interactions with O(2)-symmetry Proc. Roy. Soc. Edinb. **113A** 315 (1989)

[9] M.Kirby, D.Armbruster: Reconstructing phase space for PDE simulations, ZAMP **43** 999 (1992)

[10] J.L.Lumley in "Atmospheric Turbulence and Radio Wave Propagation", eds. A.M.Yaglom, V.I.Tatarski (Nauka, Moscow), 166 (1967)

[11] B.Nicolaenko, Z.S. She: Symmetry-breaking homoclinic chaos in the Kolmogorov flows, in: Nonlinear World, Int. workshop on nonlinear and turbulent processes in physics, Kiev 1989, eds. V.G. Baryakhtar et al., 602 (1990)
Z.S. She, B.Nicolaenko: Temporal intermittency and turbulence production in the Kolmogorov flow, In: "Topological Fluid Mechanics" Eds. H.K. Moffatt, Cambridge University Press, 256 (1990)

B.Nicolaenko, Z.S. She: Symmetry breaking homoclinic chaos and vorticity bursts in periodic Navier-Stokes flows, Eur. J. Mech. B/Fluids **10**(2), 67 (1991)

[12] L.Sirovich: Turbulence and the dynamics of coherent structures, Parts I - III, Quarterly of Appl. Math. Vol XLV (3), 561 (1987)

[13] N.Smaoui, D.Armbruster, P.Chossat, B.Nicolaenko: in preparation 1994

OSCILLATOR NETWORKS WITH THE SYMMETRY
OF THE UNIT QUATERNION GROUP

PETER ASHWIN,* GERHARD DANGELMAYR,† IAN STEWART*
MICHAEL WEGELIN†

ABSTRACT In this note we consider oscillatory dynamics of a group of eight oscillators, coupled in such a way that their symmetry group is that of the unit quaternions. We investigate Hopf bifurcation and the isotropy subgroups for the weakly coupled limit. We find a generic scenario where structurally stable heteroclinic cycles connecting limit cycles arise; these may be asymptotically or essentially asymptotically stable, in the terminology of Krupa and Melbourne.

1 Introduction

Networks of coupled oscillators can possess many possible types of symmetries, typically coming from spatial symmetries of the system, but there may be more subtle symmetries coming from the coupling arrangements. In particular, one is not confined just to solid-body rotational symmetries (subgroups of $O(3)$): for example the full permutation group S_n on n objects is commonly met as the symmetry group of an oscillator network. This arises naturally in a group of n oscillators coupled in such a way that there is identical coupling from every cell to every other cell as, for example, in a series array of Josephson junctions.

In fact, we can consider the symmetry of a network of n couped oscillators to be a subgroup of S_n by examining the different types of coupling between oscillators, provided there are no internal symmetries of the oscillators, i.e. as long as they are generic. This approach has been investigated for the subgroups D_n and Z_n as well as S_n in [2] in the weak coupling limit, where the dynamics of the network are

*Maths Institute, University of Warwick, Coventry CV4 7AL, UK
†Institut für Informationsverarbeitung der Universität Tübingen, Köstlinstrasse 7, 72074 Tübingen, Germany

P. Chossat (ed.), Dynamics, Bifurcation and Symmetry, 35–48.

governed only by the phases of the oscillators. If we assume that the oscillators are near Hopf bifurcation as opposed to being weakly coupled, we can apply generic Hopf bifurcation theory with symmetry (see [11]). Symmetric networks where the oscillators also have internal symmetries have been studied in [6, 7, 8, 12].

We can realise many different subgroups of S_n by use of a little graph theory. The automorphism group of the coloured graph is defined to be its symmetry group. It is possible to design oscillator networks with symmetry group $\Gamma \leq S_n$ by constructing the Cayley graph of Γ (more precisely, the Cayley graph given a generating set of permutations). This allows us to investigate a class of physically realisable systems which have unusual mathematical properties; Stork [14] has used this to investigate variational steady-state problems.

In this paper, we consider systems with symmetry group Q, the unit quaternions. As pointed out by Golubitsky *et al.* [11], with such symmetries, one can have centre eigenspaces where the representation of the symmetry group commutes with a set of linear maps isomorphic to the field of Quaternions H (also known as the Hamiltonians).

We examine this system firstly in the limit of Hopf bifurcation and secondly, with weak coupling (using the setting of Ashwin and Swift [2]). The generic Hopf bifurcations with symmetry Q are exactly the same as Hopf bifurcations with D_4 investigated by Golubitsky and Stewart [9] and Swift [15]; the interpretation of the branches is however different. In the weakly couped case, it is possible to reduce the asymptotic dynamics to a flow on an eight-dimensional torus T^8 on which an assumption of weak coupling allows us to average and to introduce an extra S^1 symmetry. We consider the various invariant subspaces in this torus and show how they can generically give rise to global heteroclinic connections which can be structurally stable, i.e. connections between periodic orbits that are not destroyed by generic perturbations of the system preserving the symmetry $Q \times S^1$. Moreover, these connections can by dynamically attracting, repelling or *essentially attracting* as defined by Krupa and Melbourne [13]. Such heteroclinic cycles have a large measure set of initial conditions in a neighbourhood which have the cycle as their ω-limit set but also have a positive measure set of points in any neighbourhood of themselves that are repelled away from the cycle. We finish with a numerical example of such a heteroclinic cycle.

2 The Quaternion group

The eight element group of the unit quaternions $\mathbf{Q} = \{1, i, j, k, -1, -i, -j, -k\}$ is defined abstractly by the algebraic relations between its elements,

$$i^2 = j^2 = k^2 = ijk = -1. \tag{1}$$

Thus, $\mathbf{Q}$ is generated by, for example, the two elements i and j. The group $\mathbf{Q}$ has some interesting group theoretical properties, in that it is the smallest dicyclic group, the smallest Hamiltonian group and the smallest group of rank 1 [5]. As the smallest Hamiltonion group it is the smallest non-Abelian group all of whose subgroups are normal. As the smallest group of rank 1 it is the smallest non-Abelian group all of whose proper subgroups are Abelian. Furthermore, $\mathbf{Q}$ is the *only* finite non-Abelian group all of whose proper subgroups are Abelian and normal.

There is a close relationship between $\mathbf{Q}$, $\mathbf{H}$ and the two-dimensional special unitary group, $\mathbf{SU}(2)$. In fact, $\mathbf{SU}(2)$ is the group of isometries of the quaternions of unit norm. Previous work on the dynamics of systems with symmetries commuting with a nonabsolutely irreducibly represented group of quaternionic type has been confined to the case $\mathbf{SU}(2)$ by Cicogna and Gaeta [3, 4] who have studied Hopf bifurcation with this symmetry, and work of Stork [14] who has considered gradient problems with symmetry $\mathbf{Q}$. The latter author has shown how one can easily construct a concrete example of a problem commuting with symmetry $\mathbf{Q}$ using a graphical technique. We shall outline this for an example with oscillatory dynamics.

2.1 The Cayley Graph

Recall that a finitely generated discrete group G has a generating set $J \subset G$ if (a) J generates G, (b) J is finite and (c) $J = J^{-1}$. For the finite group that we shall consider, we can drop assumptions (b) and (c).

The Cayley graph of a group G with given generating set $J \subset G$ is a directed coloured graph whose vertices are the elements of the group. The generating set J is a set of "colours" of the directed edges, and there is an edge from a to b, $(a, b \in G)$ of colour $c \in J$ if and only if

$$a = cb.$$

At any given vertex there are two edges for each generator, one pointing to the vertex; the other pointing away. The graph is connected because J generates G.

The Cayley graph for $G = \mathbf{Q}$ is shown in Fig. 1. Thus, we may identify the vertices as cells with a certain dynamics and the edges as couplings between the cells, as discussed by Stork [14]. In this way we may construct a network which has precisely the symmetry of the group represented by the Cayley graph. If we number the elements $(1, i, -1, -i, k, j, -k, -j)$ sequentially by one up to eight, the action of $\mathbf{Q}$ on the cells can be written as the permutation group generated by

$$i := (1234)(5876), \tag{2}$$
$$j := (1638)(2547), \tag{3}$$
$$k := (1735)(2846). \tag{4}$$

Thus by assigning coupling g between cells related by i on the Cayley graph and h between cells related by j, we can construct the following pairwise coupled system in $\mathbf{R}^8$ with quaternionic symmetry.

$$
\begin{aligned}
\dot{x}_1 &= f(x_1) + g(x_4, x_1) + h(x_8, x_1) \\
\dot{x}_2 &= f(x_2) + g(x_1, x_2) + h(x_7, x_2) \\
\dot{x}_3 &= f(x_3) + g(x_2, x_3) + h(x_6, x_3) \\
\dot{x}_4 &= f(x_4) + g(x_3, x_4) + h(x_5, x_4) \\
\dot{x}_5 &= f(x_5) + g(x_6, x_5) + h(x_2, x_5) \\
\dot{x}_6 &= f(x_6) + g(x_7, x_6) + h(x_1, x_6) \\
\dot{x}_7 &= f(x_7) + g(x_8, x_7) + h(x_4, x_7) \\
\dot{x}_8 &= f(x_8) + g(x_5, x_8) + h(x_3, x_8),
\end{aligned}
\tag{5}
$$

where $f : \mathbf{R} \to \mathbf{R}$, and $g, h : \mathbf{R}^2 \to \mathbf{R}$. As discussed by Ashwin and Stork [1], this network has the symmetry of the quaternions for generic f, g, h in the sense that the isotropy of this vector field under the action of $\mathbf{O}(8)$ is generically $\mathbf{Q}$.

3 Hopf Bifurcation

By the work of Golubitsky and Stewart [9, 10], we can consider generic one-parameter Hopf bifurcations with symmetry $\mathbf{Q}$ by looking at the complex irreducible representations of $\mathbf{Q}$. These are of one or two complex dimensions.

Recall that the linear orthogonal representation of a group Γ

$$\alpha_\Gamma : \Gamma \times W \to W$$

on the complex vector space W is irreducible if and only if the only Γ-invariant subspaces of W are trivial; i.e. $\{0\}$ or W itself. Note that (a) there need be no faithful irreducible representation, and (b) this is typical. The "amount by which the representation fails to be faithful" is the kernel of the action α_Γ.

The group $\mathbf{Q}$ has five complex irreducible representations: four of them are one-dimensional and the other is two-dimensional. The one-dimensional cases can

isotropy in $\mathbf{Q} \times \mathbf{S}^1$	isotropy in $\mathbf{D}_4 \times \mathbf{S}^1$	Fix	$\dim_{\mathbf{C}}$ Fix	Name ($\mathbf{D}_4$)
$\mathbf{Q} \times \mathbf{S}^1$	$\mathbf{D}_4 \times \mathbf{S}^1$	$(0,0)$	0	Trivial solution
$\tilde{\mathbf{Z}}_4^i$	$\tilde{\mathbf{Z}}_4(\rho)$	$(z,0)$	1	Rotating Wave
$\tilde{\mathbf{Z}}_4^j$	$\tilde{\mathbf{Z}}_2(\rho^2) \times \mathbf{Z}_2(\kappa)$	(z,z)	1	Edge Oscillation
$\tilde{\mathbf{Z}}_4^k$	$\tilde{\mathbf{Z}}_2(\rho^2) \times \mathbf{Z}_2(\rho\kappa)$	(z,iz)	1	Vertex Oscillation
$\tilde{\mathbf{Z}}_2$	$\tilde{\mathbf{Z}}_2(\rho^2)$	(w,z)	2	Submaximal

Table 1: This table relates the isotropies of points in $\mathbf{C}^2$ for the identical actions of $\mathbf{Q} \times \mathbf{S}^1$ and $\mathbf{D}_4 \times \mathbf{S}^1$.

be interpreted as Hopf bifurcation with trivial or $\mathbf{Z}_2$ symmetry, corresponding to a quotient group of $\mathbf{Q}$, and will not be discussed further.

The standard irreducible action of $\mathbf{Q}$ on $\mathbf{C}^2$ is given by

$$
\begin{aligned}
i(z_+, z_-) &= (iz_+, -iz_-), \\
j(z_+, z_-) &= (iz_-, iz_+) \text{ and} \\
k(z_+, z_-) &= (-z_-, z_+).
\end{aligned}
$$

(careful; i has two meanings here!) and there is a natural "phase shift" action of $\mathbf{S}^1$ given by

$$
R_\phi(z_+, z_-) = (e^{i\phi}z_+, e^{i\phi}z_-).
$$

for $\phi \in \mathbf{S}^1$. In fact, the action of $\mathbf{Q} \times \mathbf{S}^1$ is identical to the action of $\mathbf{D}_4 \times \mathbf{S}^1$ investigated by [9, 15]; this action is generated by

$$
\begin{aligned}
\kappa(z_+, z_-) &= (z_-, z_+) \\
\rho(z_+, z_-) &= (iz_+, -iz_-),
\end{aligned}
$$

where $\rho^4 = \kappa^2 = 1$ and $\rho\kappa = \kappa\rho^3$. The kernel of this action is the 2–cycle in $\mathbf{D}_4 \times \mathbf{S}^1$ generated by (ρ^2, π) whereas the kernel of the action of $\mathbf{Q} \times \mathbf{S}^1$ is generated by $(-1 = i^2 = j^2 = k^2, \pi)$. It is then possible to check that

$$
\mathbf{Q} \times \mathbf{S}^1 / \ker \alpha_{\mathbf{Q} \times \mathbf{S}^1} \equiv \mathbf{D}_4 \times \mathbf{S}^1 / \ker \alpha_{\mathbf{D}_4 \times \mathbf{S}^1}.
$$

This means that we can apply the results for $\mathbf{D}_4$ Hopf bifurcation with merely a re-interpretation of the branches. Table 1 compares the different maximal and submaximal symmetries of possible solutions. We write $\tilde{H}$ to mean that the subgroup H in $\mathbf{Q}$ or $\mathbf{D}_4$ is twisted by a homeomorphism into $\mathbf{S}^1$ (see [11]). The normal form to third order giving the behaviour of generic branching at such Hopf bifurcations is that for $\mathbf{D}_4$ symmetry, i.e.

$$\begin{aligned}
\dot{z}_+ &= (\lambda + i\omega)z_+ + (A(|z_+|^2 + |z_-|^2) + B|z_+|^2)z_+ + c\bar{z}_+z_-^2 \\
\dot{z}_- &= (\lambda + i\omega)z_- + (A(|z_+|^2 + |z_-|^2) + B|z_-|^2)z_- + c\bar{z}_-z_+^2.
\end{aligned}$$

and generic bifurcation theory implies that this will determine the branching behaviour if the nondegeneracy conditions on A, B and C listed in §3.10 of [15] are satisfied (λ is a bifurcation parameter and ω gives the perturbation of the period from the linear frequency at Hopf bifurcation).

4 Weak Coupling

Now we consider another setting where our system is governed by an ODE of the form

$$\dot{x}_i = f(x_i) + \epsilon g_i(x_1, \cdots, x_8)$$

for $i = 1, \ldots, 8$, $x_i \in X$ and g commuting with the permutation action of $\mathbf{Q}$ on X^8. We assume that f and g_i are smooth (infinitely differentiable). The parameter $\epsilon \geq 0$ represents the coupling strength. We assume that $\dot{x} = f(x)$ has an hyperbolic stable limit cycle.

In the case of weak coupling, this means that there is a natural reason why we should not just look at irreducible representations of $\mathbf{Q}$. The fact that there are 8 stable hyperbolic limit cycles in the limit of $\epsilon = 0$ means that the asymptotic dynamics of the system factors into the asymptotic dynamics of eight limit cycles, and so it can be expressed as a flow on an eight-dimensional torus, $\mathbf{T}^8$.

The assumption of hyperbolicity of the individual limit cycles means that $\mathbf{T}^8$ is normally hyperbolic and so will persist for small enough values of $\epsilon > 0$ (although it may progressively lose differentiability), and this justifies expressing the dynamics of the system as an ODE in terms of eight phases, i.e. an ODE on $\mathbf{T}^8$ which is equivariant under the action of $\mathbf{Q}$ by permutation.

Moreover, for small values of ϵ, it is possible to *average* the equations and introduce an approximate decoupling between the fast variation of the phases and the slow variation of the phase differences. This can be seen as introducing an $\mathbf{S}^1$ phase shift symmetry which acts on $\mathbf{T}^8$ by translation along the diagonal;

$$R_\theta(\phi_1, \cdots, \phi_8) := (\phi_1 + \theta, \cdots, \phi_8 + \theta),$$

for $\theta \in \mathbf{S}^1$. Figure 2 shows schematically the regions of validity of the Hopf bifurcation and weak coupling approximations.

4.1 Isotropy types for weak coupling

We recall that we now have an ODE on $\mathbf{T}^8$ that is equivariant under an action of $\mathbf{Q} \times \mathbf{S}^1$; it is of importance to classify the isotropy types of points under this action, and we list this in Table 2.

Σ	Fix(Σ)	Generators	dim Fix(Σ)
$\mathbf{Q}$	$(0,0,0,0,0,0,0,0)$	$(i,0), (j,0)$	0
$\tilde{Q}^i$	$(0,0,0,0,\pi,\pi,\pi,\pi)$	$(j,\pi), (k,\pi)$	0
$\tilde{Q}^j$	$(0,\pi,0,\pi,\pi,0,\pi,0)$	$(i,\pi), (k,\pi)$	0
$\tilde{Q}^k$	$(0,\pi,0,\pi,0,\pi,0,\pi)$	$(i,\pi), (j,\pi)$	0
$\mathbf{Z}_4^i$	$(0,0,0,0,\phi,\phi,\phi,\phi)$	$(i,0)$	1
$\mathbf{Z}_4^j$	$(0,\phi,0,\phi,\phi,0,\phi,0)$	$(j,0)$	1
$\mathbf{Z}_4^k$	$(0,\phi,0,\phi,0,\phi,0,\phi)$	$(k,0)$	1
$\tilde{\mathbf{Z}}_4^i$	$(0,\pi,0,\pi,\phi,\phi+\pi,\phi,\phi+\pi)$	(i,π)	1
$\tilde{\mathbf{Z}}_4^j$	$(0,\phi,0,\phi,\phi+\pi,\pi,\phi+\pi,\pi)$	(j,π)	1
$\tilde{\mathbf{Z}}_4^k$	$(0,\phi,0,\phi,\pi,\phi+\pi,\pi,\phi+\pi)$	(k,π)	1
$\tilde{\mathbf{Z}}_4^{i/2}$	$(0,\frac{3\pi}{2},\pi,\frac{\pi}{2},\phi,\phi+\frac{\pi}{2},\phi+\pi,\phi+\frac{3\pi}{2})$	$(i,\frac{\pi}{2})$	1
$\tilde{\mathbf{Z}}_4^{j/2}$	$(0,\phi,\pi,\phi+\pi,\phi+\frac{3\pi}{2},\frac{3\pi}{2},\phi+\frac{\pi}{2},\frac{\pi}{2})$	$(j,\frac{\pi}{2})$	1
$\tilde{\mathbf{Z}}_4^{k/2}$	$(0,\phi,\pi,\phi+\pi,\frac{\pi}{2},\phi+\frac{\pi}{2},\frac{3\pi}{2},\phi+\frac{3\pi}{2})$	(k,π)	1
$\mathbf{Z}_2$	$(0,\phi_1,0,\phi_1,\phi_2,\phi_3,\phi_2,\phi_3)$	$(-1,0)$	3
$\tilde{\mathbf{Z}}_2$	$(0,\phi_1,\pi,\phi_1+\pi,\phi_2,\phi_3,\phi_2+\pi,\phi_3+\pi)$	$(-1,\pi)$	3

Table 2: Isotropy subgroups and fixed point subspaces for the $\mathbf{Q} \times \mathbf{S}^1$ action on $\mathbf{T}^8$. Note that points in Fix(Σ) are fibred by circles in the $(1,\cdots,1)$ direction on $\mathbf{T}^8$; we show where the fixed point space intersects the codimension one torus $\phi_1 = 0$.

We shall be interested in particular in the three-dimensional space Fix($\mathbf{Z}_2$); note that there are several fixed point subspaces of one- and zero-dimensions contained in this space; these are fixed by elements in the normaliser of Fix($\mathbf{Z}_2$) rather than in $\mathbf{Z}_2$ itself.

We can define coordinates in Fix($\mathbf{Z}_2$) by taking a basis

$$
\begin{aligned}
e_1 &= \tfrac{1}{2}(1,1,1,1,-1,-1,-1,-1), \\
e_2 &= \tfrac{1}{2}(1,-1,1,-1,-1,1,-1,1), \\
e_3 &= \tfrac{1}{2}(1,-1,1,-1,1,-1,1,-1),
\end{aligned}
$$

and consider the invariant space spanned by $\{e_1, e_2, e_3\}$, parameterised by $\{\theta_1, \theta_2, \theta_3\}$:

$$\sum_{n=1}^{3} \theta_n e_n. \tag{6}$$

4.2 An example vector field on Fix($\mathbf{Z}_2$)

Using the coordinates (6) in Fix($\mathbf{Z}_2$), an example vector field on this space is given by

$$
\begin{aligned}
\dot{\theta}_1 &= \sin\theta_1 \cos\theta_2 + \epsilon \sin 2\theta_1 \cos 2\theta_2 \\
\dot{\theta}_2 &= \sin\theta_2 \cos\theta_3 + \epsilon \sin 2\theta_2 \cos 2\theta_3 \\
\dot{\theta}_3 &= \sin\theta_3 \cos\theta_1 + \epsilon \sin 2\theta_3 \cos 2\theta_1 + q(1 + \cos(\theta_1 - \theta_2))\sin 2\theta_3,
\end{aligned}
\tag{7}
$$

with ϵ and q real constants. We claim that this vector field contains examples of structurally stable, attracting heteroclinic cycles which may be asymptotically stable or essentially asymptotically stable, depending on the values of ϵ and q.

Firstly, note that the planes $\theta_i = 0 \pmod{\pi}$ are invariant under this flow. In fact, this is not necessary for a vector field with this symmetry; there are no two-dimensional fixed point spaces in Fix($\mathbf{Z}_2$), only the one-dimensional Fix($\mathbf{Z}_4$), Fix($\tilde{\mathbf{Z}}_4$) and the zero-dimensional Fix($\mathbf{Q}$) and Fix($\tilde{\mathbf{Q}}$). As mentioned already, these symmetries are not in $\mathbf{Z}_2$, though they are in the normalizer of Fix($\mathbf{Z}_2$). We can assume without loss of genericity that the space Fix($\mathbf{Z}_2$) is normally attracting for the dynamics, and so the stabilities or otherwise of the dynamics within the fixed point space determine the stabilities in the full system.

The eigenvalues of the flow at the four different (non-conjugate) zero-dimensional fixed points are given by

Fix	$(\theta_1, \theta_2, \theta_3)$	λ_1	λ_2	λ_3
$\mathbf{Q}$	$(0,0,0)$	$1-2\epsilon$	$1-2\epsilon$	$1-2\epsilon$
$\tilde{\mathbf{Q}}^i$	$(\pi,0,0)$	$-1+2\epsilon$	$1+2\epsilon$	$-1+2\epsilon$
$\tilde{\mathbf{Q}}^j$	$(0,\pi,0)$	$-1+2\epsilon$	$-1+2\epsilon$	$1+2\epsilon$
$\tilde{\mathbf{Q}}^k$	$(0,0,\pi)$	$1+2\epsilon$	$-1+2\epsilon$	$-1+2\epsilon+4q.$

What is more, there are two-dimensional fixed point spaces Fix($\mathbf{Z}_2$) and Fix($\tilde{\mathbf{Z}}_2$) which connect these fixed points as shown in figure 3. On these fixed point spaces, there are no extra (dynamically) fixed points provided

$$
\begin{aligned}
&|\epsilon| < \tfrac{1}{2} \text{ and} \\
&|\epsilon + 2q| < \tfrac{1}{2}.
\end{aligned}
$$

Thus, if these conditions are satisfied, we can have a heteroclinic connection along one-dimensional subspaces in the following way:

$$\ldots \xrightarrow{\check{Z}_4^k} \tilde{Q}^i \xrightarrow{\check{Z}_4^j} \tilde{Q}^k \xrightarrow{\check{Z}_4^i} \tilde{Q}^j \xrightarrow{\check{Z}_4^k} \ldots$$

where the connection between $\tilde{Q}^i$ and $\tilde{Q}^j$ goes along the fixed point space of $\check{Z}^k$ etc.

The stability or otherwise of this cycle (which is clearly structurally stable, as it restricts to structurally stable dynamics *within* the fixed point spaces) is determined using the criteria of Krupa and Melbourne [13]. This allows us to state that the stability of the heteroclinic cycle is governed by ρ, where $\rho = \rho_1 \rho_2 \rho_3$, and

$$\rho_i = \min\{c_i/e_i, 1 - t_i/e_i\}$$

with e_i the expanding eigenvector at the ith point on the cycle, $-c_i$ is the contracting eigenvector and t_i is the tangential eigenvector of the linearisation. For the cycle above, we have

$$\begin{aligned}
\rho_1 &= \rho_2 = \frac{1-2\epsilon}{1+2\epsilon}, \\
\rho_3 &= \begin{cases} \frac{1-2\epsilon}{1+2\epsilon} & \text{for } q < \frac{1}{4} + \frac{\epsilon}{2} \\ \frac{2-4q}{1+2\epsilon} & \text{for } q > \frac{1}{4} + \frac{\epsilon}{2}. \end{cases}
\end{aligned} \tag{8}$$

¿From Theorem 2.4 of [13], we can characterise the stability of the heteroclinic cycle depending on the parameters ϵ and q in the following way.

- asymptotically stable for $-\frac{1}{2} < \epsilon < 0$ and $q < \frac{1}{4} + \frac{\epsilon}{2}$,

- unstable but essentially asymptotically stable for $-\frac{1}{2} < \epsilon < 0$ and $\frac{1}{4} + \frac{\epsilon}{2} < q < \frac{1}{2} - \frac{(1+2\epsilon)^3}{4(1-2\epsilon)^2}$; this means that a large measure set of points in a neighbourhood of the cycle are attracted to it, while a positive measure set of points are repelled,

- unstable for $\frac{1}{2} > \epsilon > 0$.

Note that for any $\frac{1}{2} < \epsilon < 0$ we have $\frac{1}{4} + \frac{\epsilon}{2} < \frac{1}{2} - \frac{(1+2\epsilon)^3}{4(1-2\epsilon)^2}$ and so there exist values of q for which there exist essentially asymptotically stable heteroclinic connections. In figure 4 we show an example time series obtained at $\epsilon = -0.1$ and $q = 0.21$, using DSTOOL. As can be seen, there is evidence of an attracting heteroclinic cycle even though the linear stability of $\text{Fix}(\tilde{Q}^k)$ has an expanding transverse eigenvalue. This is typical of essential asymptotically stability.

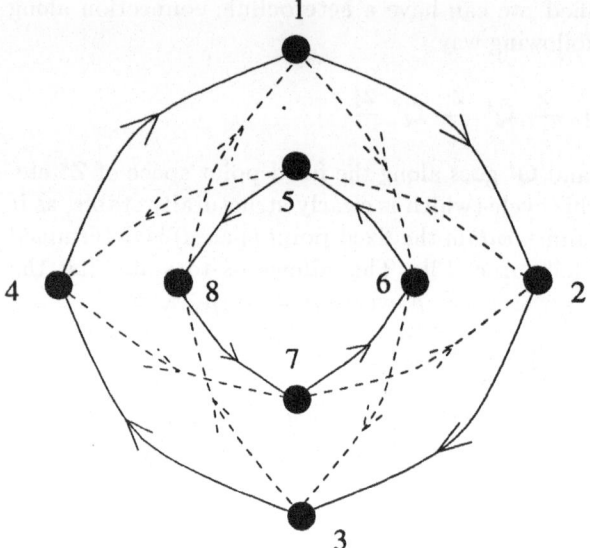

Figure 1: A Cayley graph of **Q** generated by i and j. Solid arrows correspond to left multiplication with i, dashed arrows to left multiplication with j.

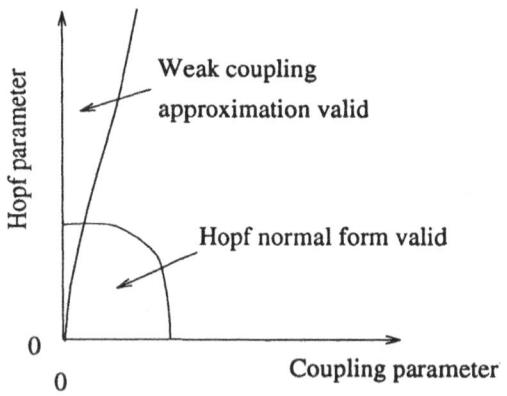

Figure 2: This figure schematically shows the regions of validity of the Hopf and the weak coupling analyses. Note that the weak coupling approximation can be valid for strongly nonlinear oscillations.

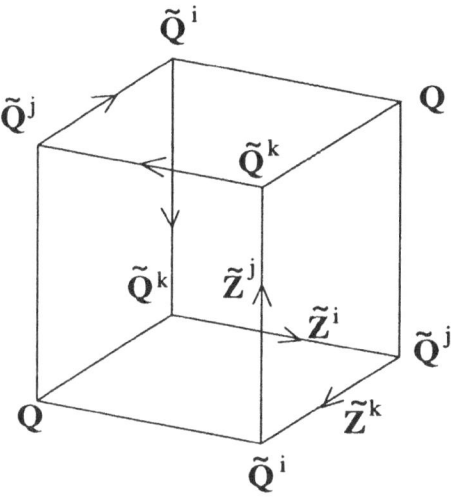

Figure 3: Within the three-dimensional space of $\mathrm{Fix}(\mathbf{Z}_2)$ for the weak coupled oscillators, there can be a structurally stable heteroclinic cycle of the form shown here.

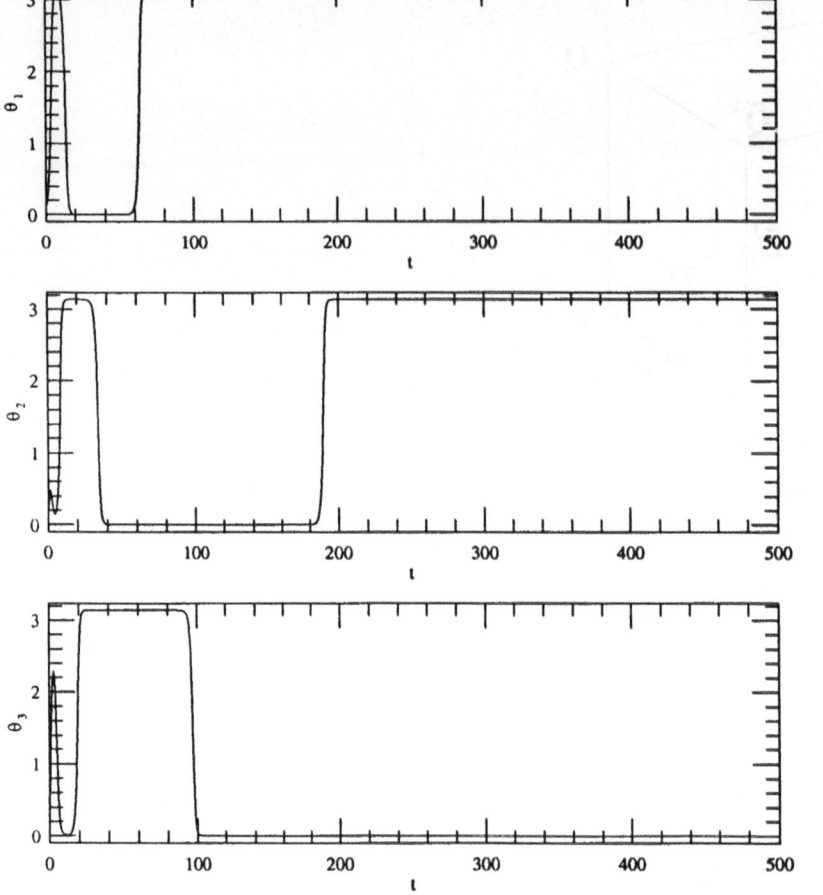

Figure 4: Equation (7) produces times series of this form for $\epsilon = -0.1$, $q = 0.21$. Although one of the fixed points on the heteroclinic chain is transversely unstable, there is a positive measure set of initial conditions that are attracted to the cycle, as evidenced here.

5 Discussion

In this paper we considered oscillator networks with a coupling scheme that induces equivariance of the governing equations under a permutation representation of the unit quaternion group. Near a Hopf bifurcation with four-(real)-dimensional centre eigenspace, the dynamics of these networks is determined by the normal form for a Hopf bifurcation with D_4 symmetry which has been already studied in detail [9, 15]. These results can be used to classify the basic periodic branches and also to predict more complicated behaviour. In the case of weak coupling we have shown by an example that there can be essentially asymptotically stable, structurally stable heteroclinic cycles for an open region in the space of ϵ, q parameters. Because we are working with a weak coupling assumption, there is the added bonus that the heteroclinic cycles pass along one-dimensional fixed point subspaces that are analytically know; other examples of structurally stable heteroclinic cycles have been found, but their analytic form in the phase space is not usually known. We are continuing our study of this system, in particular, relating the theory to electronic simulations of a network of neurons coupled with this symmetry.

This paper was written with the support of a British Council/DAAD Anglo-German Research Collaboration grant and the Science and Engineering Research Council of the UK.

References

[1] P. Ashwin and P. Stork. Permissible symmetries of coupled cell networks. *Math. Proc. Camb. Phil. Soc.*, 1994. (to appear).

[2] P. Ashwin and J.W. Swift. The dynamics of n identical oscillators with symmetric coupling. *Journal of Nonlinear Science*, 2:69–108, 1992.

[3] G. Cicogna, G. Gaeta and P. Rossi. Remarks on bifurcation with symmetry, gradient property, and reducible representations. *J. Math. Phys.* **27**:447–450, 1984.

[4] G. Cicogna. On the 'quaternionic' bifurcation. *J. Phys. A: Math. Gen.* **18**:L829–L832, 1985.

[5] H.S.M. Coxeter and W.O.J. Moser. *Generators and relations for discrete groups.* Ergebnisse der Mathematik und ihrere Grenzgebiete **14**, Springer, Berlin, 1965.

[6] G. Dangelmayr, W. Güttinger and M. Wegelin. Hopf bifurcation with $\mathbf{D_3 \times D_3}$ symmetry, *Z.A.M.P.* **44** (to appear) 1993.

[7] B. Dionne, M. Golubitsky and I.N Stewart. Coupled cells with internal symmetry, Part I: Wreath products, Preprint, University of Houston, 1994.

[8] B. Dionne, M. Golubitsky and I.N Stewart. Coupled cells with internal symmetry, Part II: Direct products, Preprint, University of Houston, 1994.

[9] M. Golubitsky and I.N. Stewart. Hopf bifurcation with dihedral group symmetry. In *Multiparameter bifurcation theory*, volume 56 of *Contemporary Maths*, pages 131–173. AMS, Providence, RI, 1985.

[10] M. Golubitsky and I.N. Stewart. An algebraic criterion for symmetric hopf bifurcation. *Proc. R. Soc. Lond. A*, **440**:727–732, 1993.

[11] M. Golubitsky, I.N. Stewart, and D. Schaeffer. *Singularities and Groups in Bifurcation Theory Volume 2*, volume 69 of *App. Math. Sci.* Springer, New York, 1988.

[12] M. Golubitsky, I.N. Stewart, and B. Dionne. Coupled cells: wreath products and direct products, *These proceedings*, 1994.

[13] M. Krupa and I. Melbourne. Nonasymptotically stable attractors in $\mathbf{O(2)}$ mode interactions. Preprint UH/MD-163, University of Houston, 1993.

[14] P. Stork. *Statische Verzweigung in Gradientenfeldern mit Symmetrien vom komplexen oder quaternionischen Typ mit numerischer Behandlung*, volume 11(7) of *Wissenschaftliche Beiträge aus europäischen Hochschulen*. Verlag an der Lottbeck, 1993.

[15] J.W. Swift. Hopf bifurcation with the symmetry of the square. *Nonlinearity*, 1:333–377, 1988.

AN INVESTIGATION OF A MODE INTERACTION INVOLVING PERIOD-DOUBLING AND SYMMETRY-BREAKING BIFURCATIONS

P. J. ASTON
Department of Mathematical
and Computing Sciences
University of Surrey
U.K.

Abstract. Mode interactions in iterated maps with a Z_2 symmetry involving a period doubling bifurcation and a symmetry breaking bifurcation of fixed points are considered. In an enlarged system of equations, a period doubling bifurcation can be considered as a Z_2 symmetry breaking bifurcation so that the standard theory for $Z_2 \times Z_2$ mode interactions can be used. However, a Lyapunov-Schmidt reduction is required to reduce the problem to two dimensions before applying the theory. In this case, only one of the two coordinate axes is invariant under the flow. However, we show that there is another flow invariant curve which is locally quadratic. An example showing all these features is then considered.

1. Introduction

In this paper, we consider a particular mode interaction in iterated maps between two different types of bifurcations. Mode interactions were first considered by Bauer, Keller and Reiss (1975) in the context of differential equations and much work has been done in this area (see for example Iooss and Langford (1980), Golubitsky and Schaeffer (1985, Ch X), Dangelmayr (1986)). More recently, bifurcation and chaos in iterated maps with symmetry has been considered (Chossat and Golubitsky (1988a,b)). We now combine these two areas by considering a mode interaction in iterated maps with a simple Z_2 symmetry which involves a period doubling bifurcation (in the fixed point space) and a symmetry breaking bifurcation of fixed points. In an appropriate framework, a period doubling bifurcation can be considered as a Z_2 symmetry breaking bifurcation and so a system can be constructed which has $Z_2 \times Z_2$ symmetry. The standard theory of Golubitsky and Schaeffer (1985) for such problems can then be applied.

P. Chossat (ed.), Dynamics, Bifurcation and Symmetry, 49–58.

2. Mode Interactions

In this section, we consider the mode interaction described in the Introduction. An enlarged system must be set up in order to realise the $Z_2 \times Z_2$ symmetry and a Lyapunov-Schmidt reduction is then performed to reduce the problem into standard form.

Consider the equations

$$
\begin{aligned}
x^{n+1} &= f(x^n, y^n, \lambda, \mu), \\
y^{n+1} &= g(x^n, y^n, \lambda, \mu).
\end{aligned}
\tag{2.1}
$$

We assume that these equations have the trivial solution $(x, y) = (0, 0)$ for all λ and μ together with a Z_2 symmetry generated by the reflection S which acts as

$$
S \begin{bmatrix} x \\ y \end{bmatrix} = \begin{bmatrix} x \\ -y \end{bmatrix}.
$$

Clearly, $\text{Fix}(Z_2) = \{(x, 0) : x \in \mathbf{R}\}$ and we will use the word 'symmetric' to refer to this fixed point space. In order to study period doubling bifurcations, steady state bifurcation theory can be applied to the algebraic equations

$$
\begin{aligned}
x &= f(f(x, y, \lambda, \mu), g(x, y, \lambda, \mu), \lambda, \mu), \\
y &= g(f(x, y, \lambda, \mu), g(x, y, \lambda, \mu), \lambda, \mu).
\end{aligned}
\tag{2.2}
$$

However, the additional Z_2 symmetry which we require, associated with period doubling bifurcations, cannot be realised in this framework. Thus, we consider a system of four equations given by

$$
\begin{aligned}
f(x_2, y_2, \lambda, \mu) - x_1 &= 0 \\
g(x_2, y_2, \lambda, \mu) - y_1 &= 0 \\
f(x_1, y_1, \lambda, \mu) - x_2 &= 0 \\
g(x_1, y_1, \lambda, \mu) - y_2 &= 0
\end{aligned}
\tag{2.3}
$$

Solutions of these equations correspond to fixed points and period 2 points of equations (2.1) and have symmetry $Z_2 \times Z_2$ generated by S_1 and S_2 which act by

$$
S_1 \begin{bmatrix} x_1 \\ y_1 \\ x_2 \\ y_2 \end{bmatrix} = \begin{bmatrix} x_1 \\ -y_1 \\ x_2 \\ -y_2 \end{bmatrix}, \qquad
S_2 \begin{bmatrix} x_1 \\ y_1 \\ x_2 \\ y_2 \end{bmatrix} = \begin{bmatrix} x_2 \\ y_2 \\ x_1 \\ y_1 \end{bmatrix}.
$$

We use the notation $\text{Fix}(S_1)$ to denote the fixed point space associated with the group generated by S_1 etc. The fixed point spaces associated with this symmetry give rise to different types of solutions of equations (2.1) and are shown in Table 1.

Fixed Point Space	Solution Type
$\text{Fix}(S_1, S_2)$	Symmetric fixed points
$\text{Fix}(S_1)$	Symmetric period 2 points
$\text{Fix}(S_2)$	Non-symmetric fixed points
$\text{Fix}(I)$	Non-symmetric period 2 points

Table 1

To put the S_2 symmetry in more standard form, we perform a change of basis and define

$$U = \begin{bmatrix} u_1 \\ u_2 \\ u_3 \\ u_4 \end{bmatrix} = \begin{bmatrix} x_1 + x_2 \\ y_1 + y_2 \\ x_1 - x_2 \\ y_1 - y_2 \end{bmatrix}.$$

The action of the S_1 symmetry on U is then unchanged but the S_2 symmetry acts on U by

$$S_2 \begin{bmatrix} u_1 \\ u_2 \\ u_3 \\ u_4 \end{bmatrix} = \begin{bmatrix} u_1 \\ u_2 \\ -u_3 \\ -u_4 \end{bmatrix}.$$

The modified equations

$$G(U, \lambda, \mu) - U = 0 \tag{2.4}$$

are then easily obtained from equations (2.3).

We now suppose that there is a two parameter path of period doubling bifurcations from the trivial symmetric fixed point to symmetric period 2 points of (2.1). The reflectional symmetry of equations (2.1) implies that the Jacobian of these equations at a symmetric fixed point is diagonal and this path is then defined by $f_x^* = -1$ where the asterisk denotes evaluation at the trivial fixed point. The mode interaction we consider occurs at a point on the two parameter path when the other diagonal entry of the Jacobian, namely g_y^* has the value $+1$. We assume that this occurs at $(\lambda, \mu) = (0, 0)$.

We now want to reduce the four equations (2.4) to a system of only two equations at a mode interaction. Now the Jacobian of equations (2.4) at a symmetric fixed point is diagonal and is given by $G_U^* - I = \text{diag}(f_x^* - 1, g_y^* - 1, -f_x^* - 1, -g_y^* - 1)$. Thus, if $f_x^* = -1$ and $g_x^* = +1$ then $G_U^* - I$ has a two-dimensional null space and so we use a Lyapunov-Schmidt reduction to reduce from four to two equations. Natural coordinates for this null space are u_2 and u_3. Thus we write

$$u_1 = h_1(u_2, u_3, \lambda, \mu)$$
$$u_4 = h_4(u_2, u_3, \lambda, \mu)$$

where h_1 and h_4 contain quadratic and higher order terms in u_2 and u_3 and satisfy the symmetry relations

$$
\begin{aligned}
h_1(u_2, -u_3, \lambda, \mu) &= h_1(-u_2, u_3, \lambda, \mu) = h_1(u_2, u_3, \lambda, \mu), \\
h_4(u_3, -u_4, \lambda, \mu) &= h_4(-u_3, u_4, \lambda, \mu) = -h_4(u_3, u_4, \lambda, \mu).
\end{aligned}
\tag{2.5}
$$

The first and last equations of (2.4) are used to find the low order terms of h_1 and h_4 and these are then substituted into the second and third equations to give a system of two equations in u_2 and u_3 with $Z_2 \times Z_2$ symmetry. In the reduced problem, the trivial solution corresponds to the branch of trivial symmetric fixed points while the two single mode branches correspond to symmetric period 2 points ($u_3 \neq 0$) and non-symmetric fixed points ($u_2 \neq 0$) and the mixed mode solutions correspond to non-symmetric period 2 points.

To be more precise, consider the equations

$$
\begin{aligned}
x^{n+1} &= -x^n + a_2(x^n)^2 + a_3 x^n \lambda + a_4 x^n \mu + a_5(y^n)^2 \\
y^{n+1} &= y^n(1 + b_1 x^n + b_2 \lambda + b_3 \mu + b_4(x^n)^2 + b_5 x^n \lambda + b_6 x^n \mu + b_7(y^n)^2)
\end{aligned}
\tag{2.6}
$$

which have the symmetry S, a trivial fixed point for all λ and μ and a mode interaction of the required type at $(\lambda, \mu) = (0, 0)$. Equations (2.4) in the U coordinates are now given by

$$
\begin{aligned}
-4u_1 + a_2(u_1^2 + u_3^2) + 2a_3 u_1 \lambda + 2a_4 u_1 \mu + a_5(u_2^2 + u_4^2) &= 0 \\
2b_1(u_1 u_2 + u_3 u_4) + 4b_2 u_2 \lambda + 4b_3 u_2 \mu + b_4(u_1^2 u_2 + u_2 u_3^2 + 2u_1 u_3 u_4) & \\
+ 2b_5(u_1 u_2 \lambda + u_3 u_4 \lambda) + 2b_6(u_1 u_2 \mu + u_3 u_4 \mu) + b_7(3u_2 u_4^2 + u_2^3) &= 0 \\
a_2 u_1 u_3 + a_3 u_3 \lambda + a_4 u_3 \mu + a_5 u_2 u_4 &= 0 \\
8u_4 + 2b_1(u_2 u_3 + u_1 u_4) + 4b_2 u_4 \lambda + 4b_3 u_4 \mu + b_4(u_1^2 u_4 + u_3^2 u_4 + 2u_1 u_2 u_3) & \\
+ 2b_5(u_2 u_3 \lambda + u_1 u_4 \lambda) + 2b_6(u_2 u_3 \mu + u_1 u_4 \mu) + b_7(3u_2^2 u_4 + u_4^3) &= 0
\end{aligned}
$$

Note that there are no linear terms in the second and third equations as anticipated. The symmetry constraints (2.5) imply that the quadratic terms of the functions h_1 and h_4 are given by

$$
\begin{aligned}
h_1(u_2, u_3, \lambda, \mu) &= c_1 u_2^2 + c_2 u_3^2, \\
h_4(u_2, u_3, \lambda, \mu) &= c_3 u_2 u_3
\end{aligned}
$$

Substituting these for u_1 and u_4 in the first and last equations, equating powers and solving for the c_i coefficients gives

$$
\begin{aligned}
u_1 &= \frac{1}{4}(a_2 u_3^2 + a_5 u_2^2), \\
u_4 &= -\frac{1}{4} b_1 u_2 u_3
\end{aligned}
$$

and substituting these into the second and third equations leads to the two bifurcation equations

$$u_2[(a_5b_1 + 2b_7)u_2^2 + (-b_1^2 + 2b_4 + a_2b_1)u_3^2 + 8b_2\lambda + 8b_3\mu] = 0,$$
$$-u_3[(a_2a_5 - b_1a_5)u_2^2 + a_2^2u_3^2 + 4a_3\lambda + 4a_4\mu] = 0, \qquad (2.7)$$

retaining only the terms required for the normal form. Note that none of the cubic terms in h_1 and h_4 affect these low order terms. The solutions of these equations give the local structure of the bifurcations involved in the mode interaction of the original equations (2.1).

We now consider stability of all the branches involved in the mode interactions. All of the bifurcations involved in the mode interaction obey the usual stability rule that if the main branch is stable subcritically, then the bifurcating branch is stable if it is supercritical. Also, Hopf bifurcations arising from the mode interaction are not possible on the branch of symmetric period 2 points, since the Jacobian is diagonal on such solutions, nor on the branch of non-symmetric fixed points, since the two eigenvalues of the Jacobian in this case are near to $+1$ and -1 and so must be real. However, Hopf bifurcations are possible on the mixed mode solutions. This behaviour is the same as for the standard $Z_2 \times Z_2$ mode interaction for vector fields and so the stability results in this case will be identical. This also implies the existence of Hopf bifurcations on branches of non-symmetric period 2 solutions in this situation which will occur in the same cases as for vector fields.

The behaviour of such systems near to the trivial solution in the neighbourhood of the mode interaction is easily described. The linearisation of equations (2.1) about the trivial solution involves the diagonal Jacobian matrix $J = \mathrm{diag}(\alpha, \beta)$ with $\alpha = f_x^0$ and $\beta = g_y^0$ where superscript zero denotes evaluation at $x = y = 0$ and for λ and μ small. Clearly the behaviour depends on the sign of α and β and their values relative to the stability boundaries ± 1. As $f_x^* = -1$, then we always have $\alpha < 0$ so that x always oscillates between positive and negative values. Also, as $g_y^* = +1$ then $\beta > 0$ and so y behaves monotonically. Typical behaviour in one particular case, showing all possible types of fixed points and period 2 points is shown in Fig. 1. Note that only one pair of non-symmetric period 2 points are shown. The diagrams can be extended by reflection in the x-axis.

In the standard mode interaction problem, the x and y axes are invariant under the flow since they correspond to fixed point spaces. In this case however, equations (2.7) are only valid for studying the fixed points and period 2 points of (2.1) and do not describe the dynamics of the problem. Due to the S symmetry of (2.1), the x axis corresponds to a fixed point space and is therefore invariant under the flow of equations (2.1). However, the y axis is not invariant in general. We now show that there is always a curve which passes through the trivial fixed point, which is invariant under the flow and which is tangent to the y axis at the origin. Moreover, if a non-symmetric fixed point exists as well for particular (small) values of λ and μ, then this curve passes through these points as well, giving rise to a heteroclinic connection.

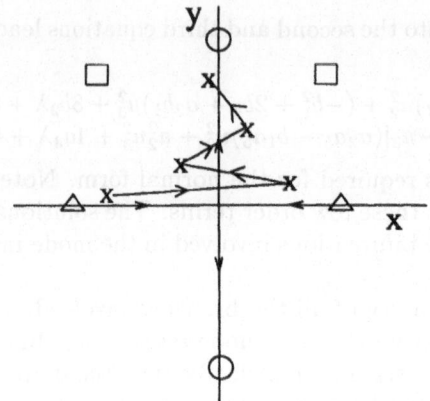

Figure 1: Typical phase diagram near to the trivial solution. $-1 < \alpha < 0$, $\beta > 1$. Symbols used : o - non-symmetric fixed points, $\triangle$ - symmetric period 2 points, $\square$ - non-symmetric period 2 points.

Suppose that the invariant curve is given by

$$x = h(y).$$

Since the curve passes through the origin and must be invariant under reflection in y, we use the approximation

$$h(y) = d_1 y^2 + d_2 y^4 + O(y^6). \tag{2.8}$$

This curve is of course tangent to the y axis at the origin. We want this curve to be flow invariant. Thus for given values (x_0, y_0), let $x_1 = f(x_0, y_0, \lambda, \mu)$ and $y_1 = g(x_0, y_0, \lambda, \mu)$. The curve is then invariant if $x_0 = h(y_0)$ implies that $x_1 = h(y_1)$ which is equivalent to

$$f(h(y_0), y_0, \lambda, \mu) = h(g(h(y_0), y_0, \lambda, \mu)). \tag{2.9}$$

It is easily verified that this equation is invariant under the transformation $y_0 \rightarrow -y_0$ and that $y_0 = 0$ is a solution. Thus, substituting for $h(y)$ using (2.8) and equating powers of y^2 enables the coefficients d_i to be calculated. The flow on this curve is then given by the single equation

$$y^{n+1} = g(h(y^n), y^n, \lambda, \mu)$$

and clearly, for any value y^{n+1}, we have, using (2.9),

$$x^{n+1} = h(y^{n+1}) = f(h(y^n), y^n, \lambda, \mu) = f(x^n, y^n, \lambda, \mu).$$

If a non-symmetric fixed point exists, then we must have

$$y = g(h(y), y, \lambda, \mu) \tag{2.10}$$

if it lies on this curve. A solution of this equation is sufficient to establish the existence of the heteroclinic connection.

We now apply this method to equations (2.6). The lowest order term in (2.9) is the y_0^2 term and equating powers of y_0^2 gives an equation which can be solved for d_1 giving

$$d_1 = \frac{a_5}{2 - a_3\lambda - a_4\mu + 2(b_2\lambda + b_3\mu) + (b_2\lambda + b_3\mu)^2}.$$

Equation (2.10) has the trivial solution $y = 0$ which confirms that the trivial solution is a fixed point on the curve $x = h(y)$. Thus, dividing (2.10) by y, the low order terms are then

$$b_2\lambda + b_3\mu + (b_7 + d_1(b_1 + b_5\lambda + b_6\mu))y^2 + O(y^4) = 0. \tag{2.11}$$

This equation has the solution $y = \lambda = \mu = 0$ and so, by the Implicit Function Theorem, can be solved for y^2 as a function of λ and μ if the y^2 coefficient with $\lambda = \mu = 0$ is non-zero. Substituting for d_1 and setting $\lambda = \mu = 0$ gives the coefficient of y^2 as

$$b_7 + \frac{1}{2}a_5 b_1$$

which we assume is non-zero. Truncating the higher order terms of (2.11), substituting for d_1 and solving for y^2 gives

$$y^2 = -2\left(\frac{b_2\lambda + b_3\mu}{2b_7 + a_5 b_1}\right) + O((\lambda, \mu)^2). \tag{2.12}$$

Clearly, this is only a valid solution if $y^2 > 0$ and if this condition is not satisfied, then the non-symmetric fixed point does not exist.

Returning to the bifurcation equations (2.7), the non-symmetric fixed points correspond to $u_3 = 0$ and $u_2 \neq 0$. These single mode solutions are given by

$$u_2^2 = -8\left(\frac{b_2\lambda + b_3\mu}{2b_7 + a_5 b_1}\right).$$

The symmetry constraints (2.5) imply that when $u_3 = 0$, then $u_4 = 0$ also and so

$$\begin{aligned} y^2 &= \frac{1}{4}u_2^2 \\ &= -2\left(\frac{b_2\lambda + b_3\mu}{2b_7 + a_5 b_1}\right) \end{aligned}$$

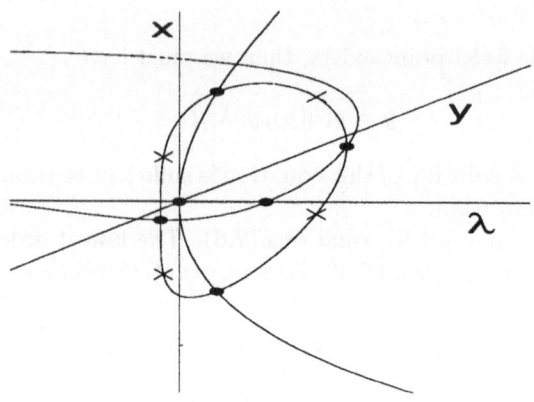

Figure 2: Bifurcation Diagram with $\mu = -0.1$. • - bifurcation points, × - Hopf bifurcation points

in agreement with (2.12). Similarly,

$$
\begin{aligned}
x &= \frac{1}{8} a_5 u_2^2 \\
&= \frac{1}{2} a_5 y^2
\end{aligned}
$$

which agrees with the approximation to $x = h(y)$ when $\lambda = \mu = 0$.

The non-degeneracy conditions associated with this procedure are

$$
a_5 \neq 0, \qquad b_7 + \frac{1}{2} a_5 b_1 \neq 0, \qquad b_2 \neq 0, \qquad b_3 \neq 0.
$$

The first condition implies that the invariant curve has a quadratic contact when it intersects the x axis while all the other terms occur as coefficients in the bifurcation equations (2.7) which must be non-zero if the problem is to be non-degenerate (Golubitsky and Schaeffer (1985)). Thus all these conditions are satisfied generically.

This proves the existence of an invariant curve passing through the trivial fixed point and which is perpendicular to the x axis at the point of intersection and which is approximately quadratic for small λ and μ. Whenever non-symmetric fixed points exist, they also lie on this curve, thus creating a heteroclinic connection. When the trivial solution is a saddle fixed point, this curve corresponds to either the stable or unstable manifold, irrespective of the existence of the non-symmetric solutions.

3. Example

In this section, we consider an example to highlight all the important features described in the previous section. In particular we take $a_1 = -1$, $a_2 = 2$, $a_3 = -1$, $a_4 = 0$, $a_5 = 2$, $b_1 = b_2 = b_3 = 1$, $b_4 = 5$, $b_5 = -10$, $b_6 = 5$, $b_7 = 1$. The bifurcation equations in this case are

$$\begin{aligned}
u_2(4u_2^2 + 11u_3^2 + 8\lambda + 8\mu) &= 0, \\
-u_3(2u_2^2 + 4u_3^2 - 4\lambda) &= 0.
\end{aligned}$$

Interchanging the order of these two equations gives rise to a bifurcation problem of Class II(d) in the classification of Langford and Iooss (1980) in which the mixed mode solutions form a closed loop connecting the two single mode branches. There are also Hopf bifurcations occurring on the mixed mode solutions. The bifurcation diagram for the problem with $\mu = -0.1$ is shown in Fig. 2 and has all the anticipated features. The trivial solution is stable for $\lambda < 0$ and loses stability at the period doubling bifurcation at $\lambda = 0$. At this point, the symmetric period 2 solutions gain stability until the secondary bifurcation at $\lambda = 0.0437$. The bifurcating loop of non-symmetric period 2 solutions is then stable up to the Hopf bifurcation points which occur at $\lambda = 0.06797$. We note that b_5 and b_6 do not occur in the bifurcation equations (2.7) but their values have been chosen to ensure that the bifurcating invariant circles are stable.

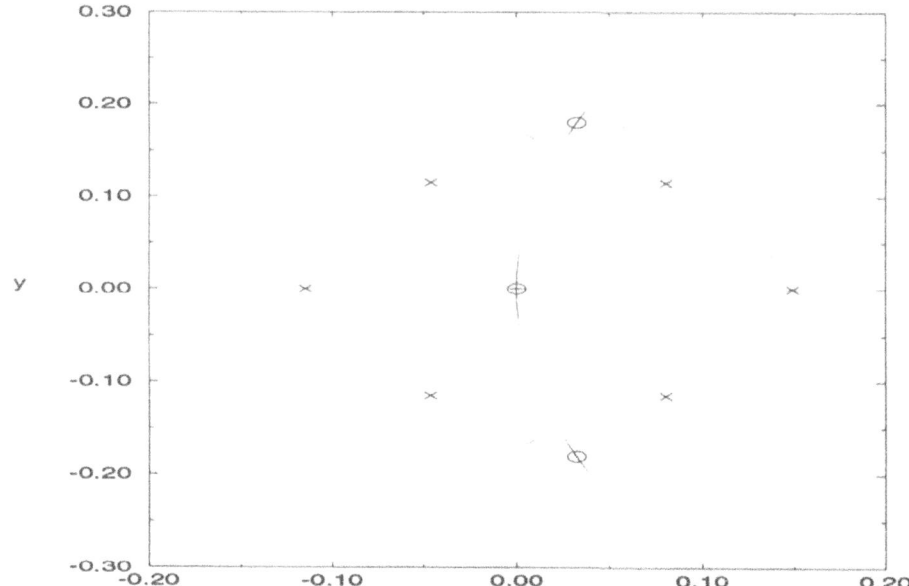

Figure 3: Phase Portrait for $\lambda = 0.0683$. o - fixed points, × - period 2 points

58

The phase diagram is shown in Fig. 3 for $\lambda = 0.0683$ which includes stable invariant circles, together with the invariant quadratic curve described in the previous section. In this case, the trivial fixed point is a saddle with the x axis as the unstable manifold. The stable manifold consists of the invariant curve $x = h(y)$ and substituting the expansion (2.8) into (2.9) gives the approximation to this manifold as

$$x = 0.99706y^2 + 0.21036y^4 - 5.2709y^6 + O(y^8).$$

We note that the y^2 coefficient is very close to $a_5/2 = 1$. Truncating the higher order terms, the resulting curve is a very good approximation to the stable manifold.

As λ is increased, the invariant circles grow in size until they touch the trivial fixed point, the non-symmetric fixed point and the symmetric period 2 points. This occurs approximately at $\lambda = 0.068429$. They break up at this point and no longer exist for larger values of λ.

References

Bauer, L., Keller, H.B. and Reiss, E.L. (1975). Multiple eigenvalues lead to secondary bifurcation. *SIAM Rev.* **17**, 101-122.

Chossat, P. and Golubitsky, M. (1988a). Symmetry increasing bifurcation of chaotic attractors. *Physica D* **32**, 423-436.

Chossat, P. and Golubitsky, M. (1988b). Iterates of maps with symmetry. *SIAM J. Math. Anal.* **19**, 1259-1270.

Dangelmayr, G. (1986). Steady-state mode interactions in the presence of O(2)-symmetry. *Dyn. Stab. Sys.* **1**, 159-185.

Golubitsky, M. and Schaeffer, D. G. (1985). *Singularities and Groups in Bifurcation Theory, Vol. I.* Appl. Math. Sci. **51**, Springer, New York.

Iooss, G. and Langford, W. (1980). Interactions of Hopf and pitchfork bifurcations, in *Bifurcation Problems and their Numerical Solution*, eds. H. Mittelmann and H. Weber, ISNM 54, Birkhauser, Basel.

SETS, LINES and ADDING MACHINES

JORGE BUESCU and IAN STEWART
Nonlinear Systems Laboratory, Mathematics Institute
University of Warwick, UK

ABSTRACT. Let X be a locally connected locally compact metric space and $f : X \rightarrow X$ a continuous map. Let A be a compact transitive set under f. If A is asymptotically stable, then it has finitely many connected components, which are cyclically permuted. If it is Liapunov stable, then A may have infinitely many connected components. Our main result states that these form a Cantor set on which f is topologically conjugate to an adding machine. A number of consequences are derived, among which is a "closing lemma" for adding machines on the interval.

This paper describes some links between dynamical stability and connectedness properties of invariant sets for discrete dynamical systems. The existence of such links was first observed by Dellnitz, Golubitsky and Melbourne [3]. Here we extend their ideas and show how they lead to strong constraints on the dynamics on such sets; the main results are presented in Theorems 2 and 3. Many corollaries that are interesting on their own right follow from these theorems. In this paper we include some new results on the accumulation of stable Cantor sets by periodic points — see Corollary 12 and the examples that follow it. The main ideas of the proofs are sketched; for a formal treatment we refer to Buescu and Stewart [2].

Let X be a metric space and $f : X \rightarrow X$ a continuous map. Suppose A is a compact topologically transitive set under f (topological transitivity means there is $x \in A$ such that $\omega(x) = A$; it then follows immediately that A is an invariant set). A is said to be *Liapunov stable* if for every open neighbourhood U of A there is a smaller open neighbourhood V of A with $f^n(V) \subset U$ for all $n \geq 0$. If in addition there exists an open neighbourhood W of A such that $\omega(x) = A$ for all $x \in W$, then A is said to be *asymptotically stable* or equivalently an *open-basin attractor*.

Adding machines are maps of the Cantor set which are best defined through symbolic dynamics. Let $\underline{k} = \{k_n\}_{n \geq 1}$ be a sequence of integers with $k_n > 1$ for all n, and let $\Sigma_{\underline{k}} = \Pi_{n=1}^{\infty}\{0, \ldots, k_n - 1\}$ be the space of semi-infinite sequences $\underline{i} = (i_n)_{n \geq 1}$ such that $0 \leq i_n < k_n$ with the product topology. $\Sigma_{\underline{k}}$ is metrizable and homeomorphic

P. Chossat (ed.), Dynamics, Bifurcation and Symmetry, 59–68.

to the Cantor set. The *adding machine on* $\Sigma_{\underline{k}}$ is a map $\alpha_{\underline{k}} : \Sigma_{\underline{k}} \to \Sigma_{\underline{k}}$ defined by the following action on elements of $\Sigma_{\underline{k}}$:

$$\alpha_{\underline{k}}(\underline{i}) = \begin{cases} (\overset{l-1}{\overbrace{0,\ldots,0}}, i_l + 1, i_{l+1}, \ldots) & \text{if } i_l < k_l - 1 \text{ and } i_j = k_j - 1 \text{ for } j < l \\ (0,0,\ldots) & \text{if } i_j = k_j - 1 \text{ for all } j. \end{cases} \quad (1)$$

It is straightforward to check that $\alpha_{\underline{k}}$ is continuous, and indeed a homeomorphism of $\Sigma_{\underline{k}}$. Moreover its action on $\Sigma_{\underline{k}}$ is *minimal*, that is, $\omega(x) = A$ for *all* $x \in \Sigma_{\underline{k}}$.

It is useful to consider the action of $\alpha_{\underline{k}}$ on cylinder sets, that is, sets of the form $C_{n;i_1,\ldots,i_n} = \{\underline{x} \in \Sigma_{\underline{k}} : x_1 = i_1, \ldots, x_n = i_n\}$, on which the first n coordinates are constant. The set of all cylinders of length n forms a finite partition of $\Sigma_{\underline{k}}$ by closed and open (abbreviated to *clopen*) sets. Moreover, this partition is invariant under $\alpha_{\underline{k}}$ (in the sense that both the image and the preimage of a cylinder of length n are cylinders of length n) and the action of $\alpha_{\underline{k}}$ on the set of cylinders of length n is just a cyclic permutation.

A useful example to keep in mind is that of the Feigenbaum map — the quadratic map $F(x) = \lambda x(1 - x)$ of the interval $I = [0,1]$ at the accumulation of the period-doubling cascade $\lambda_\infty = 3.57\ldots$. The following properties of this map are well-known, see Milnor [5]. First of all, there is an invariant transitive Cantor set K which is Liapunov stable. In fact, almost all points in I (with respect to Lebesgue measure) are asymptotic to K, the only exceptions being the countably many periodic points of period 2^n as well as their (countably many) preimages. Thus the only ω-limit sets apart from K are the periodic points. These accumulate on K, and therefore K is not an open-basin attractor. Finally, $F_{|A}$ is topologically conjugate to the binary (i. e. $\underline{k} = 2$) adding machine. Our work will show that all these apparently different structures are in fact interdependent, and that indeed this is an archetypical example of a much more general situation.

The following construction is instrumental throughout. Let X be a locally compact metric space, f a continuous map, and suppose as before that A is a compact transitive set under f. Let $\sim$ be the equivalence relation determined by the connected components of A, and let $K = A/\sim$ be the quotient space endowed with the identification topology. Then the diagram

$$\begin{array}{ccc} A & \overset{f}{\longrightarrow} & A \\ \downarrow \pi & & \downarrow \pi \\ K & \overset{\tilde{f}}{\longrightarrow} & K \end{array}$$

commutes, π being the identification map and $\tilde{f}$ the induced map on K. Our first result is a 'folklore' characterization of K.

Theorem 1 *With $X, A, f, \tilde{f}$ as above, one of the following holds.*

1. *K is finite and $\tilde{f}$ is a cyclic permutation on K.*

2. *K is a Cantor set and $\tilde{f}$ is transitive on K.*

Sketch of Proof: The Hausdorff metric on the components of A projects downstairs via π to a metric which generates the identification topology, showing metrisability of K. K is also compact because A is compact, and totally disconnected by definition. Transitivity of f upstairs implies transitivity of $\tilde{f}$ downstairs. If K is finite, this implies $\tilde{f}$ must be a cyclic permutation. If K is infinite, this means K is perfect and therefore homeomorphic to the Cantor set. □

It is clear that at this level of generality not much more can be said. The present setting includes for instance the case where X itself is the Cantor set, $A = X$ and π is the identity, in which case we are really talking about *arbitrary* (continuous) transitive maps of the Cantor set. It is known that these have a very rich structure; in particular they have not been classified. In order to produce more interesting statements some extra structure is needed. We introduce this extra structure in two places: the space X and the set A. Specifically, from now on we shall make the following additional requirements: (i) X will be *locally connected* and (ii) A will be Liapunov stable. This is the setting for our main result, the hypotheses of which we make explicit.

Theorem 2 *Suppose that X is a locally connected locally compact metric space, $f : X \to X$ is a continuous map and A is a compact transitive set. Assume A is Liapunov stable and has infinitely many components. Then $\tilde{f} : K \to K$ is topologically conjugate to an adding machine.*

Sketch of Proof: As a first step we prove that transitivity and Liapunov stability of A upstairs imply minimality of K under $\tilde{f}$. Secondly, we note that the hypotheses on X imply that A has a basis of neighbourhoods consisting of finitely many disjoint connected open subsets with compact closure. Take U and V as in the definition of Liapunov stability to be of this type. The main observation is that each component of V must map inside a single component of U. This statement translates downstairs to K in the form of a finite clopen invariant partition of K, say with k_1 elements. Moreover, the action of $\tilde{f}$ on these sets is a cyclic permutation, so that they can be considered as cylinders of length 1 in an adding machine. Refining this procedure infinitely many times by looking at suitable iterates of f in succession produces an invariant partition of K whose dynamics are conjugate to an adding machine. Finally, minimality of $\tilde{f}$ ensures that each element of the limit partition reduces to a single point. $\square$

We shall call the sequence of U, V as constructed above *tame neighbourhoods* of A. The theorem then states that tame neighbourhoods of A exist and project downstairs on K to a nested sequence of clopen invariant partitions, which we abbreviate to *invariant nest*. The order of the elements of the invariant nest is $k_1, k_1 k_2, \ldots, \Pi_{j=1}^n k_j \ldots$, where $\underline{k}$ is the sequence of integers constructed above; we call this a *tame sequence*. Different choices of tame neighbourhoods may give rise to different tame sequences; these are subject to the restrictions on Theorem 4.

Theorem 2 tells us that the mild requirement of Liapunov stability of A upstairs implies a tightly constrained structured for the map between components of A — specifically, the downstairs map $\tilde{f}$ is an adding machine. Further strengthening the stability requirement on A from Liapunov stability to asymptotic stability has the effect of destroying the Cantor structure of K altogether, and by Theorem 1 it must reduce to a finite set. In short, open-basin attractors have finitely many components which are cyclically permuted. This is the subject of Theorem 3, which strengthens a previous result of Hirsch [4].

Theorem 3 *Let X, f be as in Theorem 2 and suppose A is an open-basin attractor. Then A has finitely many components.*

Sketch of Proof: If not, K is a Cantor set. It is easy to show that, if U is a sufficiently small open neighbourhood of A with compact closure, then given any open neighbourhood V of A with $V \subset U$ there is an $N(V)$ such that $n \geq N \Rightarrow f^n(\overline{U}) \subset V$. Due to the Cantor structure of the fibres of K we may take V to have strictly more components than U, all of them intersecting A. It would then follow that $f_{|A}^n$ is not onto, contradicting transitivity. $\square$

Although the proof of Theorem 2 is constructive, there are a number of choices involved — at each step in the adding machine construction the corresponding k_n depends on how the choice of neighbourhood is performed. This means that the sequence $\underline{k} = (k_1, k_2, \ldots)$ for the adding machine in Theorem 2 is not uniquely defined. These choices, however, are not arbitrary — the statement of the Theorem itself shows that two different adding machines arising from the same A must be topologically conjugate. Thus Theorem 2 does define a *unique conjugacy class* of adding machines, giving rise to the problem of classification of adding machines up to topological conjugacy. This classification problem is solved in the following way.

Given a sequence of integers $\underline{k}$ with $k_n \geq 2$ for all n, call a prime p a *prime factor* of $\underline{k}$ if p is a prime factor of some k_i. Define the *multiplicity* $m(p)$ of a prime factor p as the sum of the powers to which p occurs as a factor in all the k_i, allowing it to be infinite if p occurs in infinitely many k_i. Let $\mathbb{P}$ denote the set of all primes; then $m : \mathbb{P} \to \mathbb{N} \cup \{\infty\}$ is the *multiplicity function* of $\underline{k}$.

Theorem 4 *Let $\underline{k}$ and $\underline{l}$ be sequences of integers ≥ 2. Then $\alpha_{\underline{k}}$ and $\alpha_{\underline{l}}$ are topologically conjugate if and only if $\underline{k}$ and $\underline{l}$ have the same multiplicity function.*

Sketch of proof: We first observe that adding machines have topologically discrete spectrum, and so the (topological) eigenvalue group is a complete invariant for their classification up to topological conjugacy (Walters [7], Buescu and Stewart [2]). Explicit computation of eigenfunctions and eigenvalues of $\alpha_{\underline{k}}$ reveals that this group is the set of all $k_1 \ldots k_n \ldots$th roots of unity. The theorem then follows by group theory. $\Box$

The preceding Theorem allows us to suppose, when dealing with properties of conjugacy classes of adding machines — as in Theorem 2 — that a reduction to a prime base $\underline{p}$, in which the p_i are primes, has been performed. This prime base may be constructed in the obvious way by factoring the k_i into primes and is unique up to reordering (which leaves the conjugacy class invariant). However, it is worth remarking that this refinement of $\underline{k}$ does not necessarily lift to corresponding tame neighbourhoods as constructed in Theorem 2. So each conjugacy class is in general larger than the class of all tame sequences.

A number of corollaries are derived more or less immediately from these results:

Corollary 5 *Let A be a compact stable transitive set under f and suppose that f admits a periodic point of period n. Then A has at most k connected components, where k divides n.*

Corollary 6 *Let A be a compact transitive set with infinitely many components. If A has a periodic point, then A is Liapunov unstable.*

In particular this corollary implies that it is impossible to imbed any symbolic subshift admitting a periodic point in an arbitrary metric space in such a way as to satisfy Liapunov stability.

The following Corollary is really a consequence of Buescu [1], but finds a very natural place here. Not that this particular result does *not* require Liapunov stability.

Corollary 7 *Let A be a compact transitive set with infinitely many components. Then A is nowhere dense in X.*

We next specialise to more specific situations.

Corollary 8 *If A is a transitive stable Cantor set, then $f_{|A}$ is conjugate to an adding machine. In particular, $f_{|A}$ is a homeomorphism.*

Corollary 9 *Let I be a compact interval and $f : I \to I$ continuous. Transitive stable sets under f are either periodic orbits, transitive cycles of intervals or Cantor sets on which f is conjugate to an adding machine.*

We say a circle homeomorphism f is a *Denjoy map* if it has an invariant Cantor set K such that $\omega(x) = K$ for all $x \in S^1$. It then follows that f has irrational rotation number and the action of f on K is minimal; see Schweitzer [6]. However, irrationality of the rotation number implies that $f_{|K}$ is not topologically conjugate to any adding machine (Buescu and Stewart [2]), proving the following corollary.

Corollary 10 *The minimal Cantor set K for any Denjoy map is Liapunov unstable.*

The next corollary is another way to state Theorem 3; in short, stable Cantor sets cannot be isolated ω-limit sets.

Corollary 11 *If A is a compact stable transitive set with infinitely many components, then A is accumulated by ω-limit sets other than itself.*

On the real line, this conclusion may be sharpened to a statement about periodic points which constitutes a "closing lemma" for stable adding machines. We provide a proof as this corollary has not been published elsewhere.

Corollary 12 *Let $f : I \to I$ be a continuous map of the interval admitting a transitive stable Cantor set K on which f is conjugate to an adding machine $\alpha_{\underline{k}}$, where $\underline{k}$ is a tame sequence. Let $x \in K$. Then there is a sequence $x_n \in I \setminus K$ with $x_n \to x$ such that x_n is periodic with period $P_n = \Pi_{j=1}^n k_j$. In particular $K \subset \overline{Per(f)}$.*

Proof. Let $h : K \to \Sigma_{\underline{k}}$ be the homeomorphism establishing conjugacy. Let $C_{1;i_1}$ be the length 1 cylinder containing $h(x)$, $K_{1;i_1} = h^{-1}(C_{1;i_1})$ be the corresponding clopen subset of K and $H_{1;i_1}$ be the closed convex hull of $K_{1;i_1}$ in I, that is, the smallest closed interval containing $K_{1;i_1}$. Following the proof of Theorem 2, we see that

$$f^n(H_{1;i_1}) \supseteq H_{1;i_1+n \pmod{k_1}}, \tag{2}$$

while Liapunov stability implies that

$$f^n(H_{1;i_1}) \cap H_{1;i_1} = \emptyset, \quad 1 \leq n \leq k_1 - 1. \tag{3}$$

As the endpoints of $H_{1;i_1}$ belong to $K_{1;i_1}$ which is invariant under f^{k_1} and $H_{1;i_1}$ is connected, it follows that $f^{k_1}(H_{1;i_1}) \supseteq H_{1;i_1}$. By the fried-egg lemma below, f^{k_1} must possess a fixed point x_1 in $H_{1;i_1}$, which must be a point of period k_1 by the cycling condition 3.

For the inductive step, suppose we have located periodic points $x_1, \ldots, x_n$ of periods $k_1, \ldots, \Pi_{j=1}^n k_j$ in a neighbourhood of x. The closed convex hull of the cylinder of length $n+1$ around x, which we denote by $H_{n+1;i_1,\ldots,i_{n+1}}$, satisfies equations similar to 2 and 3 with k_1 replaced by $\Pi_{j=0}^{n+1} k_j$. So there is a fixed point x_{n+1} for $f^{k_1 \cdots k_{n+1}}$ in $H_{n+1;i_1,\ldots,i_{n+1}}$. The cycling condition 3 again ensures x_{n+1} is not fixed by any previous iterate of f, and therefore x_{n+1} has prime period $\Pi_{j=0}^{n+1} k_j$. As $\bigcap_{n=1}^{\infty} H_{n;i_1,\ldots,i_n} = \{x\}$, it follows that $x_n \to x$, as asserted. $\square$

Remark. Application of Corollary 12 to the Feigenbaum map shows not only that F admits all periods that are powers of 2 but the stronger fact that any point on the invariant Cantor set is the limit point of a sequence of periodic points with period 2^n.

Lemma 13 (The real fried-egg) *Let $f : \mathbb{R} \to \mathbb{R}$ be a continuous map. Suppose I is a compact interval such that $f(I) \supseteq I$. Then f has a fixed point in I.*

Proof. If not, then for all $x \in I$ either $f(x) > x$ or $f(x) < x$. In the first case continuity and compactness imply there exists $\epsilon > 0$ such that, for all $x \in I$, $f(x) > x + \epsilon$. It follows that $f(I)$ cannot cover I, contradicting the assumption. The second case is identical. $\square$

We call the covering property $f(I) \supseteq I$ the *fried-egg property*. It is specific to $\mathbb{R}$ that the fried egg-property for continuous maps on compact balls implies existence of fixed points; counterexamples may be given in $\mathbb{R}^2$ — see figure 1. Here Q is the unit square, $f(Q) \supseteq Q$ but f has obviously no fixed points. It is a simple matter to modify this example in such a way that $f(Q)$ is topologically the same as Q (say, by adding a twist which covers the hole). It is also clear that f may be chosen so that Q is Liapunov stable. Therefore Corollary 12 does not generalise to other spaces with the present methods.

Example 1. The purpose of this example is to show that every adding machine can occur as a stable Cantor set for a dynamical system. Given a sequence of integers $\underline{k} = (k_1, k_2, \ldots)$, consider a smooth flow φ^t in $\mathbb{R}^2$ with a compact invariant disk D, where it has k_1 sinks and k_1 saddles, all hyperbolic. Around each sink construct a small disk D_i, $i = 0, \ldots k_1 - 1$. Define a map $\psi_1 = R_{k_1} \circ \varphi^1$, where φ^1 is the time-1 map of φ^t and R_{k_1} is just a rotation by $2\pi/k_1$, which permutes the k_1 disks $\varphi^1(D_i)$

cyclically (for clarity we assume them to be identical and symmetrically placed). See figure 1.

One can now iterate the procedure replacing D with some D_i, f by f^{k_1} and k_1 by k_2. Proceeding inductively, we get a diffeomorphism of $D \setminus K$, where K is a Cantor set; this can be extended by continuity to a homeomorphism ψ of D by making ψ act on K as the adding machine $\alpha_{\underline{k}}$. By construction K is Liapunov stable; indeed, a point in D is either attracted to K or lies in the stable manifold of the countably many periodic saddles (incidentally proving also that K attracts a set of full Lebesgue measure).

J. Guckenheimer pointed out to us that indeed every stable adding machine occurs in the bifurcation diagram of any full family of C^3-unimodal maps. A sketch of the construction is given by Milnor [5].

Remark. No general statement about periodic points is possible if the fibres over the adding machine — i. e. the components of A — are nontrivial. We give two examples to illustrate the situation; in the first one $\overline{Per(f)} = \emptyset$, in the second one $A \subset \overline{Per(f)}$. For simplicity we work with the Feigenbaum map of the interval $I = [0, 1]$ to which corresponds the binary adding machine; that the corresponding results are valid for *arbitrary* adding machines is easily seen by replacing the Feigenbaum map by example 1 above.

Example 2. Let $f : I \times S^1 \to I \times S^1$ be defined by $f(x, y) = (F(x), R_\omega(y))$ where $F : I \to I$ is the Feigenbaum map and $R_\omega(y) = y + \omega \pmod 1$ is an irrational rotation. Let K be the invariant Cantor set for F; it is easily shown that $K \times S^1$ is a stable transitive set under f (sketch: stability is trivial. For transitivity, remark that every open set in $K \times S^1$ must contain a cylinder cross an interval in S^1. Then for every open U, V there is $n \geq 0$ such that $f^n(U) \cap V \neq \emptyset$, which is equivalent to topological transitivity). The only other ω-limit sets under f, which are the ones alluded to by Corollary 11, are those of the form $A_n = \{x_n, F(x_n), \ldots, F^{2^n-1}(x_n)\} \times S^1$, x_n being the (unique) period 2^n point of F. As $\omega \in \mathbb{R} \setminus \mathbb{Q}$ it is clear that $f_{|A_n}$ is minimal; in particular f has no periodic points.

Example 3. Let $f : I \times I \to I \times I$ be defined by $f(x, y) = (F(x), \varphi(y))$, where again F is the Feigenbaum map and $\varphi(y) = 4y(1 - y)$. Here $K \times I$ is a stable transitive set and is accumulated by invariant sets of the form $A_n = \{x_n, F(x_n), \ldots, F^{2^n-1}(x_n)\} \times I$. As periodic points are dense in each A_n, it follows that $K \times I$ lies in the closure of the periodic points.

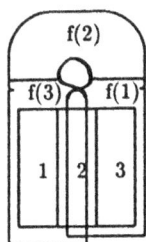

Figure 1: The unit square Q is divided into 3 vertical strips which map in the indicated way under f.

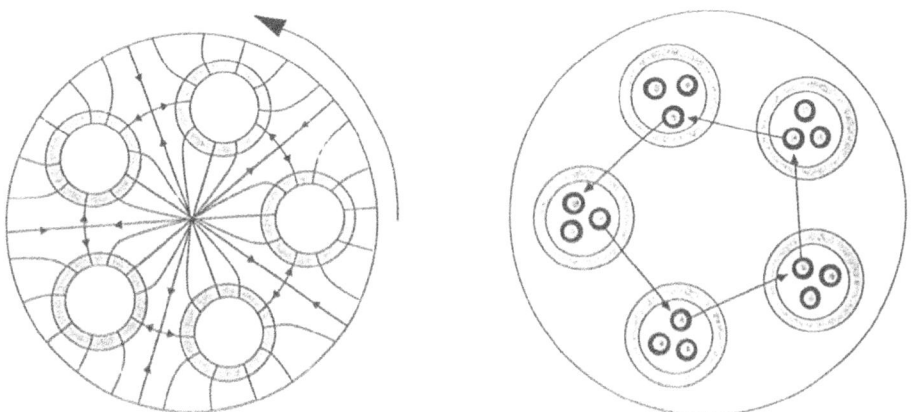

Figure 2: The first two stages in the definition of ψ.

68

Acknowledgments

We are grateful to Peter Ashwin, Jim Denvir, John Guckenheimer, Robert MacKay, Anthony Manning, Ian Melbourne, Pat McSwiggen, Bill Parry, Mark Pollicott, and especially to Peter Walters for many helpful discussions. INS thanks the University of Cincinatti and the Fields Institute, University of Waterloo for hospitality and financial support during visits in which some of the work reported here began to crystallize. The work of JB was supported in part by a JNICT/CIENCIA grant BD-1073-90/RM. INS acknowledges partial support by a grant from the Science and Engineering Research Council of the UK and by a European Community Laboratory Twinning grant. This research was carried out under the auspices of the European Bifurcation Theory Group.

References

[1] J. Buescu, Transitive sets with non-empty interior, Preprint, University of Warwick, 1993.

[2] J. Buescu and I. Stewart, Liapunov stability and adding machines, Preprint, University of Warwick, 1993; *Erg. Th. Dyn. Sys.*, to appear.

[3] M. Dellnitz, M. Golubitsky, and I. Melbourne, The structure of symmetric attractors, Preprint, University of Houston, 1992; *Arch. Rational. Mech. Anal.*, to appear.

[4] M. Hirsch. Components of attractors, University of California at Berkeley, Preprint, 1992.

[5] J. Milnor, On the concept of attractor, *Commun. Math. Phys.*, **99** (1985) : 177–195.

[6] P.A. Schweitzer, Counterexamples to the Seifert conjecture and opening closed leaves of foliations, *Ann. Math.*, **100** (1974) : 368–400.

[7] P. Walters, *An Introduction to Ergodic Theory*, Springer-Verlag, New York, 1982.

MIXED-MODE SOLUTIONS IN MODE INTERACTION PROBLEMS WITH SYMMETRY

SOFIA CASTRO
Faculdade de Economica do Porto
Portugal

ABSTRACT. We study symmetric mode interaction problems of codimension 1. We show that some features existing in each single-mode problem are preserved in some way in the mode interaction. For example, single-mode branches become mixed-mode but keep the same symmetry. We also refer to some other types of behaviour which are not present in the single-mode problems, like the existence of secondary Hopf bifurcations and of heteroclinic connections between equilibria.

1 Introduction

Assume that a centre manifold reduction has been performed and define a steady-state mode interaction bifurcation problem with symmetry group Γ as follows:

$$
\begin{aligned}
\dot{u} + g_u(u, v, \lambda) &= 0 \\
\dot{v} + g_v(u, v, \lambda) &= 0,
\end{aligned}
$$

where $g \equiv (g_u, g_v) : U \times V \times \mathbb{R} \to U \times V$ is Γ-equivariant under the Γ-action given by

$$
\gamma.(u, v) = (\gamma u, \gamma v) \ \forall (u, v) \in U \times V, \ \gamma \in \Gamma.
$$

For the equations above to define a bifurcation problem, we assume further that

$$
g(0) = 0, \quad Dg(0) = 0
$$

and the 0-eigenspace of $Dg(0)$ decomposes as the direct sum of U and V, which we assume to be Γ-absolutely irreducible in the following. Under these hypothesis, the bifurcation problem is a *mode interaction* (Golubitsky et al [3], chapter XX, section

69

P. Chossat (ed.), Dynamics, Bifurcation and Symmetry, 69–77.

0). If these two subspaces U and V are of dimension n and m, say, we call the bifurcation problem a (n, m)-*mode interaction*.

The question we want to answer is, assuming the problem is the least degenerate possible, how much of each single-mode problem appears in the mode interaction, and how much is new.

2   Existence of mixed-mode solutions

Here we prove that, under a few additional assumptions, a mode interaction bifurcation problem always has solutions involving both modes. They are called *mixed-mode solutions*. These assumptions are very similar to those of the *equivariant branching lemma* (Golubitsky et al. [3], Theorem XIII). First remark that a least degenerate mode interaction problem is one of codimension 1, which means that the unfolding parameter appears at the linear level to deal with the linear degeneracy introduced by the condition $Dg(0) = 0$. The main result in this section is

Proposition 1 *Let* Γ *act absolutely irreducibly on* U *and* V, *and such that* U *and* V *are non isomorphic representations of* Γ. *Consider the generic mode interaction bifurcation problem on* $U \times V$ *and assume that there exist subgroups* Σ *and* Σ' *such that:*

(i) $Fix(\Sigma')_{|V} = \{0\}$ *and* $dimFix(\Sigma')_{|U} = 1$, *so that* Σ' *is a maximal isotropy subgroup and that there exists a bifurcated branch* $\lambda \equiv h(u)$ *with isotropy* Σ'

(ii) $\Sigma \subset \Sigma'$, $Fix(\Sigma)_{|U} = Fix(\Sigma')$ *and* $dimFix(\Sigma)_{|V} = 1$. *Hence*

$$Fix(\Sigma) = \{(u, v) \mid u \in Fix(\Sigma'), \ v \in Fix(\Sigma)_{|V}\}.$$

(iii) *Let* $l(u, \lambda)$ *be the coefficient of the term of order 1 with respect to* v *in* g_v. *In other words,* $g_v(u, v, \lambda) = l(u, \lambda)v + \bar{g}_v(u, v, \lambda)$ *where* $\bar{g}_v(u, v, \lambda) = o(\|v\|)$.

Assume there exists u^* *such that* $l(u^*, \lambda^*) = 0$, *with* $\lambda^* = h(u^*)$. *Assume further that*

$$\frac{\partial}{\partial u}(dg_v)(u, 0, h(u))_{|u=u^*}(v_0) \neq 0$$

for $v_0 \in Fix(\Sigma)_{|V}$.

Then a mixed-mode branch with isotropy Σ *bifurcates from the* Σ'-*symmetric branch at* $(u^*, 0, \lambda^*)$.

We emphasize the fact that this result proves the existence of a branch involving the two modes and not that of a branch with smaller isotropy subgroup. In fact the latter is not always true, (see Castro [1], chapter 5).

Proof. Let us write $l(u, \lambda) = f(u, \lambda) - \alpha$, where α is the unfolding parameter. We restrict the equations to $Fix(\Sigma)$, which is two dimensional, and compute the derivative L of g at the point $(u^*, 0, \lambda^*, \alpha)$:

$$L = \begin{pmatrix} \frac{\partial g_u}{\partial u} & \frac{\partial g_u}{\partial v} \\ \frac{\partial g_v}{\partial u} & (f(u, \lambda) - \alpha) + \frac{\partial \bar{g}_v}{\partial v} \end{pmatrix}_{|(u^*, 0, \lambda^*, \alpha)}$$

Note that:

(a) $f(u, \lambda) - \alpha = 0$ by assumption, (b) $\frac{\partial g_u}{\partial u}(u^*, 0, \lambda^*, \alpha) \neq 0$ because of the existence of a branch with isotropy Σ', and (c) $\frac{\partial \bar{g}_v}{\partial v}(u^*, 0, \lambda^*, \alpha) = 0$ because there are no linear terms in v in $\bar{g}_v$.

Moreover, we claim that $\frac{\partial g_u}{\partial v}(u^*, 0, \lambda^*, \alpha) = 0$. Indeed, U and V being non isomorphic representations of Γ, any non-zero equivariant maps E from U to V are non-linear. It follows that $\frac{\partial E}{\partial v} = 0$ at $v = 0$, which proves the claim.

It follows that $\text{rank}(L) = \dim Fix(\Sigma) - 1$. Hence a Lyapunov-Schmidt reduction with simple 0 eigenvalue can be performed at the point $((u^*, 0, \lambda^*, 0)$ in $Fix(\Sigma)$. Condition (iii) is then just the usual non degeneracy condition which ensures the existence of a bifurcated branch, and this branch has a non zero component in V since $kerL = V$. ¶

Remark. In the above proposition we can assume $\dim Fix(\Sigma') = k$, $k > 1$, as long as we know the existence of a branch of solutions $u = k(\lambda)$ with symmetry Σ'. The proof carries on exactly in the same way, the assumption (iii) being reformulated as

$$\frac{\partial}{\partial \lambda}(dg_v)(k(\lambda), 0, \lambda)_{\lambda=\lambda^*(v_0)} \neq 0 \quad for v_0 \in Fix(\Sigma)_{|V}.$$

When $U = \mathbb{R}$ is the trivial representation of Γ, more can be said, although we can no longer use the equivariant branching lemma to prove the existence of a branch in U. However we shall make use of the Birkhoff normal form theory in order to considerably simplify the equations.

3   Problems involving one trivial mode

We now assume $U = \mathbb{R}$, we identify V with $\mathbb{R}^n$ for some $n > 1$ and we consider a system of equations

$$\dot{x} + g_x(x, y, \lambda) = 0$$

$$\dot{y} + g_y(x, y, \lambda) = 0,$$

where $g \equiv (g_x, g_y) : \mathbb{R} \times \mathbb{R}^n \times \mathbb{R} \to \mathbb{R} \times \mathbb{R}^n$ is Γ-equivariant under the Γ-action given by

$$\gamma.(x, y) = (x, \gamma y)$$

(hence g_x is Γ-invariant). We also assume that $g(0) = 0$ and $Dg(0) = 0$.

Lemma 1 *Let g be as above, then a Γ-equivariant Birkhoff normal form for the equations for g is*

$$\dot{x} + a\lambda + p(x, y) = 0$$
$$\dot{y} + q(x, y) = 0,$$

where p and q are Γ-equivariant polynomials of degree ≥ 2 in x and y.

Proof. Extend the equations to a system in $\mathbb{R} \times \mathbb{R}^n \times \mathbb{R}$ as follows:

$$\dot{x} + g_x(x, y, \lambda) = 0$$
$$\dot{y} + g_y(x, y, \lambda) = 0$$
$$\dot{\lambda} = 0.$$

Define $g^* = (g_x, g_y, 0)$ and $L \equiv Dg^*(0, 0, 0)$. Notice that from our hypothesis, $Dg(0) \equiv 0$ and $\frac{\partial g_y}{\partial \lambda} = 0$ (the latter thanks to the absolute irreducibility of the action of Γ in $\mathbb{R}^n$). On the other hand, since Γ acts trivially on $\mathbb{R}$, it is generic that $\frac{\partial g_x}{\partial \lambda}(0) = a$ be non-zero. Hence L is nilpotent, of the form

$$L = \begin{pmatrix} 0_{(n+1) \times (n+1)} & a \\ 0_{1 \times (n+1)} & 0_{(n+1) \times 1} \end{pmatrix},$$

where the indices correspond to the dimension of the submatrices. One can therefore apply the Birkhoff normal form theory to this system and use theorems XVI, 5.8 and 5.9 in [3] in order to conclude that this normal form is given by

$$\begin{pmatrix} \dot{x} \\ \dot{y} \\ \dot{\lambda} \end{pmatrix} + L \begin{pmatrix} x \\ y \\ \lambda \end{pmatrix} + \begin{pmatrix} p_x(x, y) \\ p_y(x, y) \\ p_\lambda(x, y, \lambda) \end{pmatrix} = 0,$$

where p_x, p_y and p_λ are Γ-equivariant and moreover are equivariant under the action of the one-parameter group

$$S = Cl\{exp(sL^t); \ s \in \mathbb{R}\}.$$

Since we are not interested in the last equation of this system, we ignore it and write $p \equiv p_x$, $q \equiv p_y$, which finishes the proof. ¶

It follows from this lemma that, by restricting the equations to $Fix(\Gamma)$, i.e. by setting $y = 0$, we obtain, for the normal form,

$$\dot{x} + a\lambda + p(x, 0) = 0$$

which we can straightforwardly solve for bifurcated equilibria: If $a \neq 0$, there exists a branch of solutions $\lambda \equiv \lambda(x)$ of solutions with full symmetry. Moving down in the isotropy lattice and restricting the equations to the fixed-point space of a maximal isotropy subgroup, we see that any symmetry-breaking branch bifurcates at a value of x depending on $\alpha \in \mathbb{R}$, from which we can obtain the value of the bifurcation parameter λ. The proof of the following lemma is then straightforward.

Lemma 2 *Let $\dot{v} + g_v(v, \lambda) = 0$ be a codimension 0 single-mode bifurcation problem. The mode interaction problem involving g_v and a trivial 1-dimensional mode is defined by the following equations*

$$\begin{aligned}
\dot{x} + a\lambda + g_x(x, v) &= 0 \\
\dot{v} + g_v(v, x) &= 0
\end{aligned}$$

where g_x is Γ invariant and x replaces λ in g_v.

This result emphasizes the fact that, in this case, the variable x acts as a bifurcation parameter for the non-trivial mode.

Next we consider what happens to the isotropy subgroups when a trivial 1-dimensional mode is added as above.

Lemma 3 *Let Γ act on $\mathbb{R} \times V$, trivially on $\mathbb{R}$. Let $\{v_\alpha : \alpha \in A\}$ be a list of orbit representatives for the action of Γ on V and let $\Sigma_\alpha \subset \Gamma$ be the isotropy subgroup of v_α. Then*

$$\{(x, v_\alpha) : \alpha \in A\}$$

is a list of orbit representatives for γ acting on $\mathbb{R} \times V$ with isotropy subgroup Σ_α.

Proof. Take $\sigma \in \Sigma_\alpha$, then $\sigma.(x, v_\alpha) = (x, v_\alpha)$. On the other hand take $\gamma \in \Gamma$ which fixes (x, v_α), then

$$\gamma.(x, v_\alpha) = (x, \gamma v_\alpha) = (x, v_\alpha).$$

Hence $\gamma \in \Sigma_\alpha$. ¶

With this result, the computation of the isotropy subgroups for the mode interaction with one trivial mode reduces to the computation of the isotropy subgroups for the non-trivial component. Note that if this involves more than one mode, one can use Proposition XX, 2.3 in [3].

Let Σ be a maximal isotropy subgroup for the action of Γ on V, with $\dim Fix(\Sigma) = 1$. Then, by Proposition 1, the branch whose existence is guaranteed by the Equivariant branching Lemma in the single-mode proble, still exists, but now as a *mixed-mode branch*. Suppose now that for the bifurcation problem in V, there exists an isotropy subgroup $\Sigma_1 \subset \Sigma$ such that there is a branch $B(\Sigma_1)$ with this isotropy. Then we have the following

Lemma 4 *In the above circumstances, there exists a branch with symmetry Σ_1 given by*

$$\{(x, v) : x \in \mathbb{R}, \ v \in B(\Sigma_1)\}$$

for the Γ-equivariant problem on $\mathbb{R} \times V$.

Proof. We use the unfolded equations in normal form

$$\dot{x} + a\lambda + p(x, v) = 0$$
$$\dot{v} + c(x - \alpha)v + q(v, x) = 0$$

where $x - \alpha = \lambda'$ acts as a second parameter. We know that there exists a solution to

$$c\lambda' v + q(v, x) = 0$$

namely, with values in $B(\Sigma_1)$. By replacing these values in

$$a\lambda + p(x, v) = 0$$

we obtain an expression for λ as a function of x and v which describes a branch with Σ_1-symmetry in the subspace of $\mathbb{R} \times V$ defined by

$$\{(x, v) : \ x \in \mathbb{R}, \ v \in Fix(\Sigma_1)_{|V}\}.$$

¶

This means that a branch which exists for a single-mode problem still exists for the mode interaction with a trivial mode but as a mixed-mode branch.

4   An example

Examples illustrating the application of the above results can be found in Castro [1]. For instance consider the (1,5)-mode interaction with spherical symmetry. The group action can be simplified to that of S_3, as in Golubitsky and Schaeffer [2], and the system then reduces to the three equations

$$
\begin{aligned}
\dot{x} + \lambda - x^2 + b_2(y^2 + z^2) &= 0 \\
\dot{y} + c_1(x - \alpha + \lambda)y + c_2(y^2 - z^2) &= 0 \\
\dot{z} + c_1(x - \alpha + \lambda)z - 2c_2yz &= 0; \quad c_1 = \pm 1
\end{aligned}
$$

We assume $c_1 = +1$. Then we have
(i) a branch with full symmetry described by

$$
\lambda = x^2; \quad y = z = 0
$$

(ii) a branch with $\mathbb{Z}_2$ symmetry given by

$$
\lambda = x^2 - b_2 y^2; \quad x = \alpha - c_2 y; \quad z = 0.
$$

The interesting feature is that the second branch was present in the corresponding single-mode problem (see [2]), and near the bifurcation point it remains transcritical (Figure 1).

The mixed-mode branch becomes stable at the turning point and loses stability at a secondary Hopf bifurcation point. So far, we have been concerned in what type of behaviour can be predicted for a mode interaction when looking at each single-mode bifurcation problem. The existence of a secondary Hopf bifurcation is not an expected feature of steady-state bifurcation although it can be proven to exist in a mode interaction problem for certain values of the coefficients in the equations. A proof of this can be found in [1]. Other interesting features occur in this mode interaction problem, which cannot be expected from looking at each single-mode separately. For example if we use KAOS [4] to integrate the equations, we find that the limit cycle created by the Hopf bifurcation disappears for a value of λ in the interval (1.3,1.4) (see figure 2).

Let $(x_+, 0)$ and $(x_-, 0)$ be the positive and negative equilibria with trivial isotropy repectively. For $\lambda = 1.3$ the phase portrait is as in figure 2(a), where the unstable manifold of $(x_-, 0)$ is closer to the x-axis than the stable manifold of $(x_+, 0)$. For $\lambda = 1.4$ the situation is reversed, as we can see on figure 2(b).

Note that in figure 2(b) the limit cycle no longer exists, The only way the two manifolds can change their order in the phase space is if they are joined together

76

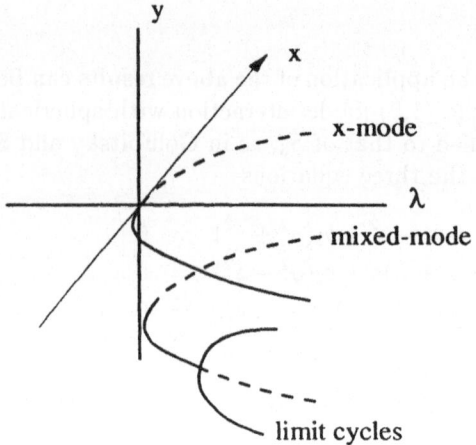

Figure 1: The bifurcation diagram

in a heteroclinic connection between $(x_-, 0)$ and $(x_+, 0)$, by which the limit cycle disappears. This heteroclinic connection is shown in figure 2(c). This in fact realizes a heteroclinic cycle in the (x, y)-plane. This cycle is not structurally stable. Hence it is of a different nature from the heteroclinic cycles studied by Armbruster and Chossat [5] in the (3,5) mode interaction with $O(3)$ symmetry.

Acknowledgments This work was done with the financial supoport of JNICT - Junta Nacional de Investiačao Científica e Tecnológica and while on leave fromm the Faculdade de Economica do Porto.

References

[1] S.B.S.D. Castro. *Mode interactions with symmetry.* PhD thesis, University of Warwick, 1993.

[2] M. Golubitsky and D. Schaeffer. Bifurcations with O(3) symmetry including applications to the Bénard problem. *Comm. Pure and Appl. Math.*, **35**, p. 81-111, 1982.

[3] M. Golubitsky, I. Stewart and D. Schaeffer. *Singularities and Groups in Bifurcation Theory, vol. 2*, Springer Verlag, 1988.

[4] J. Guckenheimer and S. Kim. *Kaos; Dynamical System Toolkit with Interactive Graphics Interface*, Cornell University, 1990.

[5] D. Armbruster and P. Chossat. *Heteroclinic cycles in a spherically invariant system.* Physica D 50, p. 155-176, 1991.

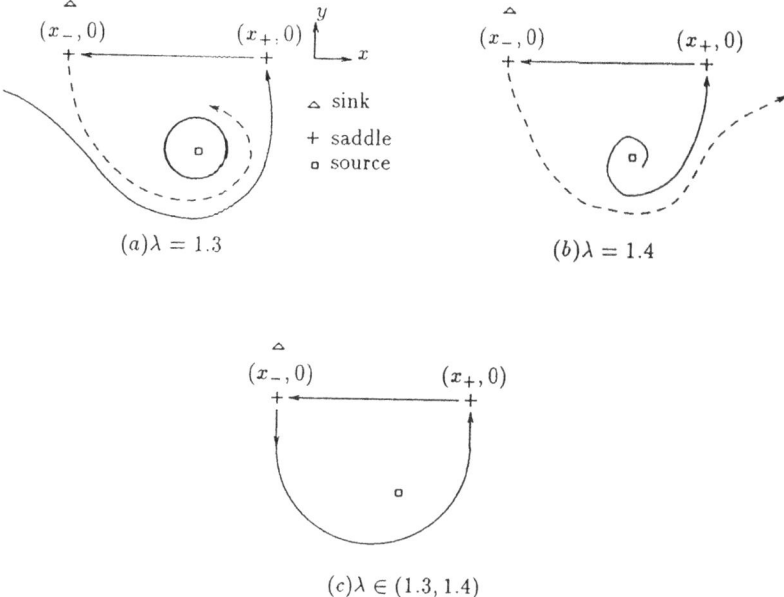

$(a)\lambda = 1.3$

$(b)\lambda = 1.4$

$(c)\lambda \in (1.3, 1.4)$

Figure 2: Existence of a heteroclinic connection

A CLASSIFICATION OF 2-MODES INTERACTIONS WITH $SO(3)$ SYMMETRY AND APPLICATIONS

PASCAL CHOSSAT and FREDERIC GUYARD
Institut Non Linéaire de Nice
CNRS & Université de Nice
Sophia Antipolis FRANCE

ABSTRACT : We give an explicit criterion for the determination of the isotropy lattices for mode interaction bifurcation problems with $SO(3)$ symmetry, which makes use of the already known classification of isotropy types in the one-parameter case. As an application, we classify isotropy lattices for 2-modes interaction with $SO(3)$ symmetry and we study three cases of physical interest.

1. Introduction

Mode interactions in bifurcation problems with spherical symmetry occur naturally in applications. Let us recall that irreducible representations of degree l of the group $O(3)$ act in spaces of spherical harmonics $Y_m^l(\theta, \phi)$ with $m = -l, \cdots, l$ and l fixed. Therefore they are $2l+1$-dimensional. In the spherical Bénard problem for instance, i.e. the onset of convection in a self-gravitating fluid spherical shell, the degree l of the critical representation depends monotoneously on the (normalized) thickness of the shell. Therefore if this thickness is taken as a parameter, there exist critical values of it at which both representations of degrees l and $l+1$ are critical. This leads to 2-parameters bifurcation problems in dimension $4l+4$... The case when $l = 1$ has been thoroughly investigated ([9], [18],[2]). To classify all $l, l+1$ mode interactions is another approach which we have undertaken, and this will be the subject of a forthcoming paper. There is also some interest in looking at mode interactions for $SO(3)$, not $O(3)$, representations. The reason is that an interaction between $O(3)$ representations with degrees l and m both even can occur in certain problems, as indicated below. Since in most physical problems such representations act essentially like the $SO(3)$ representations of same degree, one can for such problems reduce to classifying $SO(3)$ mode interactions. The examples which we have in mind are the following ones:

1. The 2,2 mode interaction for the onset of convection in binary fluids. It is a known fact that in the case of a planar symmetry, the onset of convection in such problems is typically a codimension two phenomenon, where two bifurcations, associated with the same irreducible representation of the symmetry group, interact. A similar phenomenon

P. Chossat (ed.), Dynamics, Bifurcation and Symmetry, 79–95.

should also occur in spherical shells but it was not our purpose in this work to perform such a (numerical) linear stability analysis.

2. The 2, 4 mode interaction seems to arise in the numerical simulation of convection of a compressible fluid in a spherical shell, as reported in [21]. A local model of such interactions can give valuable informations on the kind of solutions which can be expected in this case.

3. The 2, 6 mode interaction seems to play a role in the pattern formation of the Earth's mantle, as indicated by geophysical measurements. A discussion of this problem is given in [6].

In this paper we classify all isotropy subgroups for the interaction of two irreducible representations of $SO(3)$. The idea for this is to exploit the knowledge of the isotropy subgroups for the single irreducible representations. Indeed, it is easy to show that isotropy subgroups for the sum of two irreducible representations are intersections of isotropy subgroups of each of these representations. We then apply our knowledge of the natural realization of these groups as symmetry groups in $I\!R^3$ to perform the classification. In the last part of the paper we show how this applies to the three examples listed above.

2. The irreducible representations of $O(3)$ and their isotropy subgroups

The group $O(3)$ can be decomposed as $O(3) = SO(3)\oplus Z_2^c$ with $Z_2^c = \{I, -I\}$. The irreducible representations of $O(3)$ are therefore easily deduced from those of $SO(3)$ (see [17], [10]). The irreducible representations of $SO(3)$ are $2l+1$ dimensional, with $l = 0, 1, \cdots$. Up to isomorphism, there is only one representation for each value of l. This representation is noted D^l or simply l and it acts on the complexified space V^l of spherical harmonics

$$\{Y_m^l(\theta, \phi), m = -l, \cdots, l, \ (\theta, \phi) \in S^2\}$$

This family of spherical harmonics (for $l = 0, 1, \cdots$) forms a Hilbert basis for the space of square integrable functions over the sphere S^2. For each D^l, there are two non isomorphic irreductible representations of $O(3)$. Indeed, Z_2^c can act on V^l as the identity or as minus identity. The natural action of $O(3)$ on the sphere S^2 leads to the relations :

$$-I.Y_m^l(\theta, \phi) = (-1)^l Y_m^l(\theta, \phi)$$

So the *natural* irreducible representations of $O(3)$ are those for which Z_2^c acts trivially if l is even and as the reflection through the origin in $I\!R^{2l+1}$ if l is odd. The natural representations of $O(3)$ are the most commonly encountered in applications, for example in the problem of onset of convection in a spherical shell. In this case, when l is even, it follows that the symmetry analysis can be reduced to the action of $SO(3)$ instead of $O(3)$. The classification of isotropy subgroups for irreducible representations of $SO(3)$ was given by L.Michel [16] and it was completed by E.Ihrig & M.Golubitsky ([15]). This classification, together with the dimension of the corresponding fixed-point subspaces, can be found in [12] and, including various useful additional informations, in [4]. Note that the trivial group $\mathbb{I}$ and the total group $SO(3)$ are always isotropy subgroups, with fixed-point subspaces $\mathbb{R}^{2l+1}$ and 0 respectively. Every other isotropy subgroup is isomorphic to one group listed in the following set :

$$C = \{\mathbf{Z}_m \ m > 0, \mathbf{D}_m \ m > 1, O(2), \mathbf{T}, \mathbf{O}, \mathbf{I}\} \tag{1}$$

A group isomorphic to $\mathbf{Z}_m$ is generated by a rotation of angle $\frac{2\pi}{m}$ around some axis a in $\mathbb{R}^3$. We shall note it $C_m(a)$ in the following. The other isomorphy classes are $\mathbf{D}_m$: the rotational symmetry group of the regular m-prism, $O(2)$: the rotational symmetry group of the cylinder, $\mathbf{T}$: the rotational symmetry group of the tetrahedron, $\mathbf{O}$: the rotational symmetry group of the cube and $\mathbf{I}$: the rotational symmetry group of the icosahedron. Complete definitions can be found in [17].

3. Isotropy subgroups for coupled representation

The following proposition characterizes the isotropy subgroups for the action of a group G on a space $V = V_1 \oplus V_2$, where V_1 and V_2 are support spaces for two irreductible representations of G.

Proposition 1 *Let G act irreducibly on the spaces V_1 and V_2 and set $V = V_1 \oplus V_2$. Then H is an isotropy subgroup for the action of G in V defined by this direct sum if and only if there exist isotropy subgroups H_1 and H_2 for the representations in V_1 and V_2 resp., such that*

$$H = H_1 \cap H_2.$$

Then $Fix(H) = Fix(H_1) \oplus Fix(H_2)$ (with $Fix(H_i) \subset V_i$).

Proof : Let (x, y) be any element of $V_1 \oplus V_2$ and let H be its isotropy subgroup. Clearly H must fix both elements x and y, so $H \subset H_x \cap H_y$ with H_x and H_y the isotropy subgroups of x and y. Conversely, any element in $H_x \cap H_y$ fixes x as well as y, hence is in H. The fact that $Fix(H) = Fix(H_x) \oplus Fix(H_y)$ follows straightforwardly. ∎

Remark that $SO(3)$ is always an isotropy subgroup with $Fix(SO(3)) = \{0\}$. Therefore if H is a subgroup of $SO(3)$ for one of the representations D^{l_1} or D^{l_2}, then H is an isotropy subgroup for the coupled representation $D^{l_1} \oplus D^{l_2}$ since $H \cap SO(3) = H$. In particular, $\mathbb{I}$ and $SO(3)$ are always isotropy subgroups of any coupled representations.

Thanks to proposition 1, our problem reduces to that of finding all possible intersections of isotropy subgroups of $SO(3)$, i.e. of conjugate groups to elements of C. Since we are interested in conjugacy classes, it is clearly enough to determine, for any groups H_1, H_2 which are isomorphic to some elements of C, the set

$$P(H_1, H_2) = \{\dot{H} / \exists g \in O(3), H = H_1 \cap g H_2 g^{-1}\}$$

Type of intersection	$P(H,K)$
$P(\mathbf{Z}_n, \mathbf{Z}_m)$	$\mathbb{I}, \mathbf{Z}_p$
$P(\mathbf{Z}_n, \mathbf{D}_m)$	$\mathbb{I}, \mathbf{Z}_p, \mathbf{Z}_2$ n even
$P(\mathbf{Z}_n, \mathbf{T})$	$\mathbb{I}, \mathbf{Z}_2$ if n even , $\mathbf{Z}_3$ if $3|n$
$P(\mathbf{Z}_n, \mathbf{O})$	$\mathbb{I}, \mathbf{Z}_4$ if $4|n$, $\mathbf{Z}_3$ if $3|n$, $\mathbf{Z}_2$ n even
$P(\mathbf{Z}_n, \mathbf{I})$	$\mathbb{I}, \mathbf{Z}_5$ if $5|n$, $\mathbf{Z}_3$ if $3|n$, $\mathbf{Z}_2$ n even
$P(\mathbf{Z}_n, O(2))$	$\mathbb{I}, \mathbf{Z}_n, \mathbf{Z}_2$ if n even
$P(\mathbf{D}_m, \mathbf{D}_n)$	$\mathbb{I}, \mathbf{D}_p, \mathbf{Z}_p, \mathbf{D}_2$ if n and m even , $\mathbf{Z}_2$
$P(\mathbf{D}_n, \mathbf{T})$	$\mathbf{D}_2$ n even , $\mathbf{Z}_3$ if $3|n$, $\mathbf{Z}_2$
$P(\mathbf{D}_n, \mathbf{O})$	$\mathbb{I}, \mathbf{Z}_2, \mathbf{D}_2$ n even
	$\mathbf{D}_3$ and $\mathbf{Z}_3$ if $3|n$
	$\mathbf{D}_4$ and $\mathbf{Z}_4$ if $4|n$
$P(\mathbf{D}_n, \mathbf{I})$	$\mathbb{I}, \mathbf{D}_2$ n even , $\mathbf{Z}_2$
	$\mathbf{D}_3$ and $\mathbf{Z}_3$ if $3|n$
	$\mathbf{D}_5$ and $\mathbf{Z}_5$ if $5|n$
$P(\mathbf{D}_n, O(2))$	$\mathbb{I}, \mathbf{D}_n, \mathbf{D}_2$ if n even , $\mathbf{Z}_2$
$P(\mathbf{T}, \mathbf{T})$	$\mathbb{I}, \mathbf{T}, \mathbf{Z}_3, \mathbf{Z}_2$
$P(\mathbf{T}, \mathbf{O})$	$\mathbb{I}, \mathbf{T}, \mathbf{D}_2, \mathbf{Z}_2$
$P(\mathbf{T}, \mathbf{I})$	$\mathbb{I}, \mathbf{T}, \mathbf{Z}_3, \mathbf{Z}_2$
$P(\mathbf{T}, O(2))$	$\mathbb{I}, \mathbf{D}_2, \mathbf{Z}_2, \mathbf{Z}_3$
$P(\mathbf{O}, \mathbf{O})$	$\mathbb{I}, \mathbf{O}, \mathbf{D}_4, \mathbf{Z}_4, \mathbf{D}_3, \mathbf{Z}_3, \mathbf{D}_2, \mathbf{Z}_2$
$P(\mathbf{O}, \mathbf{I})$	$\mathbb{I}, \mathbf{T}, \mathbf{D}_3, \mathbf{Z}_3, \mathbf{D}_2, \mathbf{Z}_2$
$P(\mathbf{O}, O(2))$	$\mathbb{I}, \mathbf{D}_4, \mathbf{D}_3, \mathbf{D}_2, \mathbf{Z}_2$
$P(\mathbf{I}, \mathbf{I})$	$\mathbb{I}, \mathbf{I}, \mathbf{T}, \mathbf{D}_5, \mathbf{Z}_5, \mathbf{D}_3, \mathbf{Z}_3, \mathbf{Z}_2$
$P(\mathbf{I}, O(2))$	$\mathbb{I}, \mathbf{D}_5, \mathbf{D}_3, \mathbf{D}_2, \mathbf{Z}_2$
$P(O(2), O(2))$	$O(2), \mathbf{D}_2, \mathbf{Z}_2$

Table 1: Type of intersections for subgroups of $SO(3)$

where $\dot{H}$ is the conjugacy class of H in $SO(3)$. From the knowledge of the lattices of isotropy types for the irred reps D^{l_1} and D^{l_2} it is then straightforward to compute the lattice of isotropy types of the coupled representation $D^{l_1} \oplus D^{l_2}$. Another possible approach would have been to apply the proof in ([15]) directly to the representations of the form $D^{l_1} \oplus D^{l_2}$, all l_1 and l_2. We choose not to do this because it would result in a very heavy listing of the isotropy subgroups with respect to the values of l_1 and l_2.

Theorem 2 *If H and K are two subgroups of $SO(3)$, then the sets $P(H,K)$ are given by the table (1).*

The proof of this theorem is geometrical. We shall not give all details here because there are many cases to check, but rather the idea of the technique employed and an example of computation. Full proof can be found in [13]. The basic remark is that any proper isotropy subgroup of $SO(3)$ is either a union of cyclic subgroups, or it is isomorphic to $O(2)$ (which group can also be considered as a union of cyclic subgroups if the group $SO(2)$ is understood as a "topological" cyclic group, i.e. as the closure of the group generated by an irrational rotation). For a cyclic subgroup we use the general notation $C_m(a)$, where a is its axis of symmetry and m is its order. Hence $C_m(a)$ is isomorphic to $\mathbf{Z}_m$ and is generated by the rotation of angle $\frac{2\pi}{m}$ around the axis a. A group isomorphic to $SO(2)$ and with axis a is noted $C_\infty(a)$. Notice that

$$C_n(a) \cap C_m(a) = C_p(a) \text{ with } p = gcd(m,n)$$

and

$$C_\infty(a) \cap C_m(a) = C_m(a)$$

Of course, if $a \neq b$, then $C_n(a) \cap C_m(b) = \mathbb{I}$ for any m and n. We shall abuse of notations by using the same symbols to designate a subgroup isomorphic to an element in C and this element. Then the decomposition in cyclic subgroups for representatives of elements in C is given by :

$$
\begin{aligned}
\mathbf{Z}_m : \quad & C_m(a) \\
\mathbf{D}_m : \quad & C_m(a) \bigcup_{i=1}^{m} C_2(b_i), \ i = 1 \cdots m \\
\mathbf{O}(2) : \quad & C_\infty(a) \bigcup_{b \in P} C_2(b) \\
\mathbf{T} : \quad & \bigcup_{i=1}^{4} C_3(a_i) \bigcup_{i=1}^{3} C_2(b_i) \\
\mathbf{O} : \quad & \bigcup_{i=1}^{4} C_4(a_i) \bigcup_{i=1}^{4} C_3(b_i) \bigcup_{i=1}^{6} C_2(c_i) \\
\mathbf{I} : \quad & \bigcup_{i=1}^{6} C_5(a_i) \bigcup_{i=1}^{3} C_3(b_i) \bigcup_{i=1}^{6} C_2(c_i)
\end{aligned}
$$

where a, a_i, b_i and c_i are axes in $I\!R^3$ which are in a given relative position to each other. Note that the cyclic subgroups are grouped by conjugacy classes, with one exception: D_2, for which there is only one conjugacy class. For a group H of the above list, let $N(H)$ be its normalizer in $SO(3)$. Any element in $N(H)$ fixes H by conjugacy, hence it permutes (possibly trivially) the cyclic subgroups of the decomposition of H within their conjugacy classes.

We call *skeleton* of H and we note $Sk(H)$ the set of axes which appear in the cyclic decomposition of H. For example

$$Sk(D_m) = \{a, b_1, \cdots, b_m\}$$

The normalizer of H operates on $Sk(H)$ by permuting axes within the same conjugacy classes of associated cyclic subgroups. In other words, there is a natural bijection between the orbit space $Sk(H)/N(H)$ and the set of conjugacy classes of associated cyclic subgroups in the cyclic decomposition of H.

Our strategy will now be the following: given two groups H and K in the list above, act on K by $g \in SO(3)$ in order to match an axis of H with an axis of gKg^{-1}. This requires two rotations, if one parametrizes $SO(3)$ with its Euler angles for example. Hence there is an axis a and two groups $C_m(a)$ and $C_n(a)$ belonging to the cyclic decomposition of H and gKg^{-1} respectively. If $p = gcd(m, n) > 1$, this implies that $H \cap gKg^{-1}$ contains the group $C_p(a)$. After this "matching" operation, there remains one rotation g_a (around the axis a) which we can freely vary in order to get all possible groups of the form $H \cap g_a gK(g_a g)^{-1}$, which contain $C_p(a)$. In order to completely determine the set $P(H, K)$, one should a priori redo this process for all different axis in $Sk(H)$ and $Sk(K)$. It is however obviously enough to take only one representative of each $N(H)$-orbit and $N(K)$-orbit in these sets. So, for $\dot{a} \in Sk(H)/N(H)$ and $\dot{b} \in Sk(K)/N(K)$ we can define the set

$$\Delta(H, a, K, b) = \{\dot{L}/ \; \exists g_a \in C_\infty(a), \; L = H \cap g_a \cdot gK \cdot g^{-1} \cdot g_a^{-1} \text{ with } g \text{ s.t. } a = g \cdot b\}$$

Then it is clear that

$$P(H, K) = \bigcup_{\substack{a \in Sk(H)/N(H) \\ b \in Sk(K)/N(K)}} \Delta(H, a, K, b)$$

This makes a large, but tractable number of cases to scan. As an example let us show that

$$P(\mathbf{D}_m, \mathbf{D}_n) = \{\mathbf{D}_p, \mathbf{Z}_p, \mathbf{D}_2 \text{ if } m \text{ and } n \text{ are even }, \mathbf{Z}_2, \mathbb{I}\}.$$

Let us set

$$D_m = \{C_m(a_1), C_2(b_1) \cdots, C_2(b_m)\}$$
$$D_n = \{C_m(a_2), C_2(c_1) \cdots, C_2(c_n)\}$$

Note that a_1 and b_i, as well as a_2 and c_1, make an angle of $\pi/2$. This is a crucial information in the following computations.

Let $p = gcd(m, n)$. We shall note L the intersection of D_m with conjugates of D_n.

Case 1. $\Delta(D_m, a_1, D_n, a_2) = \{\mathbf{D}_p, \mathbf{Z}_p\}$. If a_1 and a_2 are made coïncident, then $C_p(a_1)$ belongs to the group L. Let us rotate $Sk(D_n)$ around a_1. When an axis b_i coïncides with an axis c_j, which is possible since $a_1 \perp b_1$ and $a_2 \perp c_1$, $C_2(b_1)$ is also a subgroup in L. Hence L is generated by $C_p(a_1)$ and $C_2(b_1)$, i.e. it is equal to D_p. Otherwise, L reduces to Z_p.

Case 2. $\Delta(D_m, a_1, D_n, c_1) = \{\mathbf{D}_2 \text{ if } m \text{ and } n \text{ are even}, \mathbf{Z}_2 \text{ if } m \text{ even }\}$. If m is odd $gcd(m, 2) = 1$ so this case is not to take into account. If m is even, $C_2(a_1)$ is in L. Now rotate $Sk(D_n)$ around a_1. The axis b_1 can then comes into coïncidence with a_2, for the reason invoked in the previous case. If however n is odd, the corresponding cyclic subgroup reduces to $\mathbb{I}$ and therefore $L = \mathbf{Z}_2$. If n is even, the corresponding cyclic subgroup is $C_2(b_1)$ and no other axes can coïncide. Hence in this case L is generated by $C_2(a_1)$ and $C_2(c_1)$, i.e. $L = \mathbf{D}_2$.

Case 3. $\Delta(D_m, b_1, D_n, c_1) = \{\mathbf{D}_2$ if m and n even, $\mathbf{D}_p\}$. Then L always contains $C_2(b_1)$. If no further axes do coïncide, then $L = \mathbf{Z}_2$. Otherwise we can either adjust a_1 and a_2, but this has already been considered in *Case 1*, or, if m and n are even, we can adjust a_1 and c_1, which has been considered in *Case 2*.

Case 4. Finally, it is easy to see that one can always find an element g in $SO(3)$ such that $D_m(a_1) \cap g \cdot D_n(a_2) \cdot g^{-1} = \mathbb{I}$. This finishes the proof.

4. Applications

In this section we study 3 examples of "even-even" interactions of irreducible representations of $O(3)$. If H is an isotropy subgroup for an even representation, then $H = H_1 \oplus \mathbf{Z}_2^c$, with H_1 a subgroup of $SO(3)$. Since :

$$H_1 \oplus \mathbf{Z}_2^c \cap H_2 \oplus \mathbf{Z}_2^c = (H_1 \cap H_2) \oplus \mathbf{Z}_2^c$$

we can apply the previous results to determine isotropy lattices of "even-even" representations.

We use the following complex coordinates :

$$X = \sum_{i=-l_1}^{l_1} x_i Y_i^{l_1}(\Theta, \phi) + \sum_{j=-l_2}^{l_2} y_j Y_j^{l_2}(\Theta, \phi) \tag{2}$$

The bifurcation equations have been determined using the algorithm of D.Sattinger ([20]). In the $2, 2$ case, a further normal form reduction can be performed because the linearized operator is nilpotent at criticallity.

The fixed point subspaces corresponding to the isotropy lattice have been determined using the results of [4] for the planar subgroups. For the exceptional subgroups we have made use of symbolic calculus (MAPLE) to compute explicitly the matricial representation of their generators (I.M.Gelfand *et al.* ([10]) and their fixed point subspaces.

We have not attempted to give an entire description of the dynamics. We have just focused our attention on specific interesting situations related to the applications we had in mind.

4.1 THE 2,2 MODES INTERACTION

For the case $l = 2$, the isotropy groups are $O(2) \oplus \mathbf{Z}_2^c, \mathbf{D}_2 \oplus \mathbf{Z}_2^c$ and $\mathbf{Z}_2^c$ ([4]). Applying theorem 2, we obtain :

$$P(O(2)\oplus Z_2^c, O(2)\oplus Z_2^c) = \{O(2)\oplus Z_2^c, D_2\oplus Z_2^c, Z_2\oplus Z_2^c\}$$
$$P(O(2)\oplus Z_2^c, D_2\oplus Z_2^c) = \{D_2\oplus Z_2^c, Z_2\oplus Z_2^c\}$$
$$P(D_2\oplus Z_2^c, D_2\oplus Z_2^c) = \{D_2\oplus Z_2^c, Z_2\oplus Z_2^c\}$$

and $P(O(2)\oplus Z_2^c, Z_2^c) = P(D_2\oplus Z_2^c, Z_2^c) = P(Z_2\oplus Z_2^c, Z_2^c) = P(Z_2^c, Z_2^c) = \{Z_2^c\}$.

Thefore the isotropy lattice and the corresponding fixed point subspaces are those given in figure (1).

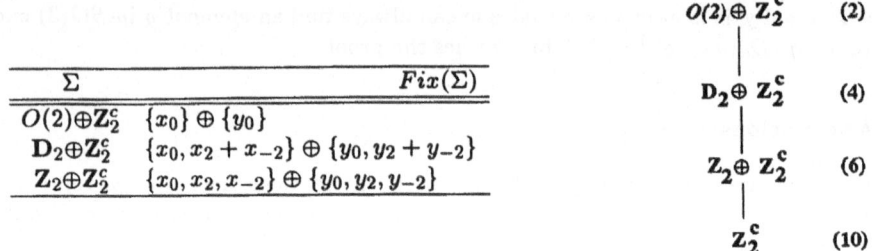

Σ	$Fix(\Sigma)$
$O(2)\oplus Z_2^c$	$\{x_0\} \oplus \{y_0\}$
$D_2\oplus Z_2^c$	$\{x_0, x_2 + x_{-2}\} \oplus \{y_0, y_2 + y_{-2}\}$
$Z_2\oplus Z_2^c$	$\{x_0, x_2, x_{-2}\} \oplus \{y_0, y_2, y_{-2}\}$

$$O(2)\oplus Z_2^c \quad (2)$$
$$D_2\oplus Z_2^c \quad (4)$$
$$Z_2\oplus Z_2^c \quad (6)$$
$$Z_2^c \quad (10)$$

Figure 1: Isotropy lattice and fixed point subspaces for the 2,2 interaction

For this representation, the support space is $V = V_1 \oplus V_2 \equiv I\!\!R^5 \oplus I\!\!R^5$, which is 10 dimensional. Since the action of $O(3)$ is the sum of two isomorphic representations, the linear part of the bifurcation equations need not be semi-simple. After normalisation, the linearized operator L in the equations takes the form :

$$L = \begin{pmatrix} O_5 & I_5 \\ O_5 & O_5 \end{pmatrix}$$

where O_5 is the 5×5 null matrix and I_5 is the 5×5 identity matrix. We note :

$$\begin{cases} \dot{X} &= \Phi(X,Y) \\ \dot{Y} &= \Psi(X,Y) \end{cases} \qquad (3)$$

the normal form for bifurcation equations, with $\Phi = (\Phi_{-2}, \cdots, \Phi_2)$ and $\Psi = (\Psi_{-2}, \cdots, \Psi_2)$. Then normal form theory (see [1] for an elementary exposition of the method) provides the following relations :

$$\begin{cases} \sum_j x_j \frac{\partial \Phi_i}{\partial y_j} &= 0 \\ \sum_j x_j \frac{\partial \Psi_i}{\partial y_j} &= \Phi_i \end{cases} \qquad (4)$$

Comparing the solutions of (4) with the general form for $O(3)$-equivariant equations yields the equations (truncated at second order) :

$$
\left\{
\begin{aligned}
\dot{x}_2 &= y_2 + A(x_0 x_2 - \sqrt{6}/4 x_1^2) \\
\dot{y}_2 &= \lambda_1 x_2 + \lambda_2 y_2 + B(x_0 x_2 - \sqrt{6}/4 x_1^2) + A/4(2x_2 y_0 - \sqrt{6} x_1 y_1 + 2x_0 y_2) \\
\dot{x}_1 &= y_1 + A/2(\sqrt{6} x_2 x_{-1} - x_0 x_1) \\
\dot{y}_1 &= \lambda_1 x_1 + \lambda_2 y_1 + B/2(x_0 x_2 - \sqrt{6}/4 x_1^2) + A/4(\sqrt{6} x_2 y_{-1} - x_1 y_0 - x_0 y_1 + \sqrt{6}/2 x_{-1} y_2) \\
\dot{x}_0 &= y_0 + A(x_2 x_{-2} + 1/2 x_1 x_{-1} - 1/2 x_0^2) \\
\dot{y}_0 &= \lambda_1 x_0 + \lambda_2 y_0 + B(x_2 x_{-2} + 1/2 x_1 x_{-1} - 1/2 x_0^2) \\
&\quad + A/4(x_1 y_{-1} + 2x_2 y_{-2} - 2x_0 y_0 + x_{-1} y_1 + 2x_{-2} y_2) \\
\dot{x}_{-i} &= (-1)^i \overline{x_{-i}} \\
\dot{y}_{-i} &= (-1)^i \overline{y_{-i}}
\end{aligned}
\right.
$$

$$(5)$$

Rescaling all the variables and the time allows to set $A = 1$ and $B = \pm 1$.

4.1.1. *Bifurcations in* $Fix(O(2) \oplus \mathbf{Z}_2^c)$. We obtain the bifurcation equation by restriction of (5) to (x_0, y_0). In this plane, the linearized operator is just the Takens-Bogdanov singularity. The bifurcation equations can be set to the equivalent form ([14]) :

$$
\left\{
\begin{aligned}
\dot{x}_0 &= y_0 \\
\dot{y}_0 &= \lambda_1 + \lambda_2 y_0 + x_0^2 \pm x_0 y_0
\end{aligned}
\right.
$$

$$(6)$$

The dynamics of this equation is well known and depends on the sign of the term $x_0 y_0$ in the second equation. We refer the reader to [14] for a complete description of the unfolding of this singularity.

4.1.2. *Bifurcations in* $Fix(\mathbf{D}_2 \oplus \mathbf{Z}_2^c)$. In the fixed point subspace of $D_2 \oplus \mathbf{Z}_2^c$, there are three planes of symmetry: $Fix(O(2) \oplus \mathbf{Z}_2^c)$ together with two of its conjugates which, in terms of coordinates, are defined as :

$$
(\pm\sqrt{\tfrac{3}{2}} x_0, 0, x_0, 0, \pm\sqrt{\tfrac{3}{2}} x_0) \oplus (\pm\sqrt{\tfrac{3}{2}} y_0, 0, y_0, 0, \pm\sqrt{\tfrac{3}{2}} y_0)
$$

Solving the bifurcation equation shows that there are no steady state solutions out of these three invariant planes.

4.1.3. *Bifurcations in* $Fix(\mathbf{Z}_2 \oplus \mathbf{Z}_2^c)$. The subspace $Fix(\mathbf{Z}_2 \oplus \mathbf{Z}_2^c)$ is six-dimensional. The normalizer of $\mathbf{Z}_2 \oplus \mathbf{Z}_2^c$ in $O(3)$ is $O(2) \oplus \mathbf{Z}_2^c$ [1]. Hence $O(2)$ acts faithfully in $Fix(\mathbf{Z}_2 \oplus \mathbf{Z}_2^c)$. Notice that $SO(2)$ acts in this space by :

$$
\forall g_\phi \in SO(2) : g_\phi.(x_{-2}, x_0, x_2, y_{-2}, y_0, y_2) \rightarrow (e^{-i\phi}.x_{-2}, x_0, e^{i\phi}.x_2, e^{-i\phi}.y_{-2}, y_0, e^{i\phi}.y_2) \quad (7)
$$

---

[1] Here, $\mathbf{Z}_2$ is the rotation by π in the $SO(2)$ component of $O(2)$

Let us therefore look for rotating wave solutions of the form :

$$x_2 = \tilde{x}_2 e^{i(\omega t + \varphi)}, y_2 = \tilde{y}_2 e^{i(\omega t + \varphi)}, x_0, y_0, x_{-2} = \overline{x_2}, y_{-2} = \overline{y_2} \tag{8}$$

where ω is close to 0 and φ is an arbitrary phase.

Thanks to relation (7), the problem is then transformed to a steady one :

$$\begin{cases} i\omega\dot{\tilde{x}}_2 &= \tilde{y}_2 + x_0\tilde{x}_2 \\ i\omega\dot{\tilde{x}}_2 &= \tilde{y}_2 + x_0\tilde{x}_2 \\ \dot{x}_0 &= y_0 + \tilde{x}_2{}^2 - 1/2x_0^2 \\ \dot{y}_0 &= \lambda_1 x_0 + \lambda_2 y_0 + \epsilon(\tilde{x}_2{}^2 - 1/2x_0^2) + 1/2(\tilde{x}_2(\tilde{y}_2 + y_{-2}) - x_0 y_0 \\ \dot{\tilde{x}}_{-2} &= \overline{\dot{\tilde{x}}_2} \\ \dot{\tilde{y}}_{-2} &= \overline{\dot{\tilde{y}}_2} \end{cases} \tag{9}$$

After resolution of this system, we find rotating waves which at first order are given by :

$$\omega^2 = 2/3\epsilon\lambda_2 - \lambda_1, \quad \tilde{x}_2 = 2/3\lambda_2(\epsilon\lambda_1 - 1/3\lambda_2), \quad x_0 = -2/3\lambda_2$$

$$\tilde{y}_2 = \tilde{x}_2(i\omega - x_0), \quad y_0 = x_0^2/2 - \tilde{x}_2{}^2$$

If we fix $\lambda_1 > 0$ and if we set $\epsilon = 1$, rotating waves bifurcate supercritically from the origin with λ_2. The branch exists in $Fix(\mathbf{Z}_2\oplus\mathbf{Z}_2^c)$ until $\lambda_2 = 3\lambda_1$ where it connect to solutions in $Fix(O(2)\oplus\mathbf{Z}_2^c)$ via a Hopf bifurcation. Since $Fix(\mathbf{Z}_2\oplus\mathbf{Z}_2^c)$ is globally invariant by the action (7) of $O(2)$, we can treat the bifurcation problem in this subspace as an usual $O(2)$-symmetric Hopf bifurcation. Rotating waves and standing waves can bifurcate in $Fix(\mathbf{Z}_2\oplus\mathbf{Z}_2^c)$ from steady states in $Fix(O(2)\oplus\mathbf{Z}_2^c)$ ([15], [7]). Moreover, standing waves must belong to $Fix(\mathbf{D}_2\oplus\mathbf{Z}_2^c)$, which therefore proves the existence of limit cycles in this subspace.

4.2. THE 2,4 MODES INTERACTION

A series of numerical simulations of M. Rieutord and L.Valdetarro for convection with a compressible fluid in a spherical shell ([21], [22]) seem to indicate such an interaction, showing in particular a combination of axisymmetric modes with $l = 2$ and $l = 4$, and a 3-fold symmetry. The isotropy lattice in this case is given in figure (2) and corresponding fixed points subspaces are listed in table (2).

Let us restrict our attention to the 3-dimensional subspace $Fix(\mathbf{D}_3\oplus\mathbf{Z}_2^c)$. This space should indeed contain the above mentionned solutions. The system in this subspace is given at second order by :

$$\begin{cases} \dot{x}_0 &= \lambda_1 x_0 + Ax_0^2 + Bx_0 y_0 + C(7z^2 - 10y_0^2) \\ \dot{y}_0 &= \lambda_2 y_0 + Dx_0^2 - 3F(7z^2 - 3y_0^2) - \frac{5}{7}Gx_0 y_0 \\ \dot{z} &= \lambda_2 z + Bx_0 y_0 - 21Fy_0 z + \frac{G}{4}x_0 z \end{cases} \tag{10}$$

Σ	$Fix(\Sigma)$
$O(2){\oplus}\mathbf{Z}_2^c$	$\{x_0\} \oplus \{y_0\}$
$\mathbf{D}_4{\oplus}\mathbf{Z}_2^c$	$\{x_0\} \oplus \{y_0, y_4 + y_{-4}\}$
$\mathbf{D}_3{\oplus}\mathbf{Z}_2^c$	$\{x_0\} \oplus \{y_0, y_3 + y_{-3}\}$
$\mathbf{D}_2{\oplus}\mathbf{Z}_2^c$	$\{x_0, x_2 + x_{-2}\} \oplus \{y_0, y_2 + y_{-2}, y_4 + y_{-4}\}$
$\mathbf{Z}_2{\oplus}\mathbf{Z}_2^c$	$\{x_0, x_{\pm2}\} \oplus \{y_0, y_{\pm2}, y_{\pm4}\}$
$O{\oplus}\mathbf{Z}_2^c$	$(0, \sqrt{\frac{10}{7}}y_0, 0, 0, y_0, 0, 0, \sqrt{\frac{10}{7}}y_0, 0)$

Table 2: Isotropy subgroups and their fixed-point subspace for l=2,4

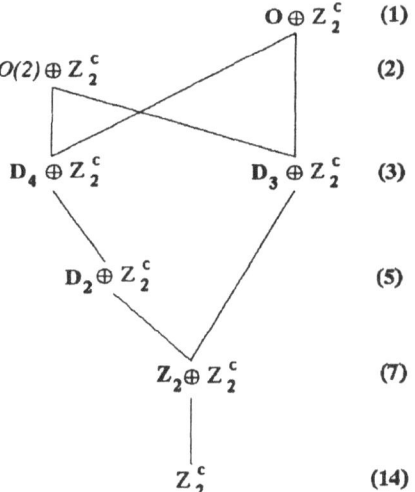

Figure 2: Isotropy lattice for the 2,4 interaction

with $z = (y_3 + y_{-3})/2$.

Notice that the structure of $Fix(\mathbf{D}_4 \oplus \mathbf{Z}_2^c)$ and the equations in it are similar to the foregoing ones.

In order to obtain bounded dynamics, we add to the system a term of the form $-HX \parallel X \parallel^2$ (which is equivariant). It turns out to be physically relevant to assume the following conditions on the coefficients :

$$H = 1, A = 0, F = 0, D = -B, G = -14C \tag{11}$$

Although this looks highly degenerate, these conditions appear "naturally" as consequences of the fact that the linear part in the equations for the classical Bénard problem is a self-adjoint operator (see [3], [9], [18] and [13] for details). In the case of bifurcation associated with a single irreducible representation, the unfolding of this kind of degeneracy has been studied by Golubitsky & Schaeffer ([11]) when $l = 2$ and by Geiger et al. ([8]) when $l = 4$.

Finally, by rescaling variables and time, we can set $H = 1, C = 1$.

4.2.1. Bifurcations in $Fix(\mathbf{O} \oplus \mathbf{Z}_2^c)$. We note $z = (0, \sqrt{\frac{10}{7}} y_0, 0, 0, y_0, 0, 0, \sqrt{\frac{10}{7}} y_0, 0)$. The bifurcated solutions are given at first order by :

$$\lambda_2 = \frac{27}{10} z^2$$

Because pure quadratic terms have a null coefficient by the above assumption, this is a supercritical bifurcation.

A typical convective flow of this kind is pictured in figure (5).

4.2.2. Bifurcations in $Fix(O(2) \oplus \mathbf{Z}_2^c)$. With the conditions (11), the equations are given

$$\begin{cases} \dot{x}_0 = \lambda_1 x_0 + B x_0 y_0 + y_0^2 - x_0(x_0^2 + y_0^2) \\ \dot{y}_0 = \lambda_2 y_0 - x_0 y_0 - B x_0^2 - y_0(x_0^2 + y_0^2) \end{cases} \tag{12}$$

In polar coordinates and after some simple algebra, we get that solutions are the intersections of two curves, defined by :

$$\Gamma_1 \; : r = (\lambda_1 \cos^2 \theta + \lambda_2 \sin^2 \theta)^{1/2}, \quad \text{and} \quad \Gamma_2 \; : r = \frac{(\lambda_2 - \lambda_1) \cos \theta \sin \theta}{B \cos \theta - \sin \theta}$$

Corresponding phase portraits are shown in figure (3). These pictures have been checked numerically (with the software DSTOOL).

We have not studied the stability of these solutions out of the invariant plane. Notice that limit cycles do bifurcate from the bifurcated foci between region I and II for example.

4.2.3 Bifurcations in $Fix(\mathbf{D}_3 \oplus \mathbf{Z}_2^c)$: In this subspace, equilibria are given by :

Figure 3: Phase portraits in the (x,y) plane for solutions with $O(2)\oplus\mathbf{Z}_2^c$ symmetry in 2,4 interaction, with the corresponding regions in the parameter plane.

$$x = \frac{22B\lambda - 2}{4(\lambda_1 - \lambda_2) - 121B^2}, y = \frac{20}{27}Bx$$

$$y = \frac{\sqrt{10}}{20}(20\lambda_2 - 7x - 20x^2 - 8000/829B^2x^2)^{1/2}$$

The domain of existence of these solutions and their stability in this subspace is given in figure (??).

Notice that in $Fix(\mathbf{D}_3\oplus\mathbf{Z}_2^c)$, the subspace $(0, -\sqrt{\frac{10}{7}}y_0, 0, 0, y_0, 0, 0, -\sqrt{\frac{10}{7}}, 0, 0)$ is conjugated to $Fix(\mathbf{O}\oplus\mathbf{Z}_2^c)$. In this subspace, the dynamics and solutions are the same as in $Fix(\mathbf{O}\oplus\mathbf{Z}_2^c)$.

Numerically, we have found stable solutions in a horn near the bifurcation curve in the $\lambda_1 > 0, \lambda_2 < 0$ quadrant. This solutions bifurcate from the solution with $O(2)\oplus\mathbf{Z}_2^c$ symmetry. They present an axial symmetry with a small $\mathbf{D}_3\oplus\mathbf{Z}_2^c$ distortion, as in the numerical simulations of M. Rieutord and L.Valdetarro ([21]). A convective flow on the sphere for such a generic stable solution is represented in figure (5).

4.3 THE 2,6 MODES INTERACTION

As we have pointed out in the introduction, this mode interaction seems to appears in geophysical measurements of convection in earth's mantle. More precisely, these measurements have shown patterns of convection with an l=2 axial symmetry and a small l=6 icosaedral

92

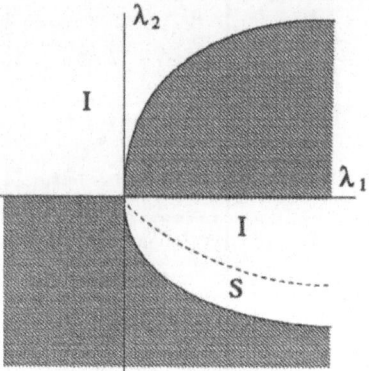

Figure 4: Domains of existence for solutions with symmetry $\mathbf{D}_3 \oplus \mathbf{Z}_2^c$ and $\mathbf{D}_4 \oplus \mathbf{Z}_2^c$ for the 2,4 interaction and for solution with $\mathbf{D}_5 \oplus \mathbf{Z}_2^c$ symmetry for the 2,6 interaction. Solutions are stables (in the fixed point subspace of their respective isotropy group) in the region S, unstable in the region I.

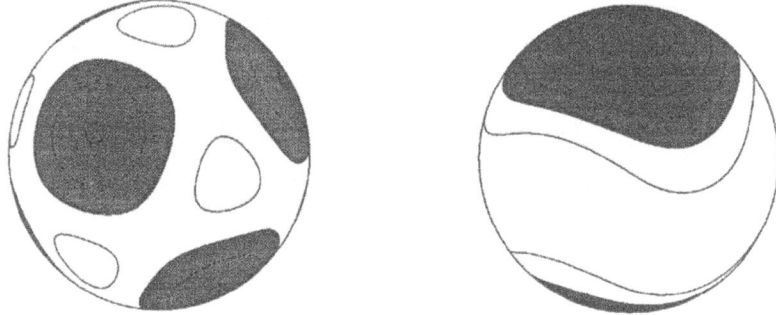

Figure 5: Convective solution with $\mathbf{O} \oplus \mathbf{Z}_2^c$ symmetry (left) and $\mathbf{D}_3 \oplus \mathbf{Z}_2^c$ symmetry (right). Lines are the radial isospeed. Grey regions represent the downwelling motion

component (P. Chossat & C.Stewart [2]).
The isotropy lattice and the corresponding fixed point subspaces are given respectively in figure (6) and in table (3)
With the same considerations that in the 2,4 case, the bifurcations equations in this fixed point subspace take the form :

$$
\begin{cases}
\dot{x_0} &= \lambda_1 x_0 + B(7y_0^2 - 11z^2) - x(x^2 + y^2 + 2z^2) \\
\dot{y_0} &= \lambda_2 y_0 - 7Bx_0 y_0 - y(x^2 + y^2 + 2z^2) + Fx^3 \\
\dot{z} &= \lambda_2 z + 11/2 Bx_0 z - z(x^2 + y^2 + 2z_2)
\end{cases}
\tag{13}
$$

Σ	$Fix(\Sigma)$
$O(2){\oplus}Z_2^c$	$\{x_0\} \oplus \{y_0\}$
$\mathbf{D_4}{\oplus}\mathbf{Z_2^c}$	$\{x_0\} \oplus \{y_0, y_4 + y_{-4}\}$
$\mathbf{D_3}{\oplus}\mathbf{Z_2^c}$	$\{x_0\} \oplus \{y_0, y_3 + y_{-3}, y_6 + y_{-6}\}$
$\mathbf{D_2}{\oplus}\mathbf{Z_2^c}$	$\{x_0, x_2 + x_{-2}\} \oplus \{y_0, y_2 + y_{-2}, y_4 + y_{-4}, y_6 + y_{-6}\}$
$\mathbf{Z_2}{\oplus}\mathbf{Z_2^c}$	$\{x_0, x_{\pm2}\} \oplus \{y_0, y_{\pm2}, y_{\pm4}, y_{\pm6}\}$
$\mathbf{O}{\oplus}\mathbf{Z_2^c}$	$(0, \sqrt{\frac{10}{7}}y_0, 0, 0, y_0, 0, 0, \sqrt{\frac{10}{7}}y_0, 0)$

Table 3: Fixed-point subspaces for l=2,6 isotropy subgroups

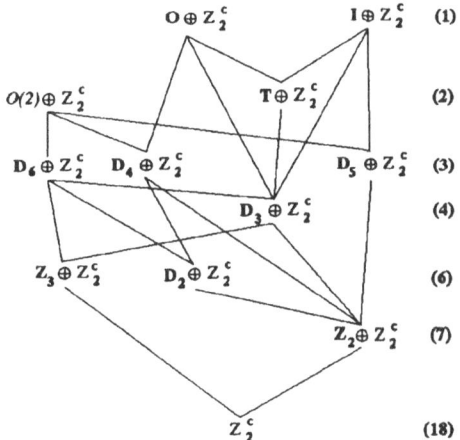

Figure 6: Isotropy lattice for the 2,6 interaction

94

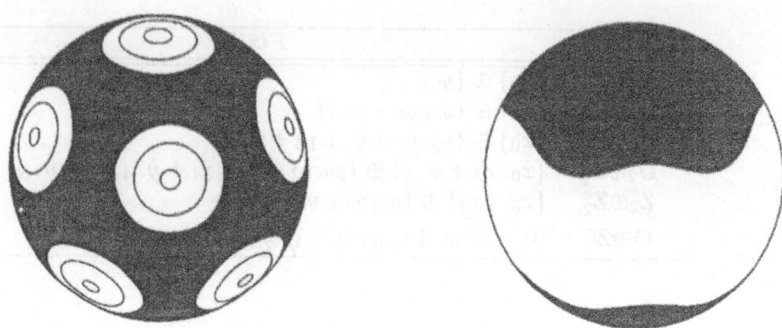

Figure 7: Convective solution with $I \oplus Z_2^c$ symmetry (left) and $D_5 \oplus Z_2^c$ symmetry (right). Lines are the radial isospeed. Grey regions represent the down welling motion

where $z = (0, -\sqrt{\frac{7}{11}}y_0, 0, 0, y_0, 0, 0, \sqrt{\frac{7}{11}}y_0, 0)$. Here we have added the term Fx^3 in order to remove a degeneracy which occurs at the quadratic level. By rescaling $x_0, y_0, z, \lambda_1, \lambda_2$ and the time, we can set $B = 1/7$.

The bifurcated equations are similar to those of the l=2,4 case. Hence, it can be shown that solutions and dynamics are of the same kind ([13]). As in the 2,4 case, we have numerically found stable solutions in $Fix(D_5 \oplus Z_2^c)$. These solutions bifurcate from solution with $O(2) \oplus Z_2^c$ symmetry. A corresponding convective flow is pictured in figure (7).

References

[1] M. Adelmeyer, G. Iooss, *Topics in Bifurcation Theory and Applications*, Advanced series in non linear dynamics, 3, World Scientific, 1992

[2] D. Armbruster , P. Chossat, *Heteroclinic orbits in a spherically invariant system* , Physica D, 50, pages 155-176, 1991

[3] F. Busse, N. Riahi, *Mixed mode patterns of bifurcation from spherically symmetric basis states*, Nonlinearity, 1, 1988, pages 379,388

[4] P. Chossat, R. Lauterbach, I. Melbourne, *Steady-State Bifurcation with O(3)-Symmetry*, Arch. Rational Mech. Anal., 113, pages 313- 376, 1990

[5] S. Chandrasekhar, *Hydrodynamic and hydromagnetic stability*, Dover, New-York, 1961

[6] P. Chossat, C.A. Stewart, *Spherical Symmetry-Breaking Bifurcations* , Chaotic Processes in Geophysical Sciences, the IMA volume of Mathematics and Applications , Vol 41, pages 11-42, 1992

[7] G. Dangelmayer, E. Knobloch, *The Takens-Bogdanov bifurcation with O(2)-symmetry*, Phi. Trans. R. Soc. Lond. A, 322, pages 243-279, 1987

[8] G. Dangelmayr, C. Geiger, J.D. Rodriguez, *On a Degenerate Bifurcation with O(3) Symmetry*, preprint of the Institut fur Informationsverarbeitung, Universitat of Tubingen, 1993

[9] R. Friedrich, H. Haken , Static, wavelike and chaotic thermal convection in spherical geometries, Phys. Rev. A, 1986 , 34, pages 2100-2120

[10] I.M. Gelfand, R.A. Milnos, Z.Y. Shapiro, *Representations of the rotation and Lorentz groups and theirs applications*, Pergamon Press, London, 1963

[11] M. Golubitsky, D.G. Schaeffer, *Bifurcations with O(3) symmetry including applications to the Bénard problem*, Comm. Pure. Appl. Math., XXXV, pages 81-111, 1982

[12] M. Golubitsky, D.G. Schaeffer, I.N. Stewart,*Singularities and Groups in Bifurcation Theory*, Vol. II, Appl. Math. Sci., Springer-Verlag, New-York, 1988

[13] F. Guyard, *Couplages de modes dans les systèmes avec symétrie sphérique*, Thèse de l'Université de Nice, in preparation, 1994

[14] P. Holmes, J. Guckenheimer, *Nonlinear Oscillations, Dynamical Systems, and Bifurcations of Vector Fields*, Appl. Math. Sci., Springer-Verlag, New-York, 1983

[15] E. Ihrig, M. Golubitsky, *Pattern selection with $O(3)$ symmetry*, Physica 13D, pages 1-33, 1984

[16] L. Michel, *Symmetry defects and broken symmetry. Confugurations. Hidden symmetry*, Reviews of Modern Physics, 52, pages 617-651, july, 1980

[17] W. Miller, *Symmetry groups and their applications*, Academic Press, New-York, 1972

[18] E. Moutrane, *Interaction de modes spheriques dans le problème de Bénard entre deux sphères*, Ph. D. Thesis, Université de Nice , 1988

[19] D.L. Turcotte, E.R. Oxburgh, *Mechanisms of Continental Drift*,Reports on Progress in Physics,1978, 41, pages 1249-1312

[20] D.H. Sattinger, *Group Theoritic Methods in Bifurcation Theory* , Lectures Notes in Mathematics, 762, Springer-Verlag, Berlin, 1979

[21] L .Valdetarro, M. Rieutord, *Numerical study of the transition to chaotic convection inside spherical shells*, In Proceeding from the Large Scale Structures in Non Linear Physics Workshop, pages 212-221, 1991, Springer-Verlag, Berlin New-York

[22] L .Valdetarro, Personnal communication

[23] E.P. Wigner,*Group Theory*, Academic Press, New-York, 1959

NON LINEAR PARABOLIC EVOLUTIONS
IN UNBOUNDED DOMAINS

P.COLLET
Centre de Physique Théorique, CNRS UPR 14
Ecole Polytechnique, F-91128 Palaiseau Cedex (France)

ABSTRACT. We describe some results on the global existence in time of the semi flow generated by some non linear parabolic equations in unbounded space domains. We also discuss some interesting solutions emerging from the bifurcation of a continuous spectrum.

1. Introduction

The time evolution of many important physical systems is given by a non linear parabolic equation (or coupled systems of such equations). These equations are in general supplemented by boundary conditions in order to provide a well defined semi-flow of time evolution in some function space. One of the most well known example corresponds to the time evolution of an incompressible viscous two dimensional fluid in a finite container. The phase space is a space of divergence-less vector fields which vanish at the boundary equipped with an adequate topology, and the time evolution is given by the Navier Stokes equation. In a bounded space domain it turns out under rather general hypothesis that the semi-flows generated by such non linear parabolic equations behave very much like a finite dimensional dynamical system although they are defined on an infinite dimensional phase space. In other words, although the system possesses an infinite number of degrees of freedom, only a finite number of them really drive the time evolution the (infinite number of) others are somehow slaves of the dominant ones (this is sometimes referred to as a finite number of degrees of freedom excited). At a Mathematical level this is reflected for example in the fact that bifurcation theory applies equally well in Banach spaces provided that tangent maps have a discrete spectrum. Other results in the same direction include the finiteness of the Hausdorff dimension of the global attractors and the recent constructions of inertial manifolds for a large class of problems.

Unfortunately, when the size of the system becomes large with respect to the size of the spatial structures (or natural length scales), the finite dimensional picture (although it remains true) is not so useful anymore. The spectra of the various interesting operators get denser and denser and bifurcations become difficult to identify (especially sequences of bifurcations occur on very small parameter ranges). Similarly the dimensions of the (strange) attractors become larger and larger and the concepts from finite dimensional dynamical systems become less and less efficient.

It is clear that for large systems one should abandon the hope of a very detailed description and concentrate on more global concepts. A natural idea borrowed from Statistical Mechanics is to consider infinite spatial domains, hoping that far away

97

P. Chossat (ed.), Dynamics, Bifurcation and Symmetry, 97–104.
© 1994 *Kluwer Academic Publishers.*

boundary conditions will not modify too much the behavior of the interesting quantities. Also for large systems the spatial behavior becomes more interesting. For systems with only few modes excited the spatial behavior is essentially characterized by the spatial structure of these few modes and it is usually quite simple. However in large systems one may observe the formation of structures whose size is much smaller than (and often independent of) the size of the entire system. In fact when we speak of large systems we really mean large with respect to the size of the natural structures (or the natural scales). A typical example is given by the clouds which have an extension much smaller than the size of the earth.

The above idea brings immediately two important problems. First of all is the time evolution well defined in infinite domain, or equivalently is there a semi-flow global in time which integrates the evolution equation. In order to solve such questions one must first fix a reasonable Banach space for the time evolution of the system, i.e. a phase space. A natural candidate is the space of uniformly bounded functions in the domain under consideration, bearing in mind that the semi-flow will have in general regularizing properties. The boundedness assumption is in general quite natural from a physical point of view (think of the Navier Stokes equation for example). Note that in this context there is no boundary conditions "at infinity" which would correspond to other types of Banach spaces (or Banach Manifolds), for example L^p spaces (for finite p) or Sobolev spaces contain functions which are essentially small at large distances. A variant of this idea would be functions which have a given limit at infinity. This is not to say that such Banach spaces are not interesting in the present context, in fact they are quite useful to study a number of particular problems (like the front solutions described below). We refer to [C] for some recent references. However what is needed is a sufficiently large Banach space that can be used as phase space and contains simultaneously all the interesting solutions. Recall in particular that the stability of stationary solutions for example depends very much on the space of allowed perturbations. Small spaces have of course a tendency to enhance stability. In the second section of this paper we describe results about the existence of parabolic non-linear semi-flows in infinite domain for an important family of equations: the complex Ginzburg Landau equations. We briefly describe a method of approach to the problem of existence of the semi-flow and some of the results. The method can be modified and extended to study many other equations. The results are in fact also valid in bounded large domain, and they are uniform in the (large) size. We also give a brief sketch of a proof that the solutions are real analytic in the space variables.

Once the problem of existence of the semi-flow has been solved, one wants as usual to study the asymptotic time evolution. In order to do so, one of course tries to borrow some intuition from the theory of finite dimensional dynamical systems. Although we are interested in non-linear phenomena, it is well known that linearized operators (or tangent maps) play a very important role (at least for the study of the local behavior of the dynamics). This brings the second major new problem in the infinite space context which is that instead of discrete spectrum for the linearized operators one has in general to deal with continuous spectrum. However a new convenient tool is also available (at least for operators with constant coefficients) which is the Fourier transform. This tool was used since a long time by physicists with quite a lot of success to analyze phenomena in large domains neglecting the boundary conditions. In the third section we will briefly present some results associated to the presence of a continuous spectrum like the amplitude equation and some interesting solutions emerging by bifurcation.

2. Existence of Semi-Flows

As we will see in the next section, an important equation in the context of bifurcation of continuous spectrum is the complex Ginzburg Landau equation which is given (in a normalized form) by

$$\partial_t u = (1 + i\alpha)\Delta u + u - (1 + i\beta)u|u|^2 \ , \qquad \text{(CGL)}$$

where the unknown $u(x,t)$ is a complex valued function of space ($x \in \mathbb{R}^n$) and time ($t \in \mathbb{R}^+$), and where α and β are real numbers. We will denote by u_t the function $u(\cdot,t)$ for fixed t. When $\alpha = \beta = 0$ this equation will be called below the real Ginzburg Landau equation (RGL). As we explained in the introduction, a natural question is wether this equation defines a global semi-flow in the Banach space $L^\infty(\mathbb{R}^n)$. The main result in this direction is the following bound (see [C]).

Theorem 1. *For $n = 1, 2$, or for $n = 3$ if $\alpha\beta > 0$ or $|\alpha| < \sqrt{3}$ or $|\beta| < \sqrt{3}$ and for any function $u_0 \in L^\infty(\mathbb{R}^n)$ the solution of (CGL) with initial condition u_0 is defined for all times in $L^\infty(\mathbb{R}^n)$. Moreover, there is a positive constant C (which depends only on α and β) and there is a positive number $T = T(\alpha, \beta, u_0)$ such that for $t > T$ the solution $u(x,t)$ of (CGL) with initial condition $u(\cdot, 0) = u_0(\cdot)$ satisfies*

$$\|u(\cdot, t)\|_\infty \le C \ .$$

Similar estimates hold for the partial derivatives of u of any given order.

This Theorem is proven in two steps. First of all, the existence of a solution for a finite time in L^∞ is easy (using for example an integral equation, see [H] or [P]). Moreover this leads to the regularity of the solution. In order to propagate this result one needs an a priori estimate on the solution. In finite domains, this is achieved by using Lyapunov functionals (also called sometimes energy estimates). In the context of infinite domain, these energy functionals diverge and it is not clear how to renormalize them in general. Under some restrictions on the function space this is however possible. For example in some L^p spaces the energy functionals can be well defined (see [BV] for some applications) or for functions with a well defined limit at infinity one can think of renormalization which are more or less equivalent to subtracting a fixed function to the solution so that the remainder goes to zero at infinity (this is for example the case of the problem of the flow around a bounded obstacle). In order to circumvent this difficulty, in the proof of Theorem 1 one uses local energy estimates. The idea is to introduce inside the Lyapunov functional a space cut-off with compact support. More precisely one uses a fixed function with compact support and it's translates. The goal is to obtain estimates for the functional which are independent of the translation. A typical example is the following uniform local L^2 a priori estimate.

Proposition 2. *For the CGL equation in any dimension, if φ is a positive function C^∞ with compact support, there is a positive constant C_1 such that for any $a \in \mathbb{R}^n$*

$$\partial_t \|u^2 \varphi_a\|_{L^1} \le -\|u^2 \varphi_a\|_{L^1} + C_1 \ , \qquad (1)$$

where φ_a denotes the function φ translated by a ($\varphi_a(x) = \varphi(x - a)$). This implies that for t large enough (depending only on $\|u_0\|_\infty$ and not on a) we have

$$\sup_{a \in \mathbb{R}^n} \|u(\cdot, t)^2 \varphi_a\|_{L^1} \leq 2C_1 .$$

The proof of this proposition is similar to the standard proof of energy estimates. One uses the evolution equation to give an expression of the left hand side of equation (1). Integration by parts give rise to several terms (more than in the usual case) which have to be estimated if they are not of constant sign. Theorem 1 follows for the restricted values of α and β ($\alpha\beta > 0$ for example) from this estimate and a similar one for the partial derivatives using also the integral form of the evolution equation (CGL).

Note that Theorem 1 gives in fact more that the global existence of the flow. It implies that there is a ball in L^∞ which is invariant and absorbs all the trajectories. A similar result also holds for the various derivatives of the solutions.

The proof of Theorem 1 for general values of α and β where the sign of some potentially dangerous terms cannot be asserted requires the use of the Gagliardo-Nirenberg inequalities. In dimension larger than 2 these inequalities do not permit to get the desired estimates. In fact one needs weighted versions of the extended Sobolev inequalities. For example, in space dimension 2, and for a suitable function ψ, C^∞ with support in the unit ball B, there is a positive constant K such that for any function in $H^2(2B)$ we have

$$\int \psi |\nabla u|^2 |u|^2 \leq K \left(\int \psi |\Delta u|^2 + \int \psi |u|^4 \int_{2B} |u|^4 + \|u\|_{H^1(2B)} \right) ,$$

where $2B$ is a ball of radius 2 and the same center as B

We also observe that for large positive values of the two parameters, after an adequate rescaling, one gets in the limit the non linear Schrödinger equation which is known to develop singularities in finite time. It is not known if such singularities can also develop for the CGL equation.

Proposition 2 is also valid in a bounded space domain provided the support of φ_a does not meet the boundary. One can prove a result similar to that of Theorem 1 which for large domains gives a uniform estimate on the time asymptotic L^∞ norm of the solution outside of a boundary layer near the border of the domain. This estimate is independent of the size and the boundary condition, and is reached in a finite time (outside the boundary layer).

It turns out that the solutions of (CGL) become real analytic (even though the initial condition is only measurable). This is well known for periodic boundary conditions (see [DT]), but the proof which relies on Fourier series does not seem to apply here. We will now give a direct proof which can also be used in bounded domains provided some analyticity of the kernel of the linearized semigroup is known. This proof applies also in higher dimension. We first formulate the result and then give a breef sketch of the proof.

Theorem 3. *Under the hypothesis of Theorem 1, there are two positive numbers η and C_ω which depend only on α and β, and a positive number $T_\omega = T_\omega(\alpha, \beta, u_0)$*

such that for $t > T_\omega$ the function u_t is analytic in the strip $|\mathrm{Im}(z)| \le \eta$ and bounded in modulus in this strip by C_ω.

It turns out that this is simply an existence theorem. We first observe that we can decompose (CGL) into a system of equations for the real and imaginary parts of u. Indeed, let $u = X + iY$, with X and Y real, we have

$$\partial_t X = \Delta X - \alpha \Delta Y + X - (X - \beta Y)(X^2 + Y^2$$
$$\partial_t Y = \Delta Y + \alpha \Delta X + Y - (Y + \beta X)(X^2 + Y^2)$$

which can be written in vectorial form

$$\partial_t \begin{pmatrix} X \\ Y \end{pmatrix} = L \begin{pmatrix} X \\ Y \end{pmatrix} + R(X, Y)$$

where L is the linear matrix valued differential operator which is the linearized part of the above system at the origin, namely

$$L = \begin{pmatrix} 1 + \Delta & -\alpha\Delta \\ \alpha\Delta & 1 + \Delta \end{pmatrix}$$

and R is the non-linear part. The first lemma concerns the kernel of the semigroup generated by L.

Lemma 4. *There is a matrix valued entire function g_t ($t > 0$) such that the kernel of the semigroup $\exp(tL)$ is given by*

$$e^{tL}(x_1, x_2) = g_t(x_1 - x_2) .$$

Moreover we have for some constant $c > 1$ and for every $x \in \mathbb{R}^d$ $y \in \mathbb{R}^d$ and $t > 0$

$$\|g_t(x + iy)\| \le c t^{-d/2} e^{-c^{-1}x^2/t} e^{cy^2/t}$$

The proof is particularly easy using Fourier transform. L becomes the multiplication by a matrix but whose eigenvectors can be taken constant independent of the wave vector. It follows that the function g_t can be computed explicitly (it is a combination of Gaussians) and the analyticity properties and estimates are obvious.

Corollary 5. *There is a positive constant K such that for any vector $U \in L^\infty$, the vector field $\exp(tL)U$ is for $t > 0$ analytic in the strip $|\mathrm{Im}(z)| \le \sqrt{t}$ continuous at the boundary and has in that strip a modulus bounded by $K\|U\|_\infty$.*

The proof is an easy consequence of the preceding Lemma. Note that the vector field is in fact entire but we will not use this result. The proof will now proceed by an existence and uniqueness theorem in a suitable Banach space $\mathcal{B}$. In order to define this Banach space we first fix a positive number T whose value will be determined later on. The Banach space $\mathcal{B}$ is the space of two dimensional vector valued functions $U(z,t)$ defined for $t \in [0,T]$ and $z = x + iy$ with $x,y \in \mathbb{R}^d$ and $|y| \leq \sqrt{t}$ which are analytic in z in that strip. It is easy to verify that equiped with the sup norm

$$\|V\| = \sup_{t \in [0,T]} \sup_{|\mathrm{Im}(z)| \leq \sqrt{t}} \|V(z,t)\|$$

the vector space $\mathcal{B}$ becomes a Banach space.

We now consider a function $u_0 \in L^\infty$ satisfying $\|u_0\|_\infty \leq C$ where C is the uniform constant which comes out from Theorem 1. We now write the evolution equation in integral form and with initial condition u_0. We have in vector form

$$V_t = \mathcal{E}(V,t) \,,$$

where the non linear map $\mathcal{E}$ is given by

$$\mathcal{E}(V,t)(x) = e^{tL}V_0(x) + \int_0^t d\tau \int g_\tau(x - \xi)R(V_{t-\tau}(\xi))d\xi \,,$$

with $V_0 = (\mathrm{Re}(u_0), \mathrm{Im}(u_0))$.

In order to be able to define this map on the space $\mathcal{B}$ we will shift the integration contour in the second integral. Let $z = x + iy$, with $|y| \leq \sqrt{t}$. We will now assume that $y \geq 0$ (the other case being treated similarly) and we will use an integration contour which is a straight line parallel to the real axis. This amounts to replacing ξ by $\xi + i\eta$ and integrating over ξ. The choice of η will depend on τ. Among many possible choices one can take

$$\eta = \frac{y\sqrt{t - \tau}}{\sqrt{t - \tau} + \sqrt{\tau}} \,.$$

Using the inequality $\sqrt{t - \tau} + \sqrt{\tau} \geq \sqrt{t}$ it is easy to verify that $|\eta| \leq \sqrt{t - \tau}$ and $|y - \eta| \leq \sqrt{\tau}$, and it follows that the map $\mathcal{E}$ is well defined on $\mathcal{B}$. It is now an easy exercise using Corollary 5 to find a positive number T which depends on u_0 only through the constant C of Theorem 1 such that the map $\mathcal{E}$ is a contraction on some ball of $\mathcal{B}$. Theorem 3 follows by the contraction mapping principle and the uniqueness in L^∞ of the solution on the real line.

3. Some Dynamical Results

We are now going to explain briefly why the CGL equation appears frequently. For simplicity we will explain the derivation on a particular equation called the Swift-Hohenberg equation which was proposed as a simpler model for convective phenomena. This equation describes the time evolution of a real scalar function in a one dimensional space and is given by

$$\partial_t u = [\alpha - (1 - \partial_x^2)^2]u - u^3 . \tag{SH}$$

By methods similar to those of section 2 one can prove that this equation defines a semi-flow in the space of bounded real functions on $\mathbb{R}$. If the parameter α is negative it is easy to verify that the stationary solution $u = 0$ is stable. However the linearized evolution equation around this solution is given by

$$\partial_t w = [\alpha - (1 - \partial_x^2)^2]w ,$$

and it is easy to see that the right hand side is an operator with continuous spectrum. This spectrum becomes unstable for $\alpha > 0$. Following the ideas of bifurcation theory one looks for solutions which point mostly in the directions which are becoming unstable and which are here the two periodic functions $e^{\pm ix}$. Note that these two functions belong to the phase space $L^\infty(\mathbb{R})$. Denoting by ϵ the positive square root of $\alpha > 0$, a simple scaling argument leads to the following ansatz for the solution u of SH

$$u(x,t) = e^{ix}A(\epsilon x, \epsilon^2 t) + c.c. , \tag{2}$$

which is a sort of amplitude approximation. To the leading approximation in ϵ one gets for the evolution of the function A the RGL equation. We refer the reader to [CE1] for a review of the rigorous results about this approximation and references. In simple words, the RGL equation appears as an amplitude equation for some bifurcations of continuous spectrum. Therefore solutions of the RGL equations will provide natural hints for the solutions of more complex equations (better Physical models), which can be constructed at least to a leading order approximation by using formula (2). Note also that for small ϵ this gives solutions which are large scale modulations of locally well defined structures which are given by the two periodic solutions $e^{\pm ix}$.

The simplest solutions of the RGL equation are the stationary solutions. A one parameter family of such solutions is given by

$$u_q(x) = \sqrt{(1 - q^2)}e^{iqx} \qquad \text{with} \quad |q| \le 1 .$$

These solutions are unstable for $|q| > 1/\sqrt{3}$ and otherwise marginally stable. This means that the operator giving the linearized time evolution around these solutions has a continuous spectrum contained in the closed let half plane but tangent to the imaginary axis.

¿From these simple solutions one can construct more interesting ones, or investigate spatial effects. A typical example of non stationary solutions is given by the front solutions. A front is a solution which interpolates between two stationary solutions one marginally stable at $-\infty$ say and the other unstable at $+\infty$ and which "moves" with a given velocity. It is of course clear that the marginally stable solution will invade the unstable one but in the front solution the interpolation between the two solutions takes place in a finite region (neglecting very small tails) which moves without deformation at a finite speed. We refer to [CE1] and [CE2] for more results on the existence and stability of front solutions.

An example of interesting spatial behavior is the asymptotic time evolution of an initial condition which interpolates between two marginally stable stationary solutions. Contrary to the problem of fronts it is not clear in this situation if one solution will take over the other. And indeed this is not what occurs. In [CE3] it was proven that if the initial condition is at every point sufficiently far away form zero, this will remain so forever. That is to say the solution evolves without going to zero (absence of the so called phase slips). It was then proved by Bricmont and Kupiainen [BK] that a new stationary solution nucleates in the region of interpolation, this new stationary solution invades the whole spatial system but in a diffusive way. Moreover the new solution may have also a diffusively growing phase.

References

[BK] J.Bricmont, A.Kupiainen. Renormalisation group and the Ginzburg-Landau equation. Commun. Math. Phys. (1992).

[C] P.Collet. Thermodynamic limit of the Ginzburg Landau equations. Preprint (1993).

[CE1] P.Collet, J.-P.Eckmann. Space-time behavior in problems of hydrodynamic type: a case study. Non linearity 5, 1265-1302 (1992).

[CE2] P.Collet, J.-P.Eckmann. *Instabilities and Fronts in Extended Systems*. Princeton University Press, Princeton 1990.

[CE3] P.Collet, J.-P.Eckmann. Solutions without phase-slip for the Ginzburg-Landau equation. Commun. Math. Phys. **145**, 345-356 (1992).

[DT] A.Doelman, E.S.Titi. Regularity of solutions and the convergence of the Galerkin method in the Ginzburg-Landau equation. Numer. Funct. Anal. and Optimiz. **14**, 299-321 (1993).

[H] D.Henry. *Geometric Theory of Semilinear Parabolic Equations*. Lecture Notes in Mathematics **840**. Springer-Verlag, Berlin, Heidelberg, New York 1981.

[P] A.Pazy. *Semigroups of Linear Operators and Applications to Partial Differential Equations*. Springer Verlag, Berlin Heidelberg New York (1983).

EIGENVALUE MOVEMENT FOR A CLASS OF REVERSIBLE HAMILTONIAN SYSTEMS WITH THREE DEGREES OF FREEDOM

MICHAEL DELLNITZ and JÜRGEN SCHEURLE
Institut für Angewandte Mathematik
Universität Hamburg, Germany

1. Introduction

It is known that in one-parameter families of Hamiltonian systems purely imaginary eigenvalues of the linearization generically split off the imaginary axis if a $1 - 1$-resonance occurs. This phenomenon is also known as the *Hamiltonian Hopf bifurcation*, see [6]. Recently it has been shown that the situation drastically changes if the underlying system possesses symmetry (see [4], [2]). In this case, generically, the eigenvalues may also pass through $1 - 1$-resonances and stay on the imaginary axis. In Hamiltonian systems without any additional structure *passing* is a phenomenon of codimension 3 (see [3], [7]).

From this point of view it is surprizing that it has been observed in [1], [5] that for the *double spherical pendulum* the passing of eigenvalues occurs in the related reduced Hamiltonian system although the reduced Hamiltonian is no longer invariant with respect to any obvious symplectic symmetry transformation. The only symmetry property that is left after reduction is given by a time reversibility.

The main purpose of this paper is to explain why passing is expected to be seen in such time reversible mechanical systems. We do this in two steps. First we show that the time reversibility indeed reduces the codimension of the passing case by 1 in the sense that just one more (nonlinear) condition on the coefficients of the underlying Hamiltonian function has identically to be satisfied so that passing is expected to be seen (Proposition 3.2). In fact, this nonlinear condition turns out to be satisfied for the reduced Hamiltonian for the double spherical pendulum. Then, in a second step, we show that non-degenerate passing is a robust phenomenon and cannot be perturbed away once it occurs in the corresponding subclass of Hamiltonian systems

105

P. Chossat (ed.), Dynamics, Bifurcation and Symmetry, 105–110.
© 1994 *Kluwer Academic Publishers.*

(Theorem 3.4).

Acknowledgment

This research has partly been supported by a European Community Laboratory Twinning Grant (European Bifurcation Theory Group).

2.   The structure of the Hamiltonian system

We consider mechanical systems with three degrees of freedom which are time reversible. Motivated by the double spherical pendulum mentioned above we assume that the time reversibility is induced by the linear transformation

$$
R = \begin{pmatrix}
-1 & 0 & 0 & 0 & 0 & 0 \\
0 & -1 & 0 & 0 & 0 & 0 \\
0 & 0 & 1 & 0 & 0 & 0 \\
0 & 0 & 0 & 1 & 0 & 0 \\
0 & 0 & 0 & 0 & 1 & 0 \\
0 & 0 & 0 & 0 & 0 & -1
\end{pmatrix},
$$

where coordinates are chosen such that the standard symplectic form on $\mathbb{R}^6$ is represented by

$$
J = \begin{pmatrix}
0 & 0 & 0 & -1 & 0 & 0 \\
0 & 0 & 0 & 0 & -1 & 0 \\
0 & 0 & 0 & 0 & 0 & -1 \\
1 & 0 & 0 & 0 & 0 & 0 \\
0 & 1 & 0 & 0 & 0 & 0 \\
0 & 0 & 1 & 0 & 0 & 0
\end{pmatrix}.
$$

We are interested in analyzing the eigenvalue behavior of the linearized system in an equilibrium, and we assume this equilibrium to be $0 \in \mathbb{R}^6$. Hence for our purposes it is sufficient to consider quadratic Lagrangians or Hamiltonians. If we denote the variables by

$$(s_1, s_2, \theta, \dot{s}_1, \dot{s}_2, \dot{\theta})$$

then the general quadratic Lagrangian possessing the type of symmetry induced by R is given by

$$
\begin{aligned}
\mathcal{L} \;=\; & a\dot{s}_1^2 + b\dot{s}_1\dot{s}_2 + c\dot{s}_2^2 + d\dot{\theta}^2 \\
& + es_1^2 + fs_1s_2 + gs_2^2 + h\theta^2 \\
& + i(\dot{\theta}s_1 - \dot{s}_1\theta) + j(\dot{\theta}s_2 - \dot{s}_2\theta),
\end{aligned}
$$

where $a, b, \ldots, j$ are arbitrary real constants.

We assume that

$$d \neq 0, \quad 4ac - b^2 \neq 0, \qquad (2.1)$$

and after a Legendre transformation we obtain the Hamiltonian $\mathcal{H}$. It follows that the matrix of the corresponding Hamiltonian system reads as follows:

$$JD^2\mathcal{H}(0) \;=\;$$

$$\begin{pmatrix} 0 & 0 & \delta(2ci - bj) & 4\delta c & -2\delta b & 0 \\ 0 & 0 & \delta(2aj - bi) & -2\delta b & 4\delta a & 0 \\ -\frac{i}{2d} & -\frac{j}{2d} & 0 & 0 & 0 & \frac{1}{d} \\ -\frac{i^2}{4d} + e & -\frac{ij}{4d} + \frac{f}{2} & 0 & 0 & 0 & \frac{i}{2d} \\ -\frac{ij}{4d} + \frac{f}{2} & -\frac{j^2}{4d} + g & 0 & 0 & 0 & \frac{j}{2d} \\ 0 & 0 & -\delta(ci^2 - bij + aj^2) + h & -\delta(2ci - bj) & -\delta(2aj - bi) & 0 \end{pmatrix},$$

where $\delta = \frac{1}{4ac-b^2}$. By construction, $JD^2\mathcal{H}(0)$ anticommutes with R, i.e.

$$R\,JD^2\mathcal{H}(0) = -JD^2\mathcal{H}(0)\,R. \qquad (2.2)$$

Lemma 2.1 *Let λ be an eigenvalue of $JD^2\mathcal{H}(0)$ with eigenvector u. Then Ru is an eigenvector corresponding to the eigenvalue $-\lambda$.*

Proof: This is immediate from (2.2) ∎

3. 1-1 resonances

As already mentioned in the introduction, in Hamiltonian systems without any additional structure passing is a phenomenon of codimension 3. Roughly speaking, with the following proposition, we show that the presence of a time reversibility reduces this codimension by one. ¿From now on we additionally assume that

$$|i| + |j| > 0. \qquad (3.1)$$

Proposition 3.2 *Inside the subclass of one-parameter families of reversible Hamiltonian systems which additionally satisfy the relation*

$$(cf - bg)i^2 + 2(ag - ce)ij + (be - af)j^2 = 0 \qquad (3.2)$$

splitting of certain eigenvalues at $1 - 1$-resonances is not possible.

Proof: Consider the following four-dimensional subspace of $\mathbb{R}^6$

$$V = \{(x, y, 0, z, w, 0)^t \in \mathbb{R}^6 : x, y, z, w \in \mathbb{R}\}.$$

In a first step we show that there is a two-dimensional subspace $\tilde{V}$ of V which is invariant under $JD^2\mathcal{H}(0)$ if and only if the relation (3.2) holds for the entries of $JD^2\mathcal{H}(0)$.

Let $v = (x, y, 0, z, w, 0)^t \in V$. Then

$$JD^2\mathcal{H}(0)v = \begin{pmatrix} 4\delta cz - 2\delta bw \\ -2\delta bz + 4\delta aw \\ -\frac{i}{2d}x - \frac{j}{2d}y \\ \left(\frac{-i^2}{4d} + e\right)x + \left(\frac{-ij}{4d} + \frac{f}{2}\right)y \\ \left(\frac{-ij}{4d} + \frac{f}{2}\right)x + \left(\frac{-j^2}{4d} + g\right)y \\ -\delta(2ci - jb)z - \delta(2aj - ib)w \end{pmatrix},$$

and $JD^2\mathcal{H}(0)v \in V$ if and only if the following two relations hold:

$$\begin{aligned} ix + jy &= 0, \\ (2ci - jb)z + (2aj - ib)w &= 0. \end{aligned} \tag{3.3}$$

By (2.1) and (3.1) we may assume that $2ci - jb \neq 0$ and $i \neq 0$, and hence those relations define the two-dimensional subspace

$$\tilde{V} = \mathbb{R}\left\{ \begin{pmatrix} -\frac{j}{i} \\ 1 \\ 0 \\ 0 \\ 0 \\ 0 \end{pmatrix}, \begin{pmatrix} 0 \\ 0 \\ 0 \\ -\frac{2aj-ib}{2ci-jb} \\ 1 \\ 0 \end{pmatrix} \right\}$$

of V. For this subspace to be invariant under $JD^2\mathcal{H}(0)$ we have to fulfill

$$4\delta cz - 2\delta bw = -\frac{j}{i}(-2\delta bz + 4\delta aw)$$

which is automatically valid, and also

$$\left(\frac{-i^2}{4d} + e\right)\left(-\frac{j}{i}\right) + \left(\frac{-ij}{4d} + \frac{f}{2}\right) = -\frac{2aj - ib}{2ci - jb}\left(\left(\frac{-ij}{4d} + \frac{f}{2}\right)\left(-\frac{j}{i}\right) + \left(\frac{-j^2}{4d} + g\right)\right).$$

This last relation can be rewritten in the form

$$(cf - bg)i^2 + 2(ag - ce)ij + (be - af)j^2 = 0$$

which is (3.2). Up to now we have proved that the subspace $\tilde{V}$ is invariant under $JD^2\mathcal{H}(0)$ if and only if (3.2) holds.

We now show that in a one-parameter family of systems in which (3.2) is additionally valid, splitting is not possible at a $1 - 1$-resonance where eigenvalues corresponding to the invariant subspace $\tilde{V}$ are involved. In that case there is always an eigenvector v of $JD^2\mathcal{H}(0)$ inside the complexified space $\tilde{V} + i\tilde{V} \subset V + iV$. Since V is R-invariant it follows from Lemma 2.1 that also the eigenvector Rv (corresponding to the eigenvalue with opposite sign) has to lie inside $V + iV$. But by (3.1) and (3.3) V itself is not an invariant subspace for $JD^2\mathcal{H}(0)$ and therefore there cannot exist a quadruplet of eigenvalues such that the corresponding real eigenspace is entirely contained in V. In particular, the eigenvalues of $JD^2\mathcal{H}(0)|_{\tilde{V}}$ cannot leave the imaginary or the real axis. ∎

Remark 3.3 Observe that the relation (3.2) is independend of d and h.

On the other hand "non-degenerate" passing involving the eigenvalues corresponding to $\tilde{V}$ is a robust phenomenon. We have the following theorem.

Theorem 3.4 *Consider a continuous one-parameter family of Hamiltonian systems as above which additionally fulfills (3.2). Assume that the eigenvalues $\pm\mu_{\tilde{V}}$ corresponding to $\tilde{V}$ and another pair $\pm\mu$ of simple eigenvalues pass each other along the real or imaginary axis as the parameter increases through a critical value. We assume that $|\mu_{\tilde{V}}| - |\mu|$ changes sign at criticality.*

Then passing also occurs in any such family of Hamiltonian systems which is sufficiently close to the given one, in terms of a matrix norm pointwise with respect to the parameter.

Proof: Remember that the invariant subspace $\tilde{V}$ is contained in V for all parameter values. Consider the eigenvalues $\mu_{\tilde{V}}^{\pm}$, $\mu^{\pm}$ in the perturbed family of systems, which are close to $\pm\mu_{\tilde{V}}$ and $\pm\mu$ respectively. Suppose for contradiction that those perturbed eigenvalues are simple near the critical parameter value. Then, by standard perturbation theory, they form nonintersecting continuous curves depending on the parameter. Of course one pair of these curves corresponds to the invariant subspace $\tilde{V}$ since $\tilde{V}$ depends continuously on the parameter by definition. In fact, those curves are close to $\pm\mu_{\tilde{V}}$, i.e. it is the pair of $\mu_{\tilde{V}}^{\pm}$ curves. Namely, strictly away from the critical parameter value they are strictly bounded away from $\pm\mu$ provided that the perturbation is sufficiently small.

Since the $\mu^{\pm}$ curves are close to the $\pm\mu$ curves in the unperturbed problem and since $|\mu_{\tilde{V}}| - |\mu|$ changes sign at criticality, continuity of the eigenvalue curves in the perturbed problem gives the desired contradiction to the above assumption of nonintersection. ∎

References

[1] M. Dellnitz, J.E. Marsden, I. Melbourne and J. Scheurle. Generic bifurcations of pendula. In: *Bifurcation and Symmetry* (E. Allgower, K. Böhmer and M. Golubitsky, eds.), ISNM **104**, 111-121, Birkhäuser, 1992.

[2] M. Dellnitz, I. Melbourne and J.E. Marsden. Generic bifurcation of Hamiltonian vector fields with symmetry. *Nonlinearity*, **5**, 979-996, 1992.

[3] D. Galin. Versal deformations of linear Hamiltonian systems. *AMS Transl.* **2** 118, 1-12, 1982 (1975 *Trudy Sem. Petrovsk.* **1** 63-74).

[4] M. Golubitsky and I. Stewart. Generic bifurcation of Hamiltonian systems with symmetry. *Physica D* **24**, 391-405, 1987.

[5] J.E. Marsden and J. Scheurle. Lagrangian reduction and the double spherical pendulum. *ZAMP* **44**, 17-43, 1993.

[6] J.C. van der Meer. *The Hamiltonian Hopf Bifurcation.* Lecture Notes in Mathematics **1160**, 1985.

[7] I. Melbourne. Versal unfoldings of equivariant linear Hamiltonian vector fields. *Math. Proc. Camb. Phil. Soc.* **114**, 559-573, 1993.

BLOWING-UP IN EQUIVARIANT BIFURCATION THEORY

MIKE FIELD
Mathematics Department
University of Houston

ABSTRACT. In these notes, our goal is to illustrate why blowing-up is a powerful technique in equivariant bifurcation theory. In addition, we shall explore some new directions that we believe may lead to more effective computational and analytical methods in the theory. Most of the work that we describe here has arisen out of a study of the effect of (high order) symmetry breaking perturbations on branches of relative equilibria [9].

1 Introduction

Let Γ be a compact non-finite Lie group. In general, Γ-equivariant bifurcations will yield branches of *relative* equilibria. That is, since Γ is not finite, branches of flow-invariant group orbits of non-zero dimension can branch off the trivial solution. Moreover, the dynamics of the flow on these group orbits may be nontrivial (see [5, 17, 8]). In recent years there has been an effort to understand this type of equivariant bifurcation. Significant applications include physically important examples, such as SO(3) and O(3) bifurcations, as well as the normal form approach to the equivariant Hopf bifurcation (see [16, 9, 10, 11]). There are also interesting questions about the effect of adding symmetry breaking terms on branches of relative equilibria (see [18] when $\Gamma = O(3)$ and many other works when $\Gamma = O(2)$).

The work of Krupa [17] on the tangent and normal decomposition for vector fields (see also [8]) gives a powerful means for analyzing bifurcations *from* relative equilibria but there are still significant technical problems in obtaining a general theory of bifurcations *to* relative equilibria. Our aim in these notes is to describe some of these problems and indicate how they may be overcome. We hope the techniques we describe may have other applications in equivariant bifurcation theory.

111

P. Chossat (ed.), Dynamics, Bifurcation and Symmetry, 111–122.
© 1994 *Kluwer Academic Publishers.*

2 Preliminaries

Throughout, we shall suppose that Γ is a compact Lie group and $\mathbf{R}^n$ is a nontrivial representation space for Γ. In the sequel, we refer to $(\mathbf{R}^n, \Gamma)$ as a Γ-*representation*. We always assume that Γ acts by isometries on $\mathbf{R}^n$ and so $\Gamma \subset O(n)$. We shall assume that $(\mathbf{R}^n, \Gamma)$ is absolutely irreducible though most of what we say applies, with minor changes in definitions, to complex irreducible representations.

For each integer $d \geq 0$, we let $P_\Gamma^d(\mathbf{R}^n, \mathbf{R}^n)$ denote the space of Γ-equivariant polynomial mappings of $\mathbf{R}^n$ which are homogeneous of degree d and $P_\Gamma^{(d)}(\mathbf{R}^n, \mathbf{R}^n)$ be the space of Γ-equivariant polynomial mappings of $\mathbf{R}^n$ which are of degree less than or equal to d.

Let Γ act on $\mathbf{R}^n \times \mathbf{R}$ as the product of the given action on $\mathbf{R}^n$ with the trivial action on $\mathbf{R}$. Let $C_\Gamma^\infty(\mathbf{R}^n \times \mathbf{R}, \mathbf{R}^n)$ denote the space of all smooth Γ-equivariant maps from $\mathbf{R}^n \times \mathbf{R}$ to $\mathbf{R}^n$. Let $f \in C_\Gamma^\infty(\mathbf{R}^n \times \mathbf{R}, \mathbf{R}^n)$. For $\lambda \in \mathbf{R}$, we define $f_\lambda : \mathbf{R}^n \to \mathbf{R}^n$ by $f_\lambda(x) = f(x, \lambda)$. By the absolute irreducibility of the action of Γ on $\mathbf{R}^n$, we may write

$$Df_\lambda(0) = a(\lambda) Id_V$$

where $a : \mathbf{R} \to \mathbf{R}$ is C^∞.

As in [13, 15], we restrict attention to *normalized* families. That is, we assume that $a(\lambda) = \lambda$, $\lambda \in \mathbf{R}$. For such families, it follows that the trivial solution $x = 0$ of $f = 0$ has a non-degenerate change of stability at $\lambda = 0$. We let $\mathcal{V}(\mathbf{R}^n, \Gamma)$ denote the space of normalized families.

3 Branches of relative equilibria

We assume known the definition of isotropy type. Let $\mathcal{O} = \mathcal{O}(\mathbf{R}^n, \Gamma)$ denote the set of isotropy types for the action of Γ on $\mathbf{R}^n$. If $\tau \in \mathcal{O}$, we let Δ_τ denote a representative Γ-orbit of isotropy type τ.

Definition 3.1 Let $f \in \mathcal{V}(\mathbf{R}^n, \Gamma)$ and $\tau \in \mathcal{O}$. A branch of *relative equilibria* (of isotropy type τ) for f at zero consists of a C^1 Γ-equivariant map

$$\phi = (\rho, \lambda) : [0, \delta] \times \Delta_\tau \to \mathbf{R}^n \times \mathbf{R}$$

such that λ is independent of $u \in \Delta_\tau$ and

1. $\phi(0, u) = (0, 0)$, all $u \in \Delta_\tau$.

2. For all $s \in (0, \delta]$, $\alpha_s = \rho(s, \Delta_\tau)$ is a relative equilibrium of $f_{\lambda(s)}$.

3. For every $u \in \Delta_\tau$, the map $\phi_u : [0, \delta] \to V \times \mathbf{R}$, $s \mapsto \phi(s, u)$ is a C^1-embedding.

If, in addition, we can choose $\delta > 0$ so that

(4) For all $s \in (0, \delta]$, α_s is a normally hyperbolic relative equilibrium $f_{\lambda(s)}$,

we refer to ϕ as a branch of *normally hyperbolic* relative equilibria for f at zero.

Remarks 3.1 (1) In our definition of branch, we do not require that ϕ is an embedding. Of course, it follows from our definitions that $\phi|(0, \delta] \times \Delta_\tau$ is an embedding. (2) In most, if not all, examples so far analyzed, ϕ can be taken to be C^∞. However, it seems to be difficult to prove that this is generically the case. $\Diamond$

3.1 Technical problems

When Γ is finite, a branch of relative equilibria consists of a finite set of embedded arcs of equilibrium points and indeed the branch is just the Γ-orbit of one of the arcs comprising the branch. If Γ is not finite and ϕ is a branch of relative equilibria with $\dim(\alpha_s) > 0$, $s > 0$, then the branch ϕ will typically have a (non-removable) singularity at the origin. In addition, the branch may have high orders of contact with other orbit strata at the origin and so certain directions (and Γ-orbits on the branch) may be distorted more than others as $\lambda \to 0$ (this is the typical situation for submaximal branches).

Suppose that $\phi = (\rho, \lambda) : [0, \delta] \times \Delta_\tau \to \mathbf{R}^n \times \mathbf{R}$ is a branch of relative equilibria. Given $t \in (0, \delta]$, $f_{\lambda(s)}$ may be nonzero on α_s. Consequently, we cannot define the hessian of $f_{\lambda(s)}$ on α_s. Nevertheless, it is possible to give a mod-Γ definition of the hessian and define eigenvalues on the relative equilibria. These 'eigenvalues' are defined modulo $\imath\mathbf{R}$ (we refer to [5, 8, 9] for details). Conceptually this approach is useful – especially when dealing with relative equilibria – but technically matters are still tricky. In addition, we must allow for zero 'eigenvalues' of high multiplicity (corresponding to directions tangent to the group orbit).

When we compute branches of relative equilibria, we have to solve a relation $f(x, \lambda) \in T_x \Gamma(x)$, rather than the simpler equation $f(x, \lambda) = 0$ that arises in the study of equilibria. Although we can use the orbit map to reduce to an equation, this may complicate rather than simplify matters. Indeed, for general non-finite compact Lie groups, it is often computationally infeasible to compute a set of generators for the $\mathbf{R}$-algebra of polynomial invariants.

Finally, suppose that ϕ is a normally hyperbolic branch of equilibria. Let Φ^s denote the time-1 map of the flow of $f_{\lambda(s)}$. For each $s \in (0, \delta]$, we have a $T\Phi^s$-invariant splitting $N^s \oplus N^u \oplus T\alpha_s$ of $T_{\alpha_s}\mathbf{R}^n$, and $N^s \oplus T\alpha_s$ is tangent to the stable manifold of α_s and $N^u \oplus T\alpha_s$ to the unstable manifold of α_s. In general, we have to allow for the angles between fibers of N^s, N^u, and $T\alpha_s$ to go to zero like a high power of $|\lambda|$ as $\lambda \to 0$. When Γ is finite, this is not serious as we can estimate the determinant of the linearization of f along the branch to prove stability and determinacy results using well-known methods in singularity theory (see [9, §8]). This avenue appears closed when Γ is not finite.

Our basic strategy is to use blowing-up transformations to ameliorate and overcome these difficulties.

4   Example I

In Field & Richardson [13, 14], (polar) blowing-up techniques were used to analyze symmetry breaking for a large class of equivariant bifurcations. The basic idea is very simple and has often been used in problems involving singular vector fields. For example, suppose that $Q \in P^3(\mathbf{R}^n, \mathbf{R}^n)$ and consider the family of ordinary differential equations

$$x' = \lambda x + Q(x) \tag{4.1}$$

Using generalized spherical polars $x = (u, R)$, $(u, R) \in S^{n-1} \times \mathbf{R}$, equation (4.1) transforms to

$$R' = \lambda R + R^3(Q(u), u) \tag{4.2}$$
$$u' = R^2 \mathcal{P}^Q(u), \tag{4.3}$$

where $\mathcal{P}^Q$ is the vector field (the 'phase vector field') on S^{n-1} defined by

$$\mathcal{P}^Q(u) = Q(u) - (Q(u), u)u$$

We have

- $R = 0$ is a flow-invariant set of (4.2,4.3) which corresponds to the trivial solution of (4.1).

- The phase portrait of (4.3) off $R = 0$ is *exactly* the same as the phase portrait of the phase vector field $\mathcal{P}^Q$ on S^{n-1} (see also [7]).

- Each zero of $\mathcal{P}^Q$ yields a branch of zeros of (4.1).

- Hyperbolic zeros of $\mathcal{P}^Q$ yield branches of hyperbolic zeros. These branches persist under perturbations of (4.1) of order three (see [13] for precise statements).

These observations are true without any assumptions about equivariance. If we assume Q is Γ-equivariant, then $\mathcal{P}^Q$ is Γ-equivariant and we may allow for relative equilibria of $\mathcal{P}^Q$ and branches of relative equilibria for (4.1) (see [9, §§5,6]).

Clearly the process of polar blowing-up leads in this instance to a major simplification of the original equations. In fact the Γ-action is also simplified. Noting that Γ lifts to $S^{n-1} \times \mathbf{R}$, we see that the action of Γ on $S^{n-1} \times \mathbf{R}$ has one less isotropy type than the action on $\mathbf{R}^n$: The Γ-isotropy type is removed under blowing-up. This simplification or 'desingularization' of the group action is particularly useful if Γ is not finite. In this case, we can expect a branch of relative equilibria of maximal isotropy type to transform to a (nonsingular) branch of Γ-orbits all of the same maximal Γ-isotropy type. Notice that the original branch will generally be singular at the origin.

5 Blowing-up

It is time to give more precise definitions of what we mean by blowing-up.

5.1 Polar blowing-up

Let $\pi_1 : \mathbf{R}^n \times S^{n-1} \to \mathbf{R}^n$ denote the projection on the first factor. Let $P_n \subset \mathbf{R}^n \times S^{n-1}$ be defined by

$$P_n = \{(x, u) \mid x \in \mathbf{R}u\}$$

Then the *polar blowing-up* of $\mathbf{R}^n$ with *center* 0 is the map $\pi : P_n \to \mathbf{R}^n$, where $\pi = \pi_1 | P_n$. It is easy to verify that π is 2:1 off the *exceptional variety* $\pi^{-1}(0)$ and that P_n is a analytic submanifold of $\mathbf{R}^n \times S^{n-1}$. Moreover, the map $\rho : S^{n-1} \times \mathbf{R} \to P_n$ defined by $\rho(u, t) = (tu, u)$ is an analytic diffeomorphism.

It is easy to extend the definition of polar blowing-up to allow for more general centers. In particular, if M is an analytic manifold and C is a closed analytic submanifold of M we can define the blowing-up $\pi : \tilde{M} \to M$ of M with center N. We may give $\tilde{M}$ the structure of an analytic manifold relative to which π is an analytic map. We refer the reader to [4, 6, 9] for details and merely note here that the process involves blowing-up M along N in transverse directions to N. These constructions work particularly well when we allow for group actions.

Example 5.1 (cf [3, 4]) Consider the $(1, 2)$-action of $O(2)$ on $\mathbf{C}^2$. That is, regard $O(2)$ as generated by $SO(2)$ and complex conjugation $\kappa(z) = \bar{z}$. For $(z_1, z_2) \in \mathbf{C}^2$ we define

$$e^{i\theta}(z_1, z_2) = (e^{i\theta}z_1, e^{2i\theta}z_2), e^{i\theta} \in SO(2)$$

$$\kappa(z_1, z_2) = (\bar{z}_1, \bar{z}_2)$$

Let $\mathbf{Z}_2$, $\mathbf{Z}_2(\kappa)$ denote the subgroups of $O(2)$ generated by $e^{i\pi}$ and κ respectively. We observe that

(a) The origin is the unique point with isotropy $O(2)$.

(b) Non-zero points on the z_2 axis have isotropy conjugate to $\mathbf{Z}_2 \times \mathbf{Z}_2(\kappa)$.

(c) Any point of the form $(x_1 e^{i\theta}, x_2 e^{2i\theta})$, with x_1, x_2 real and $x_1 \neq 0$ has isotropy conjugate to $\mathbf{Z}_2(\kappa)$.

(d) All other points in $\mathbf{C}^2$ have trivial isotropy.

If we polar blow-up $\mathbf{C}^2$ with center C_1 the origin, we obtain the analytic $O(2)$-manifold $M_1 \cong S^3 \times \mathbf{R}$. Now M_1 has three isotropy types conjugate to $\mathbf{Z}_2 \times \mathbf{Z}_2(\kappa)$, $\mathbf{Z}_2(\kappa)$ and the trivial group. The set of points $C_2 \subset S^3 \times \mathbf{R}$ with isotropy conjugate to $\mathbf{Z}_2 \times \mathbf{Z}_2(\kappa)$, is closed and of codimension 2. Blowing-up M_1 along C_2, we obtain a new $O(2)$-manifold M_2 with just two isotropy types. It is easy to see that M_2 is analytically diffeomorphic to $S^2 \times S^1 \times \mathbf{R}$.

Repeating the process again, we arrive at a free $O(2)$-manifold M_3. In this case, since the set of points in M_2 with isotropy conjugate to $\mathbf{Z}_2(\kappa)$ has codimension 1, M_3 just consists of two copies of M_2.

It is important to note that a smooth $O(2)$-equivariant vector field on $\mathbf{C}^2$ will lift to each of the spaces M_1, M_2 and M_3 as a smooth $O(2)$-equivariant vector field.

This process of blowing-up is exploited in Chossat & Field [4] to give a geometric analysis of the effect of symmetry breaking perturbations from $O(2)$ to $O(2)$ and $\mathbf{Z}_2(\kappa)$ on a homoclinic cycle arising in the study by [1]. It is shown in [1] that there is a normally hyperbolic $O(2)$-orbit α of equilibria contained in the 2-component of the $(1, 2)$-representation of $O(2)$. The unstable manifold $W^u(\alpha)$ of α is 2-dimensional; the stable manifold $W^s(\alpha)$ is 3-dimensional (α is of index 2) and $W^u(\alpha) \subset W^s(\alpha)$. The homoclinic cycle is the set $\Sigma = W^u(\alpha)$ and, in the case of greatest interest, Σ is an attractor. Topologically, Σ is an immersed Klein bottle with self intersection along α. If we blow up successively at the origin and along the set C_2, we are able to obtain a considerable simplification the dynamics of the flow near the cycle that allows a transparent geometric description of the effect of symmetry breaking perturbations of the vector field. If we ignore the radial component, the cycle transforms under blowing up into a 2-torus on which $SO(2)$ acts freely. We refer the reader to [4] for more details. ♡

Example 5.1 shows the typical features of successive blowing-ups of representations along orbit strata. One important advantage of this type of blowing-up is that it is very geometric. In particular, it is often easy to give a geometric description of the results of a succession of polar blowings-up. However, the geometric transparency of polar blowing-up is not without cost. Although we stay within the analytic framework, the blowing-up procedure has the fatal flaw that it is *not* algebraic in character[1]. The problem is that the process depends on a choice of metric and corresponding norm function $R = \sqrt{x_1^2 + \ldots + x_n^2}$. In case $n = 2$, this is usually not a problem as the norm is typically associated to a complex structure – that is, it is just the modulus of complex numbers[2]. However, for $n > 2, n \neq 4$ it is well-known that the construction and use of general spherical polar coordinates is often technically complicated and lacks naturality. Basically, this is because there is no natural choice of coordinates on the $(n-1)$-sphere for $n \neq 2, 4$. Of course, none of this matters if, for example, we only want to develop geometric insight or we can use general arguments that avoid explicit coordinate computations (as was done in [13]). This is often the case if we blow-up (once) at the origin. On the other hand, in Example 5.1, it does not seem so clear how to choose a 'natural' system of local coordinates on the blow-up M_2. More seriously, polar blowing-up of itself is unlikely to lead to effective computational algorithms.

5.2 Blowing-up

Formally, the definition of blowing-up at a point is very similar to that of polar blowing-up at a point. Let $P^{n-1}(\mathbf{R})$ denote $(n-1)$-dimensional real projective space and $\pi_1 : \mathbf{R}^n \times P^{n-1} \rightarrow \mathbf{R}^n$ denote the projection on the first factor. Let $B_n \subset \mathbf{R}^n \times P^{n-1}$ be defined by

$$B_n = \{(x, u) \mid x \in u\}$$

Then the *blowing-up* of $\mathbf{R}^n$ with *center* 0 is the map $\pi : B_n \rightarrow \mathbf{R}^n$, where $\pi = \pi_1|B_n$. It is easy to verify that π is 1:1 off the *exceptional variety* $\pi^{-1}(0)$. It can also be shown that B_n has the structure of a smooth affine algebraic variety (see [2],[9, §§9,10] and note that B_n is affine because we are working over the reals).

Obviously, the polar blowing-up and blowing-up are closely related geometrically and the natural 2:1 map $S^{n-1} \rightarrow P^{n-1}$ induces a 2:1 map of P_n onto B_n.

The definition of blowing-up may be extended to the blowing-up of smooth (affine) algebraic variety along a smooth algebraic subvariety. We give an illustration of this process in the next example and refer the reader to [2, Chapter 3, §3] and [9, §9] for further details and definitions.

---

[1]Hence the reason it is not used in desingularization in algebraic geometry.

[2]Phase amplitude equations often work well for equations written in complex coordinates... Part of the reason is that the modulus is canonically defined on $\mathbf{C}$. That is, the complex numbers of unit modulus are uniquely determined by the multiplicative structure of the complex numbers.

Example 5.2 Let $X \subset \mathbf{R}^n$ be an algebraic set. Suppose that the ideal $I(X)$ of polynomial vanishing on X has generators $f_0, \ldots, f_m$ and that X is smooth as a real algebraic set. In this context, smoothness imples that at each point of X, $df_0(x), \ldots, df_m(x)$ has rank equal to the codimension of X. We describe the blowing-up of $\mathbf{R}^n$ with center X. Let $G_X : \mathbf{R}^n \setminus X \to P^m(\mathbf{R})$ be the graph map defined by

$$G_X(v) = (f_0(v), \ldots, f_m(v)), \quad (v \in \mathbf{R}^n \setminus X),$$

where we have taken homogeneous coordinates on $P^m(\mathbf{R})$. The blow-up $\hat{\mathbf{R}}^n$ of $\mathbf{R}^n$ along X is then defined to be the Zariski closure of the graph of G_X in $\mathbf{R}^n \times P^m(\mathbf{R})$. That is, $\hat{\mathbf{R}}^n$ is the smallest (closed) algebraic subset of $\mathbf{R}^n \times P^m(\mathbf{R})$ which contains the graph of G_X. It may be shown that $\hat{\mathbf{R}}^n$ has the structure of a smooth affine algebraic variety. $\heartsuit$

In general, if we blow-up affine algebraic varieties along smooth centers, we obtain new affine algebraic varieties. The important point here is the assumption that the center has the structure of a smooth *algebraic* variety. Unlike the situation for complex algebraic varieties, there are some subtle difficulties involved in the description of non-singular real algebraic varieties.

Example 5.3 ([2, Exemples 3.3.11(b)]) Consider the irreducible algebraic sub-variety X of $\mathbf{R}^2$ defined by $y^3 + 2x^2y - x^4 = 0$. As an algebraic set, X has an isolated singularity at the origin. However, viewed as an analytic subset of $\mathbf{R}^2$, X is non-singular! Indeed, X is precisely the set of points satisfying the equation $x^2 = y(1 + \sqrt{1+y})$ and so X is a smooth analytic submanifold of $\mathbf{R}^2$. $\heartsuit$

Just as in the case of polar blowing-up, Γ-actions lift when we blow-up. However, we no longer simplify the isotropy structure.

Example 5.4 Let $\Gamma = SO(2)$ act on $\mathbf{R}^2$ in the standard way. Thus, all non-zero points have trivial isotropy. Let $\pi : \hat{\mathbf{R}}^2 \to \mathbf{R}^2$ denote the blowing-up of $\mathbf{R}^2$, center the origin. Topologically, $\hat{\mathbf{R}}^2$ is the Möbius band. The action of $SO(2)$ on $\hat{\mathbf{R}}^2$ has two isotropy types: The trivial isotropy type and $\mathbf{Z}_2$ (corresponding to the center circle of the Möbius band). Generally, we can expect to *increase* the number of isotropy types. $\heartsuit$

In comparison with polar blowing-up, we see that blowing-up typically leads to spaces with a more complicated topology and Γ-action. On the other hand, because of the algebraic character of the process and its naturality, coordinate computations are greatly simplified. Unlike the case of S^{n-1}, the projective space $P^{n-1}(\mathbf{R})$ carries a natural system of (homogeneous) coordinates. (For some explicit computations relevant to bifurcation theory see [15, §4] and Section 7.) In essence, this simplification results because we can work wholly within the category of smooth affine algebraic varieties.

6   Algebraic Geometric structure of group actions

Suppose that (V, Γ) is a Γ-representation. Let η be an isotropy type and $V_\eta \subset V$ denote the set of points of isotropy type η. A proof of the following lemma may be found in [9, §9].

Lemma 6.1 $\overline{V_\eta}$ *is a real algebraic subset of V. Provided that η is not the principal isotropy type, the set of non-singular points of $\overline{V_\eta}$ is equal to V_η.*

Remarks 6.1 (1) Note that the lemma is *trivial* if Γ is finite or abelian. However, if Γ is not finite and non-abelian the result depends on complexification arguments and the Luna slice theorem for algebraic groups.
(2) Applications of Lemma 6.1 are many and include, for example, a proof that the sets $\Sigma_\tau \subset \Sigma$ defined in [12, 9] and [7, Appendix] have the structure of smooth affine algebraic varieties and not merely semi-algebraic sets. ◊

If we give the set of isotropy types a total order, we may attempt to blow-up along the strict transforms of orbit strata starting with the fixed point space of the action and working up the order. It is shown in [9] that this process can be carried out. Specifically, at each stage the appropriate center will be a *smooth affine algebraic* variety. This result has a number of interesting implications.

First of all, it suggests the possibility that effective computer algorithms can be developed for the study of branches of relative equilibria. In view of recent developments in resolution of singularities (constructive methods developed by Bierstone, Milman and others) this is potentially an exciting area of development. In particular, the orbit strata can often be described precisely *without* detailed knowledge of all of the invariants and equivariants. A simple example is given by any of the finite reflection groups, a less trivial example by the irreducible representations of $SO(3)$.

From the analytical point of view these methods, based on blowing-up, give powerful techniques for the study of branches of relative equilibria. There is not the space here to give more than one brief indication of the possible applications of this approach (detailed applications are given in [9]). We have already indicated in Section 4 that if we blow-up a branch of relative equilibria, we can remove the geometric singularity at the origin. We can then do a tangential and normal decomposition to obtain a branch of equilibria (at the blown-up level). Because of the rationality of blowing-up, each finite jet of the blown-up field may be expressed as a rational function of a finite jet of the original field. Moreover, for generic fields, it can be shown that coefficients of the parametrization are analytic functions of finite jets of the original field. Using this type of argument, one can then prove estimates on the growth of eigenvalues that allow one to prove stability results under high-order non-equivariant perturbations.

7 Extensions

Rather remarkably, everything we have done so far extends to complex representations and the Hopf bifurcation. Instead of looking at static bifurcations, we work with $\Gamma \times S^1$-equivariant bifurcations on a complex representation $(\mathbf{C}^n, \Gamma)$. The presence of the S^1-action allows one to lift S^1-equivariant vector fields under complex blowing-up[3]. The analysis of the Hopf bifurcation can thus be reduced to the study of (complex) blowing-ups of families of vector fields at the origin. The interest of this approach lies in its algebraic character, the possibility of development of effective computer algebra methods and the suggestion that some of the singularity methods used in the static theory can be extended to the equivariant Hopf bifurcation. We refer to [15, 11] for further details.

References

[1] D Armbruster, J Guckenheimer and P Holmes. 'Heteroclinic cycles and modulated travelling waves in systems with $O(2)$ symmetry', *Physica D*, (1988), 257–282.

[2] J Bochnak, M Coste and M-F Roy. *Géométrie algébrique réelle* (Ergebnisse der Mathematik und ihrer Grenzgebiete; Folge 3, Bd. 2, Springer-Verlag, Berlin, Heidelberg, New York, 1987.)

[3] P Chossat. 'Forced reflectional symmetry breaking of an $O(2)$-symmetric homoclinic cycle', *Nonlinearity*, 6(5), (1993), 723–732.

[4] P Chossat and M J Field. 'Geometric analysis of the effect of symmetry breaking perturbations on an $O(2)$ invariant homoclinic cycle', In: Normal forms and Homoclinic chaos (W. F. Langford and W. Nagata, eds.) Fields Institute Communications, AMS, to appear.

---

[3]Even though the vector fields are not holomorphic.

[5] M J Field. 'Equivariant Dynamical Systems', *Trans. Amer. Math. Soc.*, **259** (1980),185–205. 26(1982), 161–180.

[6] M J Field. 'Resolving actions of compact Lie groups', *Bull. Austral. Math. Soc.* **18** (1978), 243–254.

[7] M J Field, 'Equivariant Bifurcation Theory and Symmetry Breaking', *J. Dynamics and Diff. Eqns.*, 1(4) (1989), 369–421.

[8] M J Field, 'Local structure of equivariant dynamics' *Singularities, Bifurcations, and Dynamics*, Proceedings of Symposium on Singularity Theory and its Applications, Warwick, 1989 (eds. R. M. Roberts and I. N. Stewart), Lecture Notes in Mathematics **1463**, Springer-Verlag, Heidelberg (1991), 168–195.

[9] M J Field, 'Symmetry breaking for compact Lie groups', preprint, University of Houston, 1994.

[10] M J Field. 'Geometric methods in bifurcation theory', In: Pattern formation and symmetry breaking in PDEs. Fields Institute Communications, AMS, to appear.

[11] M J Field. 'Determinacy and branching patterns for the equivariant Hopf bifurcation', to appear in *Nonlinearity* **7**, (1994).

[12] M J Field and R W Richardson. 'Symmetry Breaking and the Maximal Isotropy Subgroup Conjecture for Reflection Groups', *Arch. for Rational Mech. and Anal.*, **105**(1) (1989), 61–94.

[13] M J Field and R W Richardson. 'Symmetry breaking and branching patterns in equivariant bifurcation theory I', *Arch. Rational Mech. Anal.*, 118 (1992), 297–348.

[14] M J Field and R W Richardson. 'Symmetry breaking and branching patterns in equivariant bifurcation theory II', *Arch. Rational Mech. Anal.*, **120** (1992), 147–190.

[15] M J Field and J W Swift. 'Hopf fibration and the Hopf bifurcation', to appear in *Nonlinearity* **7**, (1994).

[16] M Golubitsky, I N Stewart and D G Schaeffer, *Singularities and Groups in Bifurcation Theory, Vol. II*, Applied Mathematical Sciences **69**, Springer-Verlag, New York, Berlin, Heidelberg, 1988.

[17] M Krupa, 'Bifurcations of relative equilibria', *Siam J. Math. Anal.*, **21**(6) (1990), 1453–1486.

[18] R Lauterbach and M Roberts. 'Heteroclinic cycles in Dynamical systems with broken spherical symmetry', *J. Diff. Eq.*, **100**(1) (1992), 22–48.

M J Field
Department of Mathematics
University of Houston
Houston
Texas 77204
USA
email: mike@math.uh.edu

A REMARK ON THE DETECTION OF SYMMETRY OF ATTRACTORS

KARIN GATERMANN
Konrad-Zuse-Zentrum Berlin
Germany

1   Detectives

In the last years chaotic attractors of dynamical systems have been investigated intensively. A lot of emphasis was put on the symmetry of attractors [5], [6], especially on the symmetry creation while a parameter is varied [2], [3].

The symmetry of the dynamical system is described by the finite group G acting as the faithful linear representation $\vartheta : G \to Gl(R^n)$. In [1] attractors are considered to be thickened to open sets $A \subset R^n$. The symmetry of A is investigated with a *physical space* W, where the linear representation $\rho : G \to Gl(W)$ is the group action on W. A function $f : R^n \to W$ is called ϑ-ρ-*equivariant* if

$$f(\vartheta(t)x) = \rho(t)f(x) \quad \forall\, t \in G, x \in R^n.$$

Definition 1.1 *([1]) A ϑ-ρ-equivariant C^∞ function $\phi : R^n \to W$ is called an ob-servable.*

For the detection of symmetry of attractors it is important that ρ distinguishes all subgroups, i.e. all subgroups H of G appear to be isotropy groups $H = G_y$ for one $y \in W$.

Let $\vartheta^i, i = 1, \ldots, s$ denote the lattice inequivalent irreducible representations of G.

Lemma 1.2 *([1] Thm. 4.3): If $\rho = \sum_{i=1}^{s} \vartheta^i + \tilde{\rho}$ then ρ distinguishes all subgroups of G.*

Let $\mathcal{A}$ be the class of all open subsets of R^n with piecewise smooth boundary that satisfy the dichotomy $\vartheta(t)A = A$ or $\vartheta(t)A \cap A = \emptyset$ for all $t \in G$. A *detective* is an observable which generically determines all symmetries of sets in $\mathcal{A}$.

P. Chossat (ed.), Dynamics, Bifurcation and Symmetry, 123–125.
© 1994 *Kluwer Academic Publishers.*

Definition 1.3 *([1] Def. 5.1) The observable ϕ is a detective if for each subset $A \in \mathcal{A}$ almost all near identity equivariant diffeomorphisms ψ the isotropy group of $\int_{\psi(A)} \phi d\mu$ equals the symmetry group $H(A) = \{t \in G| \vartheta(t)A = A\}$, where μ is a Lebesgue measure.*

Theorem 1.4 *([1] Thm. 5.2): Let $\phi^i, i = 1, \ldots, s$ be ϑ-ϑ^i-equivariant observables which are polynomial and $\phi^i \not\equiv 0, i = 1, \ldots, s$. Then $\phi = (\phi^1, \ldots, \phi^s)$ is a detective.*

2   Attractors in fixed point spaces

The concept of detectives was designed for attractors with points having trivial isotropy. But the following example shows that it has a shortcoming for attractors within a fixed point space.

$D_2 = Z_2(\kappa_1) \times Z_2(\kappa_2) = \{id, \kappa_1, \kappa_2, \kappa_1 \cdot \kappa_2\}$ has 4 irreducible representations $\vartheta^1, \vartheta^2, \vartheta^3, \vartheta^4$. Consider a dynamical system with the symmetry of D_2 acting as $\vartheta = \vartheta^1 + \vartheta^2 + \vartheta^3 + \vartheta^4$,

$$\vartheta(\kappa_1) = \begin{pmatrix} 1 & & & 0 \\ & -1 & & \\ & & -1 & \\ 0 & & & 1 \end{pmatrix}, \quad \vartheta(\kappa_2) = \begin{pmatrix} 1 & & & 0 \\ & -1 & & \\ & & 1 & \\ 0 & & & -1 \end{pmatrix}.$$

Observe that the fixed point space of $Z_2(\kappa_1\kappa_2) = \{id, \kappa_1\kappa_2\}$ is $\{(x_1, x_2, 0, 0)|x_1, x_2 \in R\}$. The equivariant function given by

$$\begin{array}{ll} \phi_1(x) = 1 + x_1 + x_2^2 + x_3^2 + x_4^2 & \phi_3(x) = x_3 \\ \phi_2(x) = x_3 \cdot x_4 & \phi_4(x) = x_4 \end{array}$$

fulfills the requirements of Theorem 1.4. But an attractor with the property that every point has isotropy $Z_2(\kappa_1\kappa_2)$ is detected to have symmetry D_2 which may happen to be not correct. This follows from the fact that ϕ_2 does not depend on x_2.

This example is no contradiction to Theorem 1.4 because there attractors are thickened to open sets including points with trivial isotropy. But it shows that one has to impose stronger conditions on ϕ.

Definition 2.5 *An observable $\phi : R^n \to W$ with respect to given linear representations ϑ and ρ is called a good detective, if $\phi_{|Fix(H)}$ distinguishes all subgroups of $N_G(H)/H$ for all isotropy groups H of the group action ϑ.*

In the example described above $\phi(x)$ is a good detective if the ϑ-ϑ^2-equivariant observable ϕ_2 is replaced by $\phi_2(x) = x_2$.

Acknowledgement. I wish to thank Stephan van Gils, Michael Dellnitz, and Ian Melbourne for discussion and especially Ian Melbourne for encouraging me to submit this note to the proceedings.

References

[1] E. Barany, M. Dellnitz, M. Golubitsky. *Detecting the Symmetry of Attractors.* Physica D 1993, to appear.

[2] P. Chossat, M. Golubitsky. *Symmetry-increasing bifurcation of chaotic attractors.* Physica D 32, 423-436, 1988.

[3] M. Dellnitz, M. Golubitsky, I. Melbourne. *Mechanisms of Symmetry Creation.* in Bifurcation and Symmetry (eds.: E. Allgower, K. Böhmer, M. Golubitsky), ISNM 104, Birkhäuser, p. 99-109, 1992.

[4] M. Dellnitz, M. Golubitsky, M. Nicol. *Symmetry of Attractors and the Karhunen-Loève Decomposition.* Preprint 69, Universität Hamburg, 1993.

[5] M. Field, M. Golubitsky. *Symmetry in Chaos.* Oxford University Press, Oxford, 1992.

[6] I. Melbourne, M. Dellnitz, M. Golubitsky. *The Structure of Symmetric Attractors.* Arch. Rational Mech. & Anal. 1993 to appear.

COUPLED CELLS: WREATH PRODUCTS AND DIRECT PRODUCTS

MARTIN GOLUBITSKY
Dept of Mathematics
University of Houston
USA

IAN STEWART
Maths Institute
University of Warwick
UK

BENOIT DIONNE
Dept of Mathematics
University of Ottawa
CANADA

1 Introduction

In this note we discuss the structure of systems of coupled cells (which we view as systems of ordinary differential equations) where symmetries of the system are obtained through the group $\mathcal{G}$ of global permutations of the cells and the group $\mathcal{L}$ of local internal symmetries of the dynamics in each cell. We show that even when the cells are assumed to be identical with identical coupling, the way that $\mathcal{G}$ and $\mathcal{L}$ combine to form the total symmetry group of the system Γ depends on properties of the coupling. We illustrate this point by showing how the combination of $\mathcal{L}$ with $\mathcal{G}$ can lead to a symmetry group Γ that is either a direct product or a wreath product. The symmetry group has strong implications for the dynamics of the system of cells, and the distinction between the two cases is substantial. This has important implications for the modeling of systems by coupled cells.

Several authors have studied abstractly systems of coupled cells *cf.* Alexander [1]. It has been noted previously that the form of the coupling can seriously affect the dynamics in systems of coupled cells [2]. More recently, Dangelmayr *et. al.* [6] have studied coupled cell systems with specific internal and global symmetries where the coupling produces direct product symmetry groups. Here we emphasize the point that the type of coupling does influence the total symmetry group and describe a few general bifurcation results for two natural types of coupling.

In the next section we discuss the form of the differential equations describing coupled cells. Section 3 develops the properties of the coupling that lead to direct and wreath products and Section 4 presents a number of examples of each type of coupling. In Section 5 we discuss the types of bifurcating branches that may occur in steady-state bifurcations for wreath product systems and in the last section, Section 6,

P. Chossat (ed.), Dynamics, Bifurcation and Symmetry, 127–138.
© 1994 *Kluwer Academic Publishers.*

we consider these bifurcations for direct product symmetries. These sections preview work that will appear in [8].

2   Identical Coupled Cells

In this section we discuss the form taken by systems of differential equations that model systems of identical cells with identical coupling. Imagine an array of N coupled cells — by which we mean a set of N cells with arrows connecting cell i to cell j when the output of cell i is coupled to cell j. Define the $N \times N$ *connection matrix* C by setting

$$C(i,j) = \begin{cases} 1 & \text{if cell } i \text{ is coupled to cell } j \\ 0 & \text{otherwise.} \end{cases}$$

Note that C need not be a symmetric matrix (the coupling may be directed).

Let $X_j \in \mathbf{R}^{k_j}$ be the state variables of cell j and let $X = (X_1, \ldots, X_N)$ be the state variables of the entire system of cells. Suppose that the dynamics of the coupled cell system is modeled by a differential equation

$$\dot{X} = F(X). \tag{2.1}$$

The structure of coupled cells allows us to write (2.1) in the form

$$\dot{X}_j = f_j(X_j) + h_j(X)$$

where f_j models the internal dynamics of the jth cell and h_j represents the coupling of all other cells to the jth cell.

For simplicity of exposition, assume that the total coupling to cell j is the *sum* of couplings from all cells i to cell j. (Similar conclusions may also be derived for more general types of coupling.) This assumption may be stated as

$$h_j(X) = \sum_{\{i:C(i,j)=1\}} h_{ij}(X_j, X_i)$$

for appropriate functions h_{ij}.

Now assume that all cells are identical. Then $k_j = k$ and $f_j = f$ for $j = 1, \ldots, N$; that is, the internal dynamics of each cell is governed by the same set of equations. The assumption that all couplings between cells are identical leads to the identity $h_{ij} = h$ for all i, j.

To summarize: the system of ODEs (2.1) has the form

$$\dot{X}_j = f(X_j) + \sum_{i=1}^{N} C(i,j)h(X_j, X_i) \tag{2.2}$$

where $X_j \in \mathbf{R}^k$ for $j = 1, \ldots, N$.

3 Symmetries in Coupled Cells

As mentioned in the introduction, the symmetries of coupled systems appear in two ways: through global permutation symmetries and through local internal symmetries. Studies of the effects of the global symmetries have been made by many authors, but studies of the effects of internal symmetries have been less frequent.

Global Permutation Symmetries $\mathcal{G}$

In coupled cell systems with identical cells and identical coupling the *global symmetries* are permutations, determined by patterns in the couplings themselves. There are three especially popular patterns in which the cells form rings, directed rings, or simplexes ('all-to-all' coupling). In these cases the permutation symmetries $\mathcal{G}$ are the dihedral groups $\mathbf{D}_N$, the cyclic group $\mathbf{Z}_N$ and the permutation group $\mathbf{S}_N$, respectively. A permutation $\sigma \in \mathbf{S}_N$ acts on state space by

$$\sigma \cdot X = (X_{\sigma^{-1}(1)}, \ldots, X_{\sigma^{-1}(N)}).$$

The permutation σ is a symmetry of the coupled cell system if

$$F(\sigma \cdot X) = \sigma \cdot F(X)$$

which happens precisely when

$$\sigma C \sigma^{-1} = C, \tag{3.3}$$

where σ is viewed as a permutation matrix.

Local Internal Symmetries $\mathcal{L}$

A $k \times k$ matrix ρ is an *internal symmetry* if it is a symmetry of the internal dynamics of each cell, that is,

$$f(\rho X_j) = \rho f(X_j).$$

When do the internal symmetries lead to symmetries of the entire coupled cell array? The answer depends on the type of coupling h. We discuss two types of coupling.

Direct Products

Suppose that the coupling is equivariant with respect to the internal symmetries. That is,

$$h(\rho X_j, \rho X_i) = \rho h(X_j, X_i)$$

for all $\rho \in \mathcal{L}$. If we let $\mathcal{L}$ act on state space by

$$\rho \cdot X = (\rho X_1, \ldots, \rho X_N),$$

then $\mathcal{L} \times \mathcal{G}$ is a group of symmetries of the full system of ODEs (2.1). An example is diagonal linear coupling

$$h(X_j, X_i) = \lambda(X_i - X_j),$$

where $\lambda \in \mathbf{R}$ is the coupling strength.

If h satisfies no other invariance or equivariance conditions, then the symmetry group of (2.2) is the direct product of the groups of local and global symmetries. When the coupling leads to a direct product, the internal symmetries are symmetries of the whole cell system only when they act 'diagonally' — that is, in the same way on each cell.

Wreath Products

In the second type of coupling *any* internal symmetry acting on any individual cell is a symmetry of the entire system (2.2). In this case the coupling of cell i to cell j must not 'feel' the effect of an internal symmetry applied to cell i alone. This invariance may be formalized as:

$$\begin{aligned} h(\rho X_j, \rho X_i) &= \rho h(X_j, X_i) \\ h(X_j, \rho X_i) &= h(X_j, X_i) \end{aligned}$$

for all $\rho \in \mathcal{L}$. Equivalently

$$h(\rho X_j, \sigma X_i) = \rho h(X_j, X_i)$$

for all $\rho, \sigma \in \mathcal{L}$. (That is, the coupling is invariant in X_i, equivariant in X_j.)

With such coupling the symmetries of (2.1) include the group $\mathcal{L}^N$ acting by

$$(\rho_1, \ldots, \rho_N) \cdot X = (\rho_1 X_1, \ldots, \rho_N X_N).$$

The *wreath product* of $\mathcal{L}$ with the permutation group $\mathcal{G}$, denoted by $\mathcal{L} \wr \mathcal{G}$, is the smallest group generated by $\mathcal{L}^N$ and $\mathcal{G}$ in the given actions of $\mathcal{L}^N$ and $\mathcal{G}$ on state space. See Robinson [19] p.18 or Scott [20] p.215 for general information on wreath products. The wreath product contains the direct product (identify $\mathcal{L}$ with the diagonal subgroup of $\mathcal{L}^n$) but it is huge in comparison. If the coupling satisfies no further group-theoretic constraints, then the wreath product will be the full symmetry group of these coupled cell systems. We note that an example of wreath product coupling is:

$$h(X_j, X_i) = |X_i|^2 X_j.$$

4 Examples

At first sight the above abstract considerations may appear rather artificial, especially as regards the wreath product. However, examples of both kinds of coupled cell system are widespread — and in several respects the wreath product is the most natural and the most interesting. In this section we discuss a number of such examples.

Wreath Product Examples

(a) *Coupled arrays of Josephson junctions.*

This example was suggested by Kurt Wiesenfeld. There is an extensive literature studying arrays of identical coupled Josephson junctions [17, 3]. Indeed, such arrays are prototypical examples of systems exhibiting all-to-all coupling, since the coupling is electrical and is felt equally by all junctions. Thus an array having k junctions is modeled by a system of differential equations with $\mathbf{S}_k$ symmetry. Josephson junction arrays are usually posed as an example of a system of coupled cells with no internal symmetry. However, we shall consider each Josephson junction array to be a single cell, with $\mathbf{S}_k$ as its internal symmetry group.

Josephson junction arrays are often used to model certain kinds of computer chip. From this point of view, it is natural to consider an array of N chips, also electrically coupled. Thus the global symmetry group is $\mathbf{S}_N$, since the system of chips may be modeled as having all-to-all coupling. When the resistances in the individual chip and in the array of chips are different — a reasonable modeling assumption — then this system of coupled chips has $\mathbf{S}_k \wr \mathbf{S}_N$ symmetry (rather than $\mathbf{S}_{kN}$ symmetry).

(b) *Discretizations of PDEs with gauge symmetry.*

In systems of PDEs with local gauge symmetry, the gauge group acts independently at each point in space. For example, in systems with an abelian gauge, such as the original complex Ginzburg-Landau equation modeling superconductivity, the local gauge symmetry is a phase shift and (except for smoothness considerations) the phase shift operates independently at each point in space. It is well known that when discretizations of systems of PDE are made, the resulting system of ODEs has the structure of a coupled system of cells with each cell representing the dynamics of the PDE at one point or in one small region of space. From this point of view it is natural for the gauge symmetries to act independently in each cell. Should the system of PDEs be posed on a symmetric domain, then the total symmetry group of the discrete system will be the wreath product of the local gauge symmetry with the global (permutation) symmetry of the domain.

(c) *Molecular dynamics.*

As suggested by John Guckenheimer, another example of wreath product symmetry should occur in molecular dynamics. Molecules are made up of atoms (the cells) and have permutation symmetries that depend on the type of atoms and the bonds (coupling) between the atoms. On the other hand, atoms themselves have internal symmetries and the application of one of these symmetries to one atom should have no effect on the bonds between that atom and another. If this description is valid, then symmetries of models for the dynamics of molecules that include internal variables from the individual atoms will have a wreath product symmetry. If it is merely an approximation, then the system can be considered as a symmetry-breaking perturbation of one with wreath product symmetry.

(d) *Heteroclinic cycles.*

Perhaps the best known example of a structurally stable heteroclinic cycle in a symmetric system is the one abstracted by Guckenheimer and Holmes [15] from a model by Busse and Heikes [4] on rotating convection. In the experiment the dynamics of the convection system passes near three rolls patterns — each rotated by 120° from the previous one. Guckenheimer and Holmes observed that the model in [4] can be abstracted using a certain 24 element symmetry group; this symmetry group is just $\mathbf{Z}_2 \wr \mathbf{Z}_3$. The system of ODEs has the form of a system of three coupled cells with one internal state variable ($k = 1$) and one nontrivial internal symmetry ($\mathbf{Z}_2$). Due to the rotation in the model, the coupling from cell i to cell j is not equal to the coupling from cell j to cell i. Thus the symmetry in this system is that of a directed ring system. One wonders whether the existence of heteroclinic cycles is related to the coupling pattern. Examples of Field and Richardson [9] on symmetry groups $\mathbf{Z}_2 \wr \mathbf{Z}_N$ substantiate this point of view. The 'instant chaos' scenario of Guckenheimer and Worfolk [16] involves a subgroup of index two in $\mathbf{Z}_2 \wr \mathbf{Z}_4$. The symmetry group of the cube is the wreath product $\mathbf{Z}_2 \wr \mathbf{D}_3$.

Direct Product Examples

(a) *Neural networks.*

Wegelin *et. al.* [21, 6, 7] study coupled systems of three cells where each cell is itself a system of three identical cells. In order to study patterns of oscillation they consider direct product couplings. In particular they consider the types of Hopf bifurcation that occur with this symmetry. They find eleven different patterns of oscillation, as well as states with more complicated dynamics, and they discuss the stability of the periodic solutions that they find.

(b) *Discretization of PDEs with range symmetries.*

Suppose that a system of PDEs in k functions u is posed on a domain with a symmetry group $\mathcal{G}$ and a group of range symmetries $\mathcal{L}$ acting on $\mathbf{R}^k$. For example, consider the reaction-diffusion system on the interval $[0,1]$ satisfying Dirichlet boundary conditions

$$u_t = \Delta u + f(u),$$

where $f(-u) = -u$. In this case the nontrivial domain symmetry is $x \mapsto 1 - x$ and the nontrivial range symmetry is $u \mapsto -u$. Other examples include elastic buckling of rods and plates with various symmetric geometries and appropriate boundary conditions — see for instance Buzano *et al.* [5] who study rods whose cross-sections are regular polygons, leading to $\mathbf{Z}_2 \times \mathbf{D}_n$ symmetry.

Discretizations of such PDEs will lead to coupled cell systems with the direct product of the domain and range symmetries as the group of symmetries. Here the range symmetries must act identically on all cells.

(c) *Direct products that occur by themselves.*

In a number of applications direct products occur in the standard models. For example, the Couette-Taylor system is posed on a circular cylinder with periodic boundary conditions in the axial direction. The group of symmetries for this model is $\mathbf{O}(2) \times \mathbf{SO}(2)$. Euclidean-invariant PDEs on rectangular domains with periodic boundary conditions have $\mathbf{O}(2) \times \mathbf{O}(2)$ symmetry, see Gomes and Stewart [13, 14] which also give connections with Neumann and Dirichlet boundary conditions.

5   Wreath Product Bifurcations

Wreath product bifurcations lead to some rather remarkable states, in which some cells are active while the remainder are quiescent. This kind of spatial differentiation has been found only rarely in bifurcation analyses, but appears to be natural in coupled cells with wreath product symmetry. In this section we summarize our knowledge of steady-state bifurcation in systems of coupled cells with wreath product symmetries. Details may be found in [8] along with a description of the corresponding Hopf bifurcation. To simplify the discussion here we assume that the global permutation symmetries $\mathcal{G}$ act transitively on the N cells.

Generically, steady-state bifurcations in systems of ODEs occur when the linearization at an equilibrium has zero eigenvalues and the kernel of the linearization is an absolutely irreducible representation of the group of symmetries of that equilibrium. In the context of coupled cells with wreath product $\mathcal{L} \wr \mathcal{G}$ symmetry we consider steady-state bifurcation from a group invariant equilibrium. The corresponding irreducible representations have local symmetries acting either trivially or nontrivially. When the local symmetries act trivially, the types of bifurcation reduce to the types considered in coupled cell systems with no internal symmetry. Here we only consider those types of bifurcation in which the internal symmetry group $\mathcal{L}$ acts nontrivially. In the case of a nontrivial action of $\mathcal{L}$ the irreducible spaces have the form

$$ W \oplus \cdots \oplus W, $$

where W is an irreducible representation of $\mathcal{L}$. Here we use the assumption that $\mathcal{G}$ acts transitively on the N cells. The irreducible representation on the kernel is absolutely irreducible precisely when the representation of $\mathcal{L}$ on W is absolutely irreducible, which we henceforth assume.

Recall that the equivariant branching lemma [12] guarantees that generically there exists a branch of bifurcating equilibria for each isotropy subgroup with a one-dimensional fixed-point subspace. We call an isotropy subgroup *axial* if it has a one-dimensional fixed-point subspace; we also call the corresponding bifurcating solutions axial. It turns out that axial subgroups of wreath product symmetry groups are relatively easy to describe.

Definition 5.1 *A subset $J \subset \{1, \ldots, N\}$ is a* block *if there exists a subgroup $Q \subset \mathcal{G}$ which leaves J invariant and acts transitively on J. Let Q_J be the largest subgroup of $\mathcal{G}$ that leaves J invariant.*

We now show how to form axial subgroups of $\mathcal{L} \wr \mathcal{G}$ from an axial subgroup A of $\mathcal{L}$ acting on W and a block J. Let

$$ \Sigma(A, J) = (H_1, \ldots, H_N) \dot{+} Q_J, $$

where $H_j = A$ if $j \in J$ and $H_j = \mathcal{L}$ otherwise.

Theorem 5.2 *The subgroup $\Sigma \subset \mathcal{L} \wr \mathcal{G}$ is axial if and only if Σ is conjugate to $\Sigma(A, J)$ for some axial subgroup $A \subset \mathcal{L}$ and some block J.*

Note that solutions $X = (X_1, \ldots, X_N)$ corresponding to the subgroup $\Sigma(A, J)$ have the property that $X_j = 0$ for all $j \notin J$ and $X_j \neq 0$ for all $j \in J$. Thus the cells j are quiescent when $j \notin J$ and active when $j \in J$. We note that similar results hold for Hopf bifurcation in the presence of wreath product symmetry.

6 Direct Product Bifurcations

Steady-state Bifurcation.

As in the previous section we begin our discussion with the irreducible representations of $\mathcal{L} \times \mathcal{G}$. Over $\mathbf{C}$ the irreducible representations of the direct product are just tensor products of irreducible representations of $\mathcal{L}$ and $\mathcal{G}$. This is not always true over $\mathbf{R}$ — but it is often true. (The precise description depends upon the commuting linear maps of irreducible representations, see [8].) In this note we consider only absolutely irreducible representations that are tensor products of absolutely irreducible representations. So assume that $\mathcal{L} \times \mathcal{G}$ acts absolutely irreducibly on the real tensor product $U \otimes V$. The following theorem identifies a class of axial solutions of systems with direct product symmetry.

Theorem 6.1 *Let A be axial for $\mathcal{L}$ acting on U and let B be axial for $\mathcal{G}$ acting on V. Then $A \times B$ is either axial or a subgroup of index two in an axial subgroup.*

Hopf Bifurcations.

Similar ideas apply to Hopf bifurcation. The equivariant Hopf bifurcation theorem [12] allows us to find branches of time periodic solutions for each isotropy subgroup $\Sigma \subset \Gamma \times \mathbf{S}^1$ which has a two-dimensional fixed-point subspace (in an appropriate representation). We say that Σ is $\mathbf{C}$-*axial* if its fixed-point subspace is two-dimensional.

In general, isotropy subgroups $\Sigma \subset \Gamma \times \mathbf{S}^1$ have the form of a *twisted* subgroup, that is, there is a subgroup $A \subset \Gamma$ and a homomorphism $\varphi : A \to \mathbf{S}^1$ such that $\Sigma = A^\varphi$ where
$$A^\varphi = \{(a, \varphi(a)) \in \Gamma \times \mathbf{S}^1 : a \in A\}.$$

In order to introduce the $\mathbf{C}$-axial subgroups of $\Gamma = \mathcal{L} \times \mathcal{G}$ we need one additional definition. There is a natural way to take the product of two twisted subgroups. Recall that the twisting reflects the fact that the general symmetry of a periodic solution is a mixture of a space symmetry a with a phase shift $\varphi(a)$. When taking the product of two twisted groups one must add the phase shifts. More precisely, let A^φ and B^ψ be twisted subgroups. Then define

$$A^\varphi \dot\times B^\psi = \{(a, b, \varphi(a) + \psi(b)) \in \mathcal{L} \times \mathcal{G} \times \mathbf{S}^1\}.$$

The following theorem shows that there are many periodic solutions whose isotropy subgroups are products.

Theorem 6.2 *If $A^\varphi \subset \mathcal{L} \times \mathbf{S}^1$ and $B^\psi \subset \mathcal{G} \times \mathbf{S}^1$ are C-axial in the representations on U and V, then $A^\varphi \dot\times B^\psi$ is C-axial in $\mathcal{L} \times \mathcal{G} \times \mathbf{S}^1$ in the representation on $U \otimes V$.*

For example, it is known that there are two **C**-axial subgroups of **O(2)** acting on $U = \mathbf{C} \otimes \mathbf{C}$ corresponding to rotating waves and standing waves (see [12]). It is also known that there are three **C**-axial subgroups of $\mathbf{D_3}$ acting on $V = U$ corresponding to a discrete rotating wave (ponies on a merry-go-round) and two standing waves (see [12]). Thus in the action of $\mathcal{L} \times \mathcal{G} = \mathbf{O(2)} \times \mathbf{D_3}$ on $U \otimes V$, there are at least six **C**-axial subgroups. This Hopf bifurcation problem is considered by Wegelin [21] when analyzing a system of three coupled lasers and these periodic solutions are there found by explicit computation. Wegelin also finds one additional **C**-axial subgroup in this bifurcation. Similarly, in the corresponding Hopf bifurcation for $\mathbf{D_3} \times \mathbf{D_3}$ symmetry Theorem 6.2 determines nine **C**-axial product subgroups. Wegelin *et al.* [6, 21] find these solutions in their study of neural nets with *macro* and *micro* symmetry. They also find two additional **C**-axial subgroups in this representation. It is possible to use representation theoretic ideas to compute the additional (non-product) **C**-axial subgroups. The details may be found in [8].

Acknowledgment: We thank Gerhard Dangelmayr and Michael Wegelin for helpful discussions concerning the nature of the symmetries in coupled cells. This research was supported in part by NSF Grant DMS-9101836 (MG), the Texas Advanced Research Program (003652037) (MG), a grant from the Science and Engineering Research Council of the UK (INS), and a European Community Laboratory Twinning grant (INS). The research of INS was carried out under the auspices of the European Bifurcation Theory Group. All three authors were supported by the Fields Institute for Research in the Mathematical Sciences, and are grateful for the hospitality extended to them.

References

[1] J.C. Alexander. Patterns at primary Hopf bifurcations of a plexus of identical oscillators, *SIAM J. Appl. Math.* **46**(2) (1986) 199-221.

[2] J.C. Alexander and B. Fiedler. Global decoupling of coupled symmetric oscillators. In: *Differential Equations* (C.M. Dafermos, G. Ladas and G. Papanicolaou, eds.), Lecture Notes in Pure and Applied Mathematics **118**, Marcel Dekker, Inc. New York, 1989, 7-16.

[3] D.G. Aronson, M. Golubitsky and M. Krupa. Large arrays of Josephson junctions and iterates of maps with S_n symmetry. *Nonlinearity* **4** (1991) 861-902.

[4] F.H. Busse and Heikes. Convection in a rotating layer: a simple case of turbulence. *Science* **208** (1980) 173-175.

[5] E.Buzano, G.Geymonat, and T.Poston. Post-buckling behavior of a nonlinearly hyperelastic thin rod with cross section invariant under the dihedral group $\mathbf{D}_n$, *Arch. Rational Mech. Anal.* **89** (1985) 307-388.

[6] G. Dangelmayr, W. Güttinger and M. Wegelin. Hopf bifurcation with $\mathbf{D}_3 \times \mathbf{D}_3$-symmetry, *ZAMP* **44** (1993) 595-638.

[7] G. Dangelmayr, W. Güttinger, J. Oppenländer, J. Tomes, and M. Wegelin. Coupled neural oscillators with $\mathbf{D}_3 \times \mathbf{D}_3$-symmetry. Preprint, 1992.

[8] B. Dionne, M. Golubitsky and I.N. Stewart. Arrays of oscillators with internal and global symmetries. In preparation.

[9] M. Field and R. Richardson. Symmetry breaking and the maximal isotropy subgroup conjecture for reflection groups. *Arch. Rational Mech. & Anal.* **105**, No. 1 (1989) 61-94.

[10] M.J. Field and R.W. Richardson. Symmetry breaking and branching patterns in equivariant bifurcation theory II, *Arch. Rational Mech. Anal.* **120** (1992) 147-190.

[11] M.J. Field and J.W. Swift. Stationary bifurcation to limit cycles and heteroclinic cycles, *Nonlinearity* **4** (1992) 1001-1043.

[12] M. Golubitsky, I.N. Stewart and D.G. Schaeffer. *Singularities and Groups in Bifurcation Theory* vol.II, Appl. Math. Sci. **69**, Springer-Verlag, New York 1988.

[13] M.G.M. Gomes and I.N.Stewart, Steady PDEs on generalized rectangles: a change of genericity in mode interactions, *Nonlinearity*. To appear.

[14] M.G.M. Gomes and I.N.Stewart, Hopf bifurcations on generalized rectangles with Neumann boundary conditions, in *Dynamics, Bifurcations, Symmetry*, Cargese 1993. To appear.

[15] J. Guckenheimer and P. Holmes. Structurally stable heteroclinic cycles, *Math. Proc. Camb. Phil. Soc.* **103** (1988) 189-192.

[16] J.Guckenheimer and P.Worfolk, Instant chaos, *Nonlinearity* **5** (1992) 1211-1222.

[17] P. Hadley, M.R. Beasley and K. Wiesenfeld. Phase locking of Josephson-junction series arrays, *Phys. Rev. B* **38** (1988) 8712-8719.

[18] J. Oppenländer. *Zur Dynamik hierarchischer Oszillatorennetze*. Diplomarbeit, Universität Tübingen, Institut für Informationsverarbeitung, Fakultät für Physik, 1992.

[19] D.J.S.Robinson, *Finiteness Conditions and Generalized Soluble Groups* vol.2, *Ergebnisse der Math.* **63**, Springer-Verlag, New York 1972.

[20] W.R.Scott, *Group Theory*. Prentice-Hall., Englewood Cliffs NJ 1964.

[21] M. Wegelin. *Nichtlineare Dynamik raumzeitlicher Muster in hierarchischen Systemen*. Dissertation, Universität Tübingen, Fakultät für Physik, Institut für Informationsverarbeitung, 1993.

HOPF BIFURCATIONS ON GENERALIZED RECTANGLES WITH NEUMANN BOUNDARY CONDITIONS

GABRIELA GOMES and IAN STEWART
Mathematics Institute
University of Warwick, UK

ABSTRACT. Bifurcation problems for PDEs posed on multidimensional rectangular sub-domains of Euclidean space often possess more symmetry than is immediately apparent. In particular, suppose that the partial differential operator is invariant under the subgroup of the Euclidean group generated by translations and by reflections in coordinate hyper-planes. Then solutions of the PDE may be extended periodically by reflecting them across the boundaries of the rectangle. These extra 'hidden' symmetries affect the generic bifurcation equations for mode interactions. In a previous paper we established the appropriate general forms of these bifurcation equations, for the interaction of two steady-state modes. Here we extend the analysis to interactions involving Hopf modes, namely a single Hopf mode, a steady-state/Hopf mode interaction, and a Hopf/Hopf mode interaction.

1   Introduction

Equivariant bifurcation theory exploits the symmetries of a dynamical system to understand its typical bifurcations. It has recently become apparent that a wide class of problems possess more symmetry than is immediately apparent, and that these 'hidden symmetries' can affect the generic bifurcations. The appropriate context is that of a PDE

$$u_t + \mathcal{P}(u) = 0 \tag{1}$$

posed on some domain $\Omega \subset \mathbf{R}^n$, where the differential operator $\mathcal{P}$ is invariant under the Euclidean group $\mathbf{E}_n$ or some subgroup that contains the group $\mathbf{E}_1^n$ generated by the translation group $\mathbf{R}^n$ and all reflections in the n coordinate hyperplanes $x_j = 0, j = 1, ..., n$. Note that

$$\mathbf{E}_1^n = \mathbf{E}_1 \times ... \times \mathbf{E}_1$$

139

P. Chossat (ed.), Dynamics, Bifurcation and Symmetry, 139–158.

where $\mathbf{E}_1$ is the Euclidean group on $\mathbf{R}$, that is, the group of transformations of $\mathbf{R}$ generated by translations $x \mapsto x + a$ and the reflection $x \mapsto -x$.

The methods of this paper apply most directly in the presence of Neumann boundary conditions (NBC), but they also apply — sometimes with minor modifications — to Dirichlet or even mixed Neumann/Dirichlet boundary conditions, see for example Impey et al. [17]. To avoid overcomplicating the argument we shall suppose for definiteness that Neumann boundary conditions (henceforth abbreviated to NBC) apply — that is, $\frac{\partial u}{\partial n} = 0$ where n indicates the normal to the boundary. Then the symmetries of (1.1) include all elements of $\mathbf{E}_n$ that fix Ω setwise. However, for some domains Ω there may be additional symmetries. The manner in which they arise was observed by Fujii et al. [9] in the special case of reaction- diffusion equations on the interval, and this problem was studied using singularity-theoretic methods by Armbruster and Dangelmayr [1, 7].

In this paper we shall concentrate on the simplest case, for which Ω is a *generalized rectangle* (or rectangular parallelepiped)

$$[0, \pi a_1] \times ... \times [0, \pi a_n] \subset \mathbf{R}^n.$$

(The factors π are included to simplify later calculations.) Then we may extend solutions $u(x)$ of 1.1 with NBC on Ω to solutions $\hat{u}(x)$ with NBC on the larger domain

$$\hat{\Omega} = [-\pi a_1, \pi a_1] \times ... \times [-\pi a_n, \pi a_n]$$

by reflection in coordinate hyperplanes:

$$\hat{u}(x_1, ..., x_n) = u(|x_1|, ..., |x_n|).$$

Having done so, we can consider $\hat{u}$ as a solution that satisfies *periodic* boundary conditions (henceforth abbreviated to PBC) on $\hat{\Omega}$, and extend periodically to a solution defined on the whole of $\mathbf{R}^n$, which we continue to denote by $\hat{u}$. This extension requires the operator $\mathcal{P}$ to possess appropriate invariance properties: invariance under $\mathbf{E}_1^n$ is sufficient. Moreover, appropriate regularity theorems must hold in order that, say, C^∞ solutions u on Ω extend to C^∞ solutions $\hat{u}$ on $\mathbf{R}^n$, see Field et al. [8] Theorem 5.18. These regularity conditions are valid, in particular, when $\mathcal{P}$ is an elliptic operator.

In this manner the space $\mathcal{X}_{\mathrm{NBC}}$ of NBC solutions on Ω is embedded in the larger space $\mathcal{X}_{\mathrm{PBC}}$ of PBC solutions on $\mathbf{R}^n$. Indeed it is the fixed-point space of a subgroup $\mathbf{Z}_2^n$ of $\mathbf{E}_1^n$ generated by the n coordinate reflections. Recall that the fixed-point space of a group Γ acting on a space X is

$$\mathrm{Fix}(\Gamma) = \{x \in X : \gamma.x = x \quad \forall \gamma \in \Gamma\}.$$

This additional structure affects the generic bifurcation equations for the interaction of finitely many steady-state modes (that is, when the kernel of the linearization is finite-dimensional). Crawford *et al.* [6] describe simple group-theoretic restrictions — for example that single-mode bifurcations may generically be pitchforks — and discuss a series of physical examples. Gomes and Stewart [14] develop a general normal form for the Liapunov- Schmidt reduced bifurcation equations for the interaction of two steady state modes in the case when Ω is a generalized rectangle; they also discuss the interaction of more than two modes. Gomes [13] develops a corresponding theory for domains such as cubes, with additional symmetries. Field *et al.* [8] analyse single modes on hemispherical domains by a similar extension to the full sphere, rendering the problem $\mathbf{O}(3)$-equivariant. A striking application to the Faraday crispation experiment, leading to predictions that have been verified experimentally, is given by Crawford [4] and Crawford *et al.* [5].

In this paper we extend these results to Hopf bifurcation. In the terminology of Golubitsky *et al.* [11] chapter XIX we study three cases: single Hopf modes (§4), the interaction of two Hopf modes (§5), and Hopf/steady-state mode interaction (§6). For a single Hopf mode and its interaction with a steady-state mode we employ the Liapunov-Schmidt reduction approach of Golubitsky *et al.* [11], which goes back to Hale [16]. This introduces a new period-scaling parameter, which we suppress because ultimately it is expressed in terms of the existing variables, and a circle group $\mathbf{S}^1$ of phase- shift symmetries. The bifurcating branches of steady states or of periodic solutions with period near that of the linearized flow are given by the zeros of an $\mathbf{S}^1$-equivariant mapping.

For the Hopf/Hopf mode interaction the Liapunov-Schmidt reduction approach leads to intractable equations, so we follow the technique of Golubitsky *et al.* Chapter XIX §4, namely reduction to Birkhoff normal form (in combination with centre manifold reduction since we are studying PDEs). This procedure is less satisfactory in that it introduces extra considerations regarding truncations of Taylor series and degrees of smoothness: we discuss these issues in § 5. The results are expressed as 'normal forms' for ODEs rather than mappings, which permit a further reduction to phase/amplitude equations.

We now summarize our main results, listing the general form of the Liapunov-Schmidt reduced bifurcation equations for cases involving only one Hopf mode, and the Birkhoff normal form equations for the Hopf/Hopf interaction. Here λ is the bifurcation parameter, the period-scaling parameter has been eliminated, x is the amplitude of a steady-state mode, and z, w are complex numbers, each representing the phase and amplitude of a Hopf mode. The integers k and l are the maximum values of the mode numbers for the appropriate modes, much as in Gomes and Stewart [14]. See § 3.1 for details. The regularity conditions that we impose on $\mathcal{P}$ imply that the functions a, b, c, d are smooth.

(a) *Single Hopf mode.*
$$a(|z|^2, \lambda)z = 0$$
where $a : \mathbf{C} \times \mathbf{R} \to \mathbf{C}$ and $(z, \lambda) \in \mathbf{C} \times \mathbf{R}$.

(b) *Hopf/Hopf mode interaction.*
$$a(|z|^2, |w|^2, \lambda)\begin{pmatrix} z \\ 0 \end{pmatrix} + b(|z|^2, |w|^2, \lambda)\begin{pmatrix} 0 \\ w \end{pmatrix} = 0$$
where $a, b : \mathbf{C}^2 \times \mathbf{R} \to \mathbf{C}$ and $(z, w, \lambda) \in \mathbf{C}^2 \times \mathbf{R}$.

(c) *Hopf/steady-state mode interaction.*
If all wave numbers k_j are even:

$$a(x^2, |w|^2, \lambda)\begin{pmatrix} x \\ 0 \end{pmatrix} + c(x^2, |w|^2, \lambda)\begin{pmatrix} x^{l-1}|w|^k \\ 0 \end{pmatrix} \tag{2}$$
$$+ b(x^2, |w|^2, \lambda)\begin{pmatrix} 0 \\ w \end{pmatrix} + d(x^2, |w|^2, \lambda)\begin{pmatrix} 0 \\ x^l|w|^{k-2}w \end{pmatrix} = 0.$$

If not,

$$a(x^2, |w|^2\lambda)\begin{pmatrix} x \\ 0 \end{pmatrix} + b(x^2, |w|^2, \lambda)\begin{pmatrix} 0 \\ w \end{pmatrix} = 0. \tag{3}$$

In both equations $a, b, c, d : \mathbf{R} \times \mathbf{C} \times \mathbf{R} \to \mathbf{R} \times \mathbf{C}$, and $(x, w, \lambda) \in \mathbf{R} \times \mathbf{C} \times \mathbf{R}$.

The most interesting case is (c), which — as for the interaction of two steady-state modes — leads to a normal form containing specific high order terms that cannot be read off directly from symmetries induced by those of the domain Ω. The other cases lead to simpler normal forms, but for certain parities of the mode numbers they again impose more restrictions than the symmetries of the domain alone, because some domain symmetries act trivially on the linearized eigenfunctions.

2 The Periodic Extension Method

Let $\mathcal{P}$ denote an elliptic operator on an appropriate function space. We are interested in the solution set of the PDE

$$u_t + \mathcal{P}(u) = 0 \tag{4}$$

where $u : \mathbf{R}^n \times \mathbf{R} \to \mathbf{R}$ and λ is a bifurcation parameter that will be needed in later sections. In particular our analysis applies to a reaction-diffusion operator $\mathcal{P}(u) \equiv u_t + \Delta u + F(u, \lambda)$ where F is a smooth real valued function and Δ is the Laplacian.

Let $u(\xi, t)$ be a solution of $u_t + \mathcal{P}(u) = 0$ on the generalized rectangle $\Omega = [0, \pi a_1] \times \cdots \times [0, \pi a_n]$, where all a_j are assumed to be distinct in order to avoid introducing extra symmetries that permute the coordinates. Suppose that $u(\xi, t)$ satisfies NBC

$$\frac{\partial u}{\partial \xi_i}(\xi) = 0 \qquad \text{when} \qquad \xi_i = 0, \pi a_i \qquad \text{for} \qquad 1 \leq i \leq n.$$

Then u may be extended to a solution of the same PDE on the whole of $\mathbf{R}^n$ that satisfies PBC on the larger rectangle $\hat{\Omega} = [-\pi a_1, \pi a_1] \times \cdots \times [-\pi a_n, \pi a_n]$. We make this extension by reflection across the boundaries

$$\hat{u}(\kappa \xi) = \hat{u}(\xi),$$

where κ belongs to the group K generated by

$$\kappa_j : \xi_j \mapsto -\xi_j \qquad \text{for} \qquad 1 \leq j \leq n,$$

which is isomorphic to $(\mathbf{Z}_2)^n$. Then we extend $\hat{u}$ periodically to the whole $\mathbf{R}^n$. By euclidean invariance of $\mathcal{P}$ it is clear that $\hat{u}$ is a solution of (2.1) on the whole of $\mathbf{R}^n$. Gomes and Stewart [14] and Field et al. [8] prove that the regularity of steady solutions is preserved by this extension procedure, and the same results hold for time-dependent solutions.

Moreover, $\hat{u}$ satisfies PBC on $\hat{\Omega}$ and is K-invariant. Gomes and Stewart [14] show that the converse is also true for steady solutions, and the same proof holds when $\hat{u}$ depends on time. More precisely, suppose that $\hat{u}(\xi, t)$ is a K-invariant solution to (2.1) satisfying PBC on $\hat{\Omega}$. Then the restriction u to Ω satisfies NBC.

Recall that our main interest is the symmetric structure of the solution set of equation (2.1) satisfying NBC on Ω. What is the practical method to make explicit the symmetries imposed by the boundary conditions? Start by formulating the associated periodic boundary value problem, which has the group $\mathbf{O}(2)^n$ as symmetries. This group may also be written as $K \dot{+} \mathbf{T}^n$, where K acts by reflecting the components of ξ as before, and $\theta = (\theta_1, ..., \theta_n) \in \mathbf{T}^n$ acts by

$$\theta.\xi = (\xi_1 + \theta_1, ..., \xi_n + \theta_n).$$

Then restrict the result to the subspace Fix(K).

In·the remaining sections the described method is applied to steady and Hopf bifurcations of codimensions one and two — that is, single modes and the interactions of two modes.

3 Steady-State Bifurcations

In this section we briefly recall the main results on steady-state bifurcations. See Gomes and Stewart [14] for details.

3.1 Steady-State Single Mode Bifurcation

Assume that the reaction-diffusion equation (2.1) admits the translation-invariant solution $u = 0$. Denote by $L = d\mathcal{P}$ the linearization about $u = 0$ at $\lambda = 0$. Under the assumption of PBC, a single mode bifurcation to another steady solution (satisfying $u_t = 0$) occurs if and only if ker L is an irreducible representation of $\mathbf{O}(2)^n$.

Denote $\theta = (\theta_1, \ldots, \theta_n)$ a generic n-torus element. We may write an action of $\mathbf{T}^n$ as

$$\theta : z_j \mapsto e^{i\epsilon_j \cdot \theta} z_j \qquad \text{for} \qquad 1 \le j \le 2^{n-1},$$

where the ϵ_j are all elements of the form $\left(\frac{k_1}{a_1}, \pm\frac{k_2}{a_2}, \ldots, \pm\frac{k_n}{a_n}\right)$ for some set of integers k_j. Without loss of generality we assume that these integers, called *mode numbers* or *wave numbers*, are nonnegative. The group $\mathbf{T}^n$ acts on the space of linearized eigenfunctions, which is

$$\hat{V}_k = \text{span}\left\{e^{i\epsilon_j \cdot \xi} | 1 \le j \le 2^{n-1}\right\}.$$

The action of $\mathbf{T}^n$ on $\hat{V}_k$ is naturally induced by the translations of ξ_j defined in section 2. There is also a well-defined action of K on $\hat{V}_k$ induced by the reflections κ_j of ξ_j; and $\hat{V}_k$ is an irreducible representation of $\mathbf{O}(2)^n$.

If NBC are imposed then ker L is isomorphic to the subspace of $\hat{V}_k$ that is fixed by K. Denoting this subspace by V_k we have

$$V_k = \text{span}\left\{\cos\left(\frac{k_1 \xi_1}{a_1}\right) \cdots \cos\left(\frac{k_n \xi_n}{a_n}\right)\right\} \simeq \mathbf{R}.$$

Recall from Golubitsky *et al.* [11] that if a group Γ acts on a space X, and if Σ is an isotropy subgroup of Γ, then the fixed-point space Fix(Σ) is invariant under the action of the normalizer quotient $N(\Sigma)/\Sigma$. Moreover, if $f : X \to X$ is Γ-equivariant, then $f|_{\text{Fix}(\Sigma)} : \text{Fix}(\Sigma) \to \text{Fix}(\Sigma)$ is $N(\Sigma)/\Sigma$-equivariant. Here the subgroup of $\mathbf{O}(2)^n$ that leaves V_k invariant is $\mathrm{N}_{\mathbf{O}(2)^n}(K)/K$, and $\mathbf{O}(2)^n$ acts non-faithfully through its quotient $\mathbf{Z}_2$. Therefore the bifurcation equations can be written as

$$a(x^2, \lambda)x = 0$$

where $x \in V_k$. The bifurcation diagrams for this case are classified by Golubitsky and Schaeffer [10]. In particular the generic (codimension 0) bifurcation is a pitchfork.

The symmetries of the domain are generated by

$$\tau_j : \xi_j \mapsto \pi a_j - \xi_j \qquad \text{for} \qquad 1 \leq j \leq n$$

and they act trivially on V_k if all k_j are even. Therefore the $\mathbf{Z}_2$ symmetry that causes the pitchfork to be generic is not induced by a domain symmetry, and thus would not be expected without observing the existence of a periodic extension.

3.2 Interaction of Two Steady-State Modes

Now assume that the operator $\mathcal{P}$ depends on the additional parameter r and that the solution $u = 0$ undergoes a simultaneous bifurcation of two steady states with mode numbers $\underline{k}, \underline{l} \in \mathbf{N}^n$ when the parameters are set to zero. Under the assumption of PBC, ker L is isomorphic to the direct sum

$$\hat{V}_k \oplus \hat{V}_l,$$

where $\hat{V}_k$ and $\hat{V}_l$ are two irreducible representations of $\mathbf{O}(2)^n$, defined as in §3.1. There is no loss of generality in assuming that k_j and l_j are coprime — otherwise we may factor out the kernel of the action of $\mathbf{T}^n$.

Now $\text{Fix}(K) = V_k \oplus V_l$, which may be identified with $\mathbf{R}^2$; and the action of $\mathbf{N}_{\mathbf{O}(2)^n}(K)/K$ is isomorphic to

- $\mathbf{Z}_2$ if all k_j have the same parity and all l_j have the same parity;

- $\mathbf{Z}_2 \oplus \mathbf{Z}_2$ otherwise.

Although the subgroup of $\mathbf{O}(2)^n$ that leaves $\text{Fix}(K)$ invariant imposes some symmetry constraints on the Neumann boundary value problem, these symmetries are not always sufficient to guarantee an extension to an $\mathbf{O}(2)^n$-invariant mode interaction. More precisely, define

$$k = \max_j k_j, \quad l = \max_j l_j.$$

Then the reduced bifurcation equations take the form

$$\begin{aligned} a(x^2, y^2, \lambda)x + c(x^2, y^2, \lambda)x^{l-1}y^k &= 0 \\ b(x^2, y^2, \lambda)y + d(x^2, y^2, \lambda)x^l y^{k-1} &= 0 \end{aligned}$$

if all k_j have the same parity and all l_j have the same parity; and they take the form

$$\begin{aligned} a(x^2, y^2, \lambda)x &= 0 \\ b(x^2, y^2, \lambda)y &= 0 \end{aligned}$$

otherwise. See Gomes and Stewart [14] for the deduction of these equations, and Armbruster and Dangelmayr [1] for a classification of the corresponding bifurcation diagrams.

4 Single Mode Hopf Bifurcation

Single mode Hopf bifurcation can be reduced to finding the zeros of a suitable reduced bifurcation equation by using the method of Liapunov-Schmidt reduction applied to an operator equation on a loop space, see Golubitsky et al. [11]. This method, originally due to Hale [16], introduces an additional group $\mathbf{S}^1$ of phase shift symmetries, and a new period-scaling parameter. The general theory shows that the reduced equations may be solved for this parameter in terms of the original variables, which allows us to suppress it in the following analysis.

We assume that $u = 0$ satisfies the equation $u_t + \mathcal{P}(u, \lambda) = 0$, impose PBC, and let L be the linearized operator about $u = 0$ at $\lambda = 0$. If a Hopf bifurcation occurs then ker L is generically an irreducible representation of $\mathbf{O}(2)^n \times \mathbf{S}^1$, where $\mathbf{O}(2)^n$ represents the spatial symmetries and $\mathbf{S}^1$ the temporal phase-shift symmetries.

Recall that K denotes the subgroup of $\mathbf{O}(2)^n$ generated by the coordinate reflections. The symmetries of the NBC problem are identified by restricting the representation of $\mathbf{O}(2)^n \times \mathbf{S}^1$ to Fix(K). Let $\theta = (\theta_1, \ldots, \theta_n)$ denote a general element of the n-torus $\mathbf{T}^n$. There is an action of $\mathbf{T}^n \times \mathbf{S}^1$ on $\mathbf{C}^n$ given by

$$\begin{aligned} \theta : z_j &\mapsto e^{i \epsilon_j \cdot \theta} z_j & (\theta \in \mathbf{S}^1) \\ \phi_k : z_j &\mapsto e^{i \phi_k} z_j & (\phi \in \mathbf{T}^n) \end{aligned}$$

for $1 \le j \le 2^n$, where the ϵ_j are all n-tuples of the form $\left(\pm \frac{k_1}{a_1}, \ldots, \pm \frac{k_n}{a_n} \right)$ for some set of nonnegative integers k_j. This action extends naturally to an irreducible representation of $\mathbf{O}(2)^n \times \mathbf{S}^1$ on the space of linearized eigenfunctions, which may be written as

$$\hat{W}_k = \operatorname{span} \left\{ e^{i(t_k + \epsilon_j \cdot \xi)} | 1 \le j \le 2^n \right\}.$$

Here $\mathbf{T}^n \times \mathbf{S}^1$ acts as above and the action of K is induced by reflecting the space variables ξ_j in the n coordinate hyperplanes. We claim that the restriction of $\hat{W}_k$ to

Fix(K) is

$$W_k \;=\; \mathrm{span}\left\{e^{it_k}\cos\left(\frac{k_1\xi_1}{a_1}\right)\cdots\cos\left(\frac{k_n\xi_n}{a_n}\right)\right\} \simeq \mathbf{C}.$$

To prove the claim we recall that $\hat{W}_k$ is generated by the eigenfunctions

$$e^{i\left(t\pm\frac{k_1\xi_1}{a_1}\pm\cdots\pm\frac{k_n\xi_n}{a_n}\right)}.$$

Let $E^{\pm}$ denote two eigenfunctions that differ only on the sign associated to the term $\frac{k_1\xi_1}{a_1}$ and let ϵ be a given vector of the form $\left(\pm\frac{k_2}{a_2},\ldots,\pm\frac{k_n}{a_n}\right)$ such that

$$E^{\pm} \;=\; e^{i\left(t\pm\frac{k_1\xi_1}{a_1}\right)}e^{i\epsilon\cdot(\xi_2,\ldots,\xi_n)}.$$

Let $z^{\pm}\in\mathbf{C}$ represent the respective amplitudes and write

$$\begin{aligned}
E &\;=\; z^+ e^+ + z^- e^-\\
&\;=\; e^{it}\left(z^+ e^{i\frac{k_1\xi_1}{a_1}} + z^- e^{-i\frac{k_1\xi_1}{a_1}}\right)e^{i\epsilon\cdot(\xi_2,\ldots,\xi_n)}.
\end{aligned}$$

Now denote $z^{\pm}=x^{\pm}+iy^{\pm}$ to get

$$\begin{aligned}
E &\;=\; e^{it}\left[(x^+ + iy^+)\left(\cos\left(\frac{k_1\xi_1}{a_1}\right)+i\sin\left(\frac{k_1\xi_1}{a_1}\right)\right)\right]e^{i\epsilon\cdot(\xi_2,\ldots,\xi_n)}\\
&\quad+e^{it}\left[(x^- + iy^-)\left(\cos\left(\frac{k_1\xi_1}{a_1}\right)-i\sin\left(\frac{k_1\xi_1}{a_1}\right)\right)\right]e^{i\epsilon\cdot(\xi_2,\ldots,\xi_n)}\\
&\;=\; e^{it}\left[(x^+ + x^-) + i(y^+ + y^-)\right]\cos\left(\frac{k_1\xi_1}{a_1}\right)e^{i\epsilon\cdot(\xi_2,\ldots,\xi_n)}\\
&\quad+e^{it}\left[(y^- - y^+) + i(x^+ - x^-)\right]\sin\left(\frac{k_1\xi_1}{a_1}\right)e^{i\epsilon\cdot(\xi_2,\ldots,\xi_n)}.
\end{aligned}$$

Recall that the element $\kappa_1\in K$ acts by reflecting ξ_1 as

$$\kappa_1 : \xi_1 \mapsto -\xi_1$$

and this induces an action on the eigenfunction E as

$$\begin{aligned}
\kappa_1 E &\;=\; e^{it}\left[(x^+ + x^-) + i(y^+ + y^-)\right]\cos\left(\frac{k_1\xi_1}{a_1}\right)e^{i\epsilon\cdot(\xi_2,\ldots,\xi_n)}\\
&\quad+e^{it}\left[(y^+ - y^-) + i(x^- - x^+)\right]\sin\left(\frac{k_1\xi_1}{a_1}\right)e^{i\epsilon\cdot(\xi_2,\ldots,\xi_n)}.
\end{aligned}$$

In order to satisfy the condition for κ_1-invariance

$$\kappa_1 E \;=\; E$$

we need

$$x^+ = x^- \qquad \text{and} \qquad y^+ = y^-.$$

Substitution on the formula for E yields

$$E \;=\; 2ze^{it}\cos\left(\frac{k_1\xi_1}{a_1}\right) e^{i\epsilon.(\xi_2,\dots,\xi_n)},$$

where $z \equiv z^+ = z^-$. We saw that $\mathrm{Fix}(\kappa_1)$ is generated by the eigenfunctions of the form

$$e^{it}\cos\left(\frac{k_1\xi_1}{a_1}\right) e^{i\left(\pm\frac{k_2\xi_2}{a_2}\pm\cdots\pm\frac{k_n\xi_n}{a_n}\right)}$$

and proceeding by induction on n we get that $\mathrm{Fix}(K)$ is generated by the single eigenfunction

$$e^{it}\cos\left(\frac{k_1\xi_1}{a_1}\right)\cdots\cos\left(\frac{k_n\xi_n}{a_n}\right),$$

which is what we wanted to prove.

The restriction of the group action to W_k induces an action of the normalizer quotient $\mathrm{N}_{O(2)^n \times S^1}(K)/K = S^1$. In addition, some spatial translations act on W_k as minus the identity. Although this is a nontrivial action, it adds nothing new to the action of S^1 — a phase shift of π acts in the same way. So the generic Liapunov-Schmidt reduced bifurcation equation for a Hopf bifurcation satisfying NBC is S^1-equivariant, hence of the form

$$a(|z|^2, \lambda)z \;=\; 0,$$

where a is a smooth complex-valued function with $a(0) = 0$. (The reduction to a mapping converts purely imaginary eigenvalues to zero eigenvalues, see Golubitsky et al. Chapter XVI §3.) Clearly all such a can arise in this context (that is, there are no further restrictions on a beyond normalizer equivariance, because $|z|^2$ is invariant for $O(2)^n \times S^1$, and z is equivariant). This is the usual reduced equation for a Hopf bifurcation with no symmetries apart from phase shift; that is, the domain symmetries impose *no* restrictions on the reduced bifurcation equations. See Golubitsky and Schaeffer [10] for a study of such Hopf bifurcations using singularity-theoretic methods. These authors consider not only the generic cases, but also some degenerate ones.

5 Interaction of Two Hopf Modes

Now assume that the operator $\mathcal{P}$ depends on an additional parameter r, and that the trivial solution $u = 0$ undergoes a simultaneous bifurcation to two time-periodic solutions with spatial mode numbers $\underline{k}, \underline{l} \in \mathbf{N}^n$ when $r = 0$ and λ is varied across zero. Under the assumption of PBC, ker L is isomorphic to

$$\hat{W}_k \oplus \hat{W}_l,$$

where $\hat{W}_k$ and $\hat{W}_l$ are as defined in the previous section. The group acting on $\hat{W}_k \oplus \hat{W}_l$ is $\mathbf{O}(2)^n \times \mathbf{T}^2$, where $\mathbf{O}(2)^n$ acts as before and $\mathbf{T}^2$ acts as

$$(\phi, \psi) : (z_j, w_j) \mapsto (e^{i\phi} z_j, e^{i\psi} w_j), \qquad \text{for} \qquad 1 \le j \le 2^n.$$

In order to introduce the group $\mathbf{T}^2$ we assume that the two Hopf modes are nonresonant. We perform a centre manifold reduction to convert the PDE to an ODE, and then transform the equations to Birkhoff normal form: see [11] chapter XIX §0. We therefore end up with an ODE, not a mapping. We discuss the truncations involved in this procedure, and their likely effect on the bifurcation behaviour, below. First, we analyse the Birkhoff normal form.

By factoring out the kernel of the group action if necessary we may assume that k_j and l_j are coprime. By imposing NBC we restrict the problem to Fix(K), which is obtained by imposing the conditions

$$z_1 = \cdots = z_{2^n}$$
$$w_1 = \cdots = w_{2^n}.$$

Therefore Fix(K) is isomorphic to $\mathbf{C}^2$ and the action of $N_{\mathbf{O}(2)^n \times \mathbf{T}^2}(K)/K$ is isomorphic to $\mathbf{T}^2$. So the bifurcation equations for a simultaneous bifurcation of two time-periodic modes with NBC are $\mathbf{T}^2$-equivariant, hence of the form

$$\dot{z} + a(|z|^2, |w|^2, \lambda) z = 0 \tag{5}$$
$$\dot{w} + b(|z|^2, |w|^2, \lambda) w = 0. \tag{6}$$

where a, b are complex valued functions such that $a(0) = i\omega_k, b(0) = i\omega_l$. Again there are no further restrictions on a, b beyond normalizer equivariance, for similar reasons. See Golubitsky *et al.* [11] and references therein for more details on such Hopf/Hopf mode interactions.

We now discuss smoothness problems and the effect of truncation when putting the problem into normal form. Centre manifold reduction involves a potential loss of smoothness, but since we subsequently truncate the normal form this causes no additional problem provided we choose a sufficiently smooth centre manifold. The Birkhoff normal form theorem states that (in this context) any nonresonant Hopf/Hopf mode interaction can be put into the form 5.1 up to any desired finite order, by a suitable vector field change of coordinates. See Golubitsky *et al.* [11] for further details. We now ask to what extent the behaviour predicted by the normal form 5.1 persists when high-order terms that break the torus symmetry are restored.

Equation 5.1 predicts that mixed-mode solution branches correspond to invariant tori, generated from a given solution by independent phase shifts on the two modes. By non-resonance such tori support quasiperiodic solutions. We may expect that generically such an invariant torus will be normally hyperbolic, see Guckenheimer and Holmes [15] or Arrowsmith and Place [2]. It will therefore perturb to an invariant torus when the symmetry-breaking high-order terms are restored. However, the dynamics on this torus is likely to change, and may be expected to frequency-lock into a finite set of invariant circles that attract or repel trajectories on the remainder of the torus. These frequency-locked states may include long-period periodic solutions having nontrivial winding numbers relative to the generators of the torus.

6 Steady-State/Hopf Mode Interaction

This section covers the last case of codimension 2 bifurcations. Assume that the trivial solution $u = 0$ undergoes a simultaneous bifurcation of a steady and a time periodic solution with spacial mode numbers $\underline{k}, \underline{l} \in \mathbf{N}^n$ respectively. Under the assumption of periodic boundary conditions we have that ker L is now isomorphic to

$$\hat{V}_k \oplus \hat{W}_l,$$

where $\hat{V}_k$ and $\hat{W}_l$ are as previously defined. The group acting on $\hat{V}_k \oplus \hat{W}_l$ is $\mathbf{O}(2)^n \times \mathbf{S}^1$ where $\mathbf{O}(2)^n$ acts as before, and $\mathbf{S}^1$ acts trivially on $\hat{V}_k$ and on $\hat{W}_l$ as

$$\phi : w_j \mapsto e^{i\phi} w_j, \qquad \text{for} \qquad 1 \leq j \leq 2^n.$$

Again we may assume that k_j and l_j are coprime, by factoring out the kernel of the group action. Now the defining conditions for a point in $\hat{V}_k \oplus \hat{W}_l$ to belong to Fix(K) are

$$z_1 = \cdots = z_{2^n-1} \in \mathbf{R}$$
$$w_1 = \cdots = w_{2^n} \in \mathbf{C}.$$

Therefore Fix(K) is isomorphic to $\mathbf{R} \times \mathbf{C}$ and the action of $N_{\mathbf{O}(2)^n \times \mathbf{S}^1}(K)/K$ is generated by

- $\mathbf{S}^1$ if all k_j are even and all l_j are odd;

- $\mathbf{Z}_2 \times \mathbf{S}^1$ otherwise.

In this case the action of $N_{O(2)^n \times S^1}(K)/K$ is not enough to give the generic form of an equivariant restricted to $\mathrm{Fix}(K)$. We need invariant theory methods. It is well known (see Golubitsky et al. [11]) that given a group Γ, the set of Γ- invariant functions is a ring and the set of Γ-equivariant mappings is a modulo over the ring of the invariants. In our case the group Γ is $O(2)^n \times \mathbf{S}^1$ acting as before and restriction to $\mathrm{Fix}(K)$ preserves the properties of ring and modulo. By some nice properties of torus actions, is this particular case the generators of the restricted equivariants can be obtained directly from a set of generators of the restricted invariants. The result is as follows:

Theorem 1 *Let* $O(2)^n \times \mathbf{S}^1$ *act diagonally on* $\hat{V}_k \oplus \hat{W}_l$ *as before and* $I_1, \dots I_r$ *generate the invariants restricted to* $\mathrm{Fix}(K)$. *Then the equivariants restricted to* $\mathrm{Fix}(K)$ *are generated by the mappings*

$$\begin{pmatrix} \frac{\partial I_g}{\partial x} \\ 0 \end{pmatrix} \qquad \begin{pmatrix} 0 \\ \frac{\partial I_g}{\partial w} \end{pmatrix},$$

for $1 \le g \le r$, *where* $x \equiv \mathrm{Re}(z_1) = \cdots = \mathrm{Re}(z_{2^{n-1}})$ *and* $w \equiv w_1 = \cdots = w_{2^n}$.

Proof It is well known in the context of torus group actions and toric varieties that given the generators of the $\mathbf{C}$-valued invariants $i_1, \dots, i_s$ under our diagonal action of $\mathbf{T}^n \times \mathbf{S}^1$, the equivariants are generated by the mappings

$$\mathrm{row}\ j \to \begin{pmatrix} 0 \\ \vdots \\ 0 \\ \frac{\partial i_g}{\partial \bar{z}_j} \\ 0 \\ \vdots \\ 0 \end{pmatrix}_{1 \le j \le 2^{n-1}} \qquad \mathrm{row}\ 2^{n-1} + j \to \begin{pmatrix} 0 \\ \vdots \\ 0 \\ \frac{\partial i_g}{\partial \bar{w}_j} \\ 0 \\ \vdots \\ 0 \end{pmatrix}_{1 \le j \le 2^n}$$

for $1 \le g \le s$. Recall that $\mathrm{Fix}(K)$ is isomorphic to $\mathbf{R} \times \mathbf{C}$ and denote a point in this subspace by (x, w) such that

$$\begin{aligned} x &\equiv \mathrm{Re}(z_1) = \cdots = \mathrm{Re}(z_{2^{n-1}}) \\ w &\equiv w_1 = \cdots = w_{2^n}. \end{aligned}$$

The result follows immediately by denoting I_g the restriction of the $\mathbf{T}^n \times \mathbf{S}^1$-invariant i_g to $\mathrm{Fix}(K)$, eliminating redundancies and reordering the labels if necessary. $\square$

A generic equivariant mapping restricted to $\mathrm{Fix}(K)$ is then completely characterized by a minimal set of generators for the restricted invariants. These are as follows:

Theorem 2 *Let the group $\mathrm{O}(2)^n \times \mathrm{S}^1$ act diagonally on $\hat{V}_k \oplus \hat{W}_l$ as before. Then the $\mathrm{O}(2)^n \times \mathrm{S}^1$-invariants restricted to $\mathrm{Fix}(K)$ are generated according to the mode numbers as follows:*

- *If all k_j are even and all l_j are odd the generators are $x^2, |w|^2$ and $x^l|w|^k$, where $k = \max_j k_j$ and $l = \max_j l_j$.*

- *Otherwise the generators are $x^2, |w|^2$.*

Before proving theorem 2 we restate Lemma 3 of Gomes and Stewart [14]:

Lemma 1 *Let M_n be the $2^{n-1} \times 2^{n-1}$ matrix constructed by induction as follows:*

$$M_1 = (1), \qquad M_n = \begin{pmatrix} M_{n-1} & M_{n-1} \\ M_{n-1} & -M_{n-1} \end{pmatrix}, \qquad n \geq 2.$$

Define the lattice

$$\mathcal{M} = \left\{ c \in \mathbf{Z}^{2^{n-1}} | M_n c \equiv 0 \pmod{2^{n-1}\mathbf{Z}^{2^{n-1}}} \right\}.$$

For $a \in \mathbf{Z}^n$ define

$$S_a = \left\{ c \in \mathbf{Z}^{2^{n-1}} | c_1 = a_1, c_{2^{j-2}+1} = a_j, 1 \leq j \leq n \right\}.$$

Then by letting $\mathcal{L}_a = \mathcal{M} \cap S_a$ we have

(a) *$\mathcal{L}_a \neq \emptyset$ if and only if $a_1 \equiv \ldots \equiv a_n \pmod 2$;*

(b) *If $a \in \mathbf{N}^n$ and $a_1 \geq \cdots \geq a_n$ then $\mathcal{L}_a \neq \emptyset \Rightarrow \frac{1}{2^{n-1}}M_n(\mathcal{L}_a) \cap \mathbf{N}^{2^{n-1}} \neq \emptyset$.*

Proof of theorem 2   The $\mathbf{T}^n \times \mathrm{S}^1$-invariants are generated by certain monomials represented in multi-index notation by

$$z^{\underline{\alpha}^1} \bar{z}^{\underline{\beta}^1} w^{\underline{\alpha}^2} \overline{w}^{\underline{\beta}^2} \tag{7}$$

where $z \in \hat{V}_k$, $w \in \hat{W}_l$ and $\underline{\alpha}^1, \underline{\beta}^1 \in \mathbf{N}^{2^{n-1}}$, $\underline{\alpha}^2, \underline{\beta}^2 \in \mathbf{N}^{2^n}$. Set $\underline{\gamma} = \underline{\alpha}^1 - \underline{\beta}^1 \in \mathbf{Z}^{2^{n-1}}$, $\underline{\delta} = \underline{\alpha}^2 - \underline{\beta}^2 \in \mathbf{Z}^{2^n}$. Then the conditions for $\mathbf{T}^n \times \mathbf{S}^1$-invariance are

$$\begin{pmatrix} k_1 & & 0 \\ & \ddots & \\ 0 & & k_n \end{pmatrix} L_n \underline{\gamma} + \begin{pmatrix} l_1 & & 0 \\ & \ddots & \\ 0 & & l_n \end{pmatrix} L_n' \underline{\delta} = 0 \tag{8}$$

together with

$$\sum_j \delta_j = 0. \tag{9}$$

Here L_n' is the $n \times 2^n$ matrix whose columns are all possible combinations of ± 1 and L_n is the $n \times 2^{n-1}$ matrix consisting of the columns of L_n' that have $+1$ for last entry, see Gomes and Stewart [14]. Now set

$$\underline{a} = L_n \underline{\gamma}, \qquad \underline{b} = L_{n+1} \underline{\delta}.$$

Then (6.2,6.3) can be replaced by

$$\begin{aligned} a_j k_j + b_j l_j &= 0, \qquad 1 \leq j \leq n \\ b_{n+1} &= 0. \end{aligned} \tag{10}$$

Let M_n be the $2^{n-1} \times 2^{n-1}$ matrix defined in lemma 1 and apply the coordinate change

$$\underline{c} = M_n \underline{\gamma}, \qquad \underline{d} = M_{n+1} \underline{\delta}, \tag{11}$$

or equivalently,

$$\underline{\gamma} = \frac{1}{2^{n-1}} M_n \underline{c}, \qquad \underline{\delta} = \frac{1}{2^n} M_{n+1} \underline{d}. \tag{12}$$

By reordering the columns of L_n and L_{n+1} if necessary we assign

$$\begin{array}{llll} a_j = c_{2^{j-1}+1}, & 1 \leq j \leq n-1 & b_j = d_{2^{j-1}+1}, & 1 \leq j \leq n \\ a_n = c_1 & & b_{n+1} = d_1. & \end{array} \tag{13}$$

By Lemma 1a a given $\underline{a} \in \mathbf{Z}^n$ may be lifted to $\underline{c} \in \mathbf{Z}^{2^{n-1}}$ such that $\underline{\gamma} \in \mathbf{Z}^{2^{n-1}}$ if and only if

$$a_1 \equiv \ldots \equiv a_n \pmod 2, \tag{14}$$

and by the same argument a given $\underline{b} \in \mathbf{Z}^{n+1}$ may be lifted to $\underline{d} \in \mathbf{Z}^{2^n}$ such that $\underline{\delta} \in \mathbf{Z}^{2^n}$ if and only if

$$b_1 \equiv \ldots \equiv b_{n+1} \quad (\mathrm{mod}\ 2). \qquad (15)$$

The trivial solution of (6.4) corresponds to elements generated by

$$z_j \bar{z}_j, \quad 1 \le j \le 2^{n-1} \quad \text{and} \quad w_j \bar{w}_j, \quad 1 \le j \le 2^n. \qquad (16)$$

Dividing out factors of these from (6.1) leads to a monomial in multi-index notation as

$$u^{|\gamma|} v^{|\delta|} \qquad (17)$$

where $|\,.\,|$ denotes the modulus of each component and u_j is z_j or $\bar{z}_j$, v_j is w_j or $\bar{w}_j$ depending on the sign of the j-component of $\underline{\gamma}$, $\underline{\delta}$ respectively.

By restricting to Fix(K), which we recall is isomorphic to $\mathbf{R} \times \mathbf{C}$, the invariants (6.10,6.11) become

$$N_1 = x^2, \qquad N_2 = |w|^2, \qquad T = x^\gamma (w\bar{w})^{\delta'},$$

where (in a slight abuse of notation) we denote

$$x \equiv \mathrm{Re}(z_1) = \cdots = \mathrm{Re}(z_{2^n-1})$$
$$w \equiv w_1 = \cdots = w_{2^n}.$$

Here γ represents the sum of the moduli of each component of $\underline{\gamma}$, and δ' represents the sum of the positive components of $\underline{\delta}$ (which is equal in modulus to the sum of the negative components by the condition $b_{n+1} = 0$). Letting $\delta = 2\delta'$ we get an equivalent form for T, namely

$$T = x^\gamma |w|^\delta.$$

Note that δ is the sum of the moduli of all components of $\underline{\delta}$. Solutions of (6.4) may be written as

$$(a_j, b_j) = (A_j l_j, -A_j k_j), \qquad 1 \le j \le n$$
$$b_{n+1} = 0,$$

where the A_j are integers. The assumption that k_j and l_j are coprime, together with (6.8,6.9), implies that all A_j must have the same parity. Given that $b_{n+1} = 0$, all other components of $\underline{b}$ must be even. If at least one of the k_j is odd, then all the A_j

must be even and all possible invariants of the form T are generated be N_1 and N_2. Therefore from now on we may assume that all k_j are even, in which case all l_j must be odd, because k_j and l_j are coprime. From (6.5,6.7) we see that γ has the same parity as all the a_j. Elements of the form T that are not generated by N_1, N_2 must be such that γ is odd, and this happens if and only if all l_j and A_j are odd.

Also from (6.5,6.7) we have

$$\gamma \geq \max_j |a_j|, \qquad \delta \geq \max_j |b_j|. \tag{18}$$

We claim that there exist $\underline{c} \in \mathbf{Z}^{2^{n-1}}$ and $\underline{d} \in \mathbf{Z}^{2^n}$ satisfying the equalities in (6.12). This will imply that

$$\min \gamma = \max_j |a_j|, \qquad \min \delta = \max_j |b_j|. \tag{19}$$

The proof of the claim follows as in the proof of theorem 6 of Gomes and Stewart [14], but we repeat it here for completeness. Assume without loss of generality that $\underline{a} \in \mathbf{N}^n$ and $-\underline{b} \in \mathbf{N}^{n+1}$. By induction on n, all the entries in the first column of M_n are $+1$; and in the other columns the number of $+1$s and -1s is the same. For simplicity we assume that $a_n \geq a_1 \geq \cdots \geq a_{n-1}$. Thus in order to get the first equality in (6.13) it is enough to show that $\underline{a}$ can be lifted to $\underline{c} \in \mathbf{Z}^{2^{n-1}}$ such that $\underline{\gamma} \in \mathbf{N}^{2^{n-1}}$, where $\underline{\gamma} = M_n^{-1}\underline{c}$. This comes immediately from Lemma 1b. Now let π be a permutation such that $-b_{\pi^{-1}(n+1)} \geq -b_{\pi^{-1}(1)} \geq \cdots \geq -b_{\pi^{-1}(n)}$, and apply a variation of Lemma 1 in which M_{n+1} is replaced by $\pi^{-1}M_{n+1}\pi$, to show that the two equalities (6.13) hold simultaneously.

It remains to minimize the absolute values of the components of $\underline{a}, \underline{b}$ when all A_j are odd. It is easy to verify that all the components achieve their minimum values simultaneously when $A_j = 1$ for $1 \leq j \leq n$, and the result follows. $\qquad \square$

Finally, we can tidy up the resulting normal form, depending on the parities of the mode numbers. The Liapunov-Schmidt reduced bifurcation equations for a Hopf/steady-state mode interaction with NBC can be transformed into

$$
\begin{aligned}
a(x^2, |w|^2, \lambda)x + c(x^2, |w|^2, \lambda)x^{l-1}|w|^k &= 0 \\
b(x^2, |w|^2, \lambda)w + d(x^2, |w|^2, \lambda)x^l|w|^{k-2}w &= 0
\end{aligned}
\tag{20}
$$

if all k_j are even and all l_j are odd; and into

$$
\begin{aligned}
a(x^2, |w|^2, \lambda)x &= 0 \\
b(x^2, |w|^2, \lambda)w &= 0
\end{aligned}
\tag{21}
$$

otherwise. In these equations, a, c are real-valued and b, d are complex-valued functions.

It remains to work out the bifurcation diagrams. To do this we transform equations (6.14,6.15) to phase/amplitude form, so that we can read off the bifurcation diagrams from known results. The functions a, c are real-valued and b, d are complex-valued. Write b, d as

$$b = b_r + ib_i$$
$$d = d_r + id_i$$

where b_r, b_i, d_r, d_i are real-valued functions of $x^2, |w|^2$, and λ. Introduce polar coordinates $w = ye^{i\phi}$; then a short calculation shows that the amplitudes x, y and the phase ϕ satisfy equations of the form

$$\dot{x} + a(x^2, y^2, \lambda)x + c(x^2, y^2, \lambda)x^{l-1}y^k = 0$$
$$\dot{y} + b_r(x^2, y^2, \lambda)y + d_r(x^2, y^2, \lambda)x^l y^{k-1} = 0 \qquad (22)$$
$$\dot{\phi} + b_i(x^2, y^2, \lambda) + d_i(x^2, y^2, \lambda)x^l y^{k-2} = 0$$

if all k_j are even and all l_j are odd, and equations of the form

$$\dot{x} + a(x^2, y^2, \lambda)x = 0$$
$$\dot{y} + b_r(x^2, y^2, \lambda)y = 0 \qquad (23)$$
$$\dot{\phi} + b_i(x^2, y^2, \lambda) = 0$$

otherwise. The first two equations of (6.16) constitute the system analysed by Armbruster and Dangelmayr [1, 7] and the first two equations of (6.17) form the $\mathbf{Z}_2 \oplus \mathbf{Z}_2$-invariant system described in chapter X of Golubitsky and Schaeffer [10]. Thus the bifurcation diagrams are known and will not be reproduced here. However, their interpretation is different, because now y is the amplitude of a standing wave rather than a steady state.

Acknowledgements

We are greateful to Edgar Knobloch for asking whether our methods could say anything useful about interactions involving Hopf modes. The research of MGMG was partially supported by a JNICT Scholarship from Portugal and a Grant from the Science and Engineering Research Council of the UK. The research of INS was supported in part by a Grant from the Science and Engineering Research Council of the UK and a European Community Laboratory Twinning Grant; and some of the work was done during a visit to the Fields Institute, which provided superb hospitality and facilities, and financial support. This research was carried out under the auspices of the European Bifurcation Theory group.

References

[1] D. Armbruster and G. Dangelmayr, Coupled Stationary Bifurcations in Non-Flux Boundary Value Problems, *Math. Proc. Camb. Phil. Soc.* **101** (1987) 167-192.

[2] D.K.Arrowsmith and C.M.Place, *An Introduction to Dynamical Systems*, Cambridge Univ. Press, Cambridge 1990.

[3] S. Castro, *Symmetry and Bifurcation of Periodic Solutions in Neumann Boundary Value Problems*, M.Sc Thesis, Mathematics Institute, U. of Warwick 1990.

[4] J.D.Crawford, Normal forms for driven surface waves: boundary conditions, symmetry, and genericity, preprint, U. of Pittsburgh 1990.

[5] J.D.Crawford, J.P.Gollub, and D.Lane, *Hidden Symmetries of Parametrically Forced Waves*, preprint, U. of Pittsburgh 1992.

[6] J.D.Crawford, M.Golubitsky, M.G.M.Gomes, E.Knobloch and I.N.Stewart, Boundary conditions as symmetry constraints, in *Singularity Theory and Its Applications, Warwick 1989*, Vol. 2, (eds. R.M. Roberts and I.N. Stewart), Lecture Notes in Mathematics **1463**, Springer-Verlag, Heidelberg (1991).

[7] G. Dangelmayr and D. Armbruster, Steady-state mode interactions in the presence of O(2) symmetry and in non-flux boundary value problems, in *Multiparameter Bifurcation Theory* (eds. J.Guckenheimer and M.Golubitsky), *Contemporary Math.* **56**, Amer. Math. Soc., Providence RI 1986, 53-68.

[8] M.J.Field, M.Golubitsky and I.N.Stewart, Bifurcations on hemispheres, *J. Nonlinear Sci.* **1** (1991) 201-223.

[9] H.Fujii, M.Mimura, and Y.Nishiura, A picture of the global bifurcation diagram in ecological interacting and diffusing systems, *Physica* **5D** (1982) 1-42.

[10] M. Golubitsky and D.G.Schaeffer, *Singularities and Groups in Bifurcation Theory* vol.1, *Appl. Math. Sci.* **51** (Springer-Verlag, New York, 1985).

[11] M. Golubitsky, I.N. Stewart and D.G. Schaeffer, *Singularities and Groups in Bifurcation Theory* vol.2,. *Appl. Math. Sci.* **69** (Springer-Verlag, New York, 1988).

[12] M. Golubitsky and I.N. Stewart, Symmetry and stability in Taylor-Couette flow, *SIAM J.Math.Anal.* **17** (1986), 249-288.

[13] M.G.M.Gomes, *Symmetries in Bifurcation Theory: the Appropriate Context*, Ph.D. Thesis, U. of Warwick 1992.

[14] M.G.M. Gomes and I.N. Stewart, *Steady PDEs on Generalized Rectangles: a Change of Genericity in Mode Interactions*, preprint 8/1993, U. of Warwick 1993; *Nonlinearity*, to appear.

[15] J.Guckenheimer and P.Holmes, *Nonlinear Oscillations, Dynamical Systems, and Bifurcations of Vector Fields, Appl. Math. Sci.* **42**, Springer, New York 1983.

[16] J.K.Hale, *Ordinary Differential Equations*, Wiley, New York 1969.

[17] M.Impey, R.M.Roberts and I.N.Stewart, *Mode Interactions in Lapwood Convection*, in preparation.

THE ROLE OF GEOMETRY IN COMPUTATIONAL DYNAMICS

JOHN GUCKENHEIMER
Mathematics Department
Cornell University
Ithaca, New York 14853

ABSTRACT. This paper is an informal discussion of how geometry and numerical analysis are intertwined in the computational study of dynamical systems and their bifurcations. We use the example of determining the phase portrait of planar vector fields to illustrate the more general and philosophical attitudes that constitute our main thesis. Few mathematical details are included.

1. Introduction

There are two fundamental aspects of dynamical systems theory that lead us to reliance upon computers. First, "most" nonlinear vector fields cannot be integrated explicitly, so numerical methods are absolutely essential to obtain quantitative information about the solutions to particular systems. This is usually done through the iterative computation of approximate solutions by numerical integration algorithms. Second, computers work well for simulating trajectories of vector fields of modest complexity. Modern workstations have an architecture that emphasizes very rapid floating point calculations on data that is stored in cache memories of moderate size. This is well matched to obtaining maximal performance from application of a numerical integration algorithm as long as the size of a system allows the data and inner loop of of an algorithm to be stored in cache memory. Even without worrying about data locality, computers have increased their speed enormously over the past decade.

Numerical integration is not a panacea for the investigation of dynamical systems that occur as "real world" examples. There are many problems involving dynamical systems that are not easily solved by numerical integration. Here we focus upon the mathematical problem of providing proofs for the qualitative structure of phase portraits of dynamical systems. Traditional computational theories have regarded questions of dynamical systems largely as analytic problems. However, there are fundamental limitations on error estimation for numerical computation of trajectories

P. Chossat (ed.), Dynamics, Bifurcation and Symmetry, 159–165.

for dynamical systems. Naive error estimation applied to an iterative method for computing trajectories yields an exponential growth in the estimates. Moreover, the finite precision with which calculations are performed results in a trade-off between truncation and round off errors that is due to the increasing length of a computation with reductions in step sizes.

There are alternate strategies to error estimation of trajectory computations for the rigorous analysis of dynamical systems properties. In particular, geometric approaches can be productively combined with computation to address some questions about dynamical systems. We shall make this argument in a specific context in which the geometry is relatively simple, namely the study of planar vector fields. We draw heavily upon the work of Salvador Malo [5],[3] in the discussion.

2. Hilbert's XVI Problem

Of the problems posed by Hilbert at the beginning of the century, the sixteenth remains of one of the few that has been resistant to solution. The problem asks for a bound on the number of limit cycles of a polynomial vector field

$$\dot{x} = P(x, y) \qquad \dot{y} = Q(x, y)$$

as a function of the degrees of the polynomials P, Q. The problem remains unsolved for polynomials of any degree $d > 1$. One of the difficulties associated with Hilbert's sixteenth problem is the lack of principles that allow one to deduce information about the phase portrait of a vector field from its analytic expression. There are perturbation methods that can be used to study vector fields that are close to integrable ones with families of periodic orbits and ones that are close to bifurcations of a specific codimension. For structurally stable systems, numerical integration and root finding frequently produce a correct phase portrait with minimal effort. On the other hand, proving that the computed pictures produce the correct phase portrait for a system has been a difficult task, usually involving lengthy analytic arguments for each example or class of examples. Our goal is to create computer algorithms that automate these analytic arguments. We illustrate how transversality is used by describing how to prove the existence of a hyperbolic periodic orbit in a planar vector field.

The strategy that we adopt is based upon transversality. For planar vector fields, the Poincaré-Bendixon Theorem can be used to prove the existence of a periodic orbit [1]. If a vector field X enters an annulus A and does not leave, there is either an equilibrium point or a periodic orbit in A. Verification of the fact that X enters A at all points of its boundary does not require integration of the vector field. Instead, one needs a representation of the boundary of A for which it is possible to prove that the vector field points into A at all points of this boundary. Location of the equilibrium points of X is a matter of solving polynomial equations, and Bezout's Theorem tells us that the number of complex equilibrium points is generically the product of the degrees of P and Q. Thus the existence of a hyperbolic periodic orbit of X can be proved by finding an annulus that contains the periodic orbit but no equilibrium points and has the property that the vector field enters (or leaves) the annulus at all points of its boundary.

3. Rotated Vector Fields and Interval Arithmetic: Malo's Thesis

Salvador Malo [5] has implemented a simple version of this process and applied it to polynomial vector fields. There are two aspects to the problem: choosing an annulus A and showing that the vector field enters or leaves the annulus at all points of its boundary. He uses annuli that have piecewise linear boundaries. The verification that the vector field enters the annulus A is reduced to proving that certain polynomials do not vanish on specific intervals. Specifically, if $\phi(s) = \mathbf{u} + s\mathbf{v}$, $s \in [a, b]$ is a segment of the boundary of the annulus, then we must prove that the polynomial $\mathbf{v} \times X(\phi(s))$ does not change sign for $s \in [a, b]$. This computation is carried out by interval arithmetic in a straightforward fashion. The application of this computation to each segment of the boundary of A is used to prove that X does not both enter and leave A.

The novel part of the implementation lies mainly in the strategy used to identify an annulus that whose boundary is transverse to the flow. to do so, the properties of *rotated* vector fields are used. Consider the one parameter family of vector fields $X_\theta = R_\theta \circ X$, where R_θ is rotation by the angle θ. The equilibrium points of X remain fixed under rotation, and X_θ is transverse to X at all other points. Duff [2] proved that hyperbolic periodic orbits of X_θ shrink and expand monotonically under rotation. In particular, if Γ is a hyperbolic periodic orbit of X, then there is an $\epsilon > 0$ and a continuous family of periodic orbits Γ_θ with $-\epsilon < \theta < \epsilon$ and $\Gamma = \Gamma_0$. The Γ_θ are disjoint and either expand or contract as θ increases. Thus we can use periodic orbits of rotated vector fields to form the boundary of an annulus that contains the periodic orbit Γ.

The use of periodic orbits of rotated vector fields to obtain an annulus with boundary transverse to X works well numerically. We compute periodic orbits of $X_{\pm\epsilon}$ with a numerical integration algorithm. Discrete versions of these trajectories are used to form the boundary of an annulus that contains the periodic orbit Γ. This construction works well on numerical examples that would clearly be difficult to work with analytically. For example, Malo [5] proves that in the system $\dot{z} = \lambda z + a|z|^2 z + b\bar{z}^3$ (with z, λ, a, b complex) arising in the study of fourth order resonant Hopf bifurcation from a periodic orbit, that there are parameter values with a pair of concentric limit cycles.

Transversality techniques are used by Malo to produce proofs for the non-existence of limit cycles as well as existence proofs. Once again appeal is made to topological arguments in the proof. A limit cycle must contain equilibrium points in its interior.

If two equilibrium points have a trajectory that connects the two equilibria, then no periodic orbit can separate the two equilibria. By using estimates of the size of a neighborhood contained in the domain of attraction of a sink, together with rotated vector fields, it is possible to prove the existence of a trajectory with a specified initial condition that reaches a sink. A region is identified with boundary a piecewise smooth triangle so that one vertex is at the given initial point, the vector field points inward on the boundaries of the two adjacent sides, and the opposite side lies in the domain of attraction of the sink. Of course, these techniques can be applied in backwards as well as in forward time. Information about the α and ω limit sets of the stable and unstable manifolds of a saddle can be obtained with similar calculations.

4. Uniqueness of Limit Cycles

The phase portrait for a structurally stable planar vector field is determined by

- the number and type of equilibrium points

- the number of periodic orbits and their stability

- the α and ω limit sets of the stable and unstable manifolds of the saddle points in the system.

The methods discussed above are sufficient to determine all aspects of this structure except the determination of the number of periodic orbits. For this purpose, we need the means of proving the uniqueness of periodic orbits in annuli. Around each periodic orbit, we not only need to find an annulus with boundary transverse to the vector field, we also need to prove that there is only one periodic orbit in the annulus. If the divergence of the vector field does not change sign in the annulus, then Dulac's criterion [1] finishes this step in the proof. However if the limit cycle intersects the zero set of the divergence of X, then additional numerical computations are required to establish uniqueness of the limit cycle. We outline an approach to these computations that appears feasible, but has not yet been implemented.

We want to describe an algorithm that gives a constructive procedure for estimating properties of the return map defined in a neighborhood of a periodic orbit. Let $\gamma(s)$, $s \in S^1$, be a C^2 curve parametrized by arc length that approximates a periodic orbit of the vector field X. (We think of γ as a spline approximation to a numerically computed trajectory of X that has been reparametrized.) Let $\nu(s)$ be the unit normal to γ and let $\Psi(r, s) = \gamma(s) + r\nu(s)$. The map Ψ^{-1} is a coordinate system on a tubular neighborhood of γ for which the "radial" curves are normal lines to γ. To make calculation with interval arithmetic easier, we may assume that the curve γ is piecewise polynomial. This assumption implies that Ψ is also a piecewise polynomial map. Expressed in terms of the (r, s) coordinates, X will be given by the expression $\tilde{X}(r, s) = (D\Psi_{(r,s)})^{-1} X(\Psi(r, s))$. This is a piecewise rational expression, with the denominator of the rational expression coming from $det((D\Psi_{(r,s)}))$ when computing the inverse of $(D\Psi_{(r,s)})$. Moreover, the s component of $\tilde{X}$ is non-zero, so we can reparametrize $\tilde{X}$ to yield a vector field Y which has the form

$$\dot{r} = g(r,s)$$
$$\dot{s} = 1$$

The function g is also piecewise rational.

If γ is a sufficiently good approximation to a periodic orbit of X, then the periodic orbit will be in the image of Ψ, and will project monotonically onto the s axis. The curves parallel to the r axis map into themselves under the flow of Y since $\dot{s} = 1$. Thus, the return map Θ of a cross-section $s = s_0$ is just the time T map of the flow of Y, and

$$\Theta(r) = \int_0^T g(\rho(s), s)ds$$

with $(\rho(s), s)$ the trajectory with initial condition r. To assess the potential existence of another periodic orbit near γ, we compute

$$I(\sigma) = \int_0^T \frac{\partial g(\sigma(s), s)}{\partial r}ds$$

over curves in a neighborhood of γ. Since

$$\int\int \frac{\partial g(\sigma(s), s)}{\partial r}dA$$

is the rate of change of area under the flow of Y, the existence of multiple periodic orbits implies that $I(\sigma)$ vanishes on some closed curve σ in an annulus bounded by periodic orbits. We use this fact to formulate a test to prove that such curves do not exist. Fix an annulus defined by $|r| < \epsilon$ where we shall test for the existence of periodic orbits. Compute rigorous upper and lower bounds for g along radial segments with constant s in this annulus. For curves σ with derivatives lying in the computed range of values of g, establish bounds for $I(\sigma)$ by using interval versions of algorithms for evaluating integrals. If these bounds maintain a constant sign, then there cannot be two periodic orbits in the annulus. For the algorithm to produce the existence of a unique limit cycle in the annulus, ϵ must be large enough that we can prove existence of a limit cycle in the annulus with rotated vector fields, but small enough that the $I(\sigma)$ maintain a constant sign. Successful implementation of this algorithm would complete the construction of a practical toolkit for the verification of correctness of phase portraits for structurally stable planar vector fields defined by expressions that can be computed with interval arithmetic.

5. Concluding Remarks

We would like to extend the rigorous analysis described above to more problems involving qualitative properties of dynamical systems. For example, consider the analysis of bifurcations in families of planar vector fields as well as establishing the structure of individual phase portraits. It seems feasible to use similar ideas to prove that a specified family is stable and does not contain degenerate bifurcations. To carry out this task, rigorous evaluation of higher derivatives along trajectories is required. We discuss the rigorous determination of saddle-nodes of periodic orbits as an example. Let X_λ be a one parameter family of vector fields. Suppose that there are parameter values $\lambda_0 < \lambda_1$ and an annulus A such that X_λ has no equilibrium points in A and is transverse to its boundary for $\lambda_0 < \lambda < \lambda_1$, X_{λ_0} has a pair of periodic orbits in A and trajectories of X_{λ_1} connect the two boundary components of A. These assumptions can all be verified using the techniques described above. They imply the existence of a bifurcation of periodic orbits occurs for some $\lambda_0 < \lambda < \lambda_1$. This bifurcation is a saddle-node if two non-degeneracy conditions are met by the return map: the first derivative with respect to the parameter and the second derivative with respect to the coordinate on the cross section should not vanish. These calculations can be performed in the coordinate system described above as integrals of the appropriate derivatives of the vector field along the periodic orbit. Though the periodic orbit is not known precisely, interval computations can provide rigorous estimates that hold on all curves in a C^1 neighborhood of the periodic orbit.

The computational complexity of determining the qualitative structure of vector fields increases rapidly with dimension. Nonetheless, we suggest that the use of more geometry in the development of algorithms for computing properties about dynamical systems will be productive and worthwhile. We point to the discussion of computation of two dimensional stable and unstable manifolds of equilibrium points for flows in [4] as an illustration of the efficacy of geometric approaches to computations of dynamical system structure.

References

[1] A. A. Andronov, E. A. Vitt and S. E. Khaiken (1966), Theory of Oscillators, Pergamon Press.

[2] G. F. D. Duff (1953), Limit-cycles and rotated vector fields, Ann. Math. 57, 15-31.

[3] J. Guckenheimer and S. Malo (1993), Computer-generated proofs of phase portraits for planar systems, preprint.

[4] J. Guckenheimer and P. Worfolk (1993) Dynamical systems: some computational problems, in Bifurcations and Periodic Orbits of Vector Fields, ed. Dana Schlomiuk, Kluwer Academic Publishers, 241-278.

[5] S. Malo (1993), Computer verification of planar vector field structure, Thesis, Cornell University.

HOPF BIFURCATION AT k-FOLD RESONANCES IN EQUIVARIANT REVERSIBLE SYSTEMS *

JÜRGEN KNOBLOCH
Department of Mathematics
Technische Universität Ilmenau
Germany

ANDRE VANDERBAUWHEDE
Dept of Pure Mathematics
and Computer Algebra
University of Gent
Belgium

ABSTRACT. We study Hopf bifurcation in families of equivariant reversible differential equations under the condition that $\pm i$ are non-semisimple eigenvalues of the linearization at a symmetric fixed point. We describe how families of periodic solutions with appropriate isotropies interact and bifurcate at such resonances. Our method is based on a combination of normal form reduction with the Liapunov-Schmidt method and forms an equivariant version of the approach presented in [4].

1. Introduction

Consider the autonomous differential equation

$$\dot{x} = f(x, \lambda), \tag{1}$$

where $f : \mathbb{R}^n \times \mathbb{R}^m \longrightarrow \mathbb{R}^n$ is a smooth vectorfield with $f(0, \lambda) = 0$ for all $\lambda \in \mathbb{R}^m$ and satisfying the following hypotheses:

(H1) there exists some $\mathcal{R} \in \mathcal{L}(\mathbb{R}^n)$ satisfying $\mathcal{R}^2 = id$ and such that

$$f(\mathcal{R}x, \lambda) = -\mathcal{R}f(x, \lambda), \tag{2}$$

i.e. the equation (1) is reversible;

---

*The research in this paper was supported by the E.E.C. Science Project on "Bifurcation Theory and its Applications".

P. Chossat (ed.), Dynamics, Bifurcation and Symmetry, 167–179.

(H2) there exists a compact group $\Gamma \subset O(n)$ such that

$$f(\gamma x, \lambda) = \gamma f(x, \lambda), \quad \forall \gamma \in \Gamma, \quad \forall (x, \lambda) \in \mathbb{R}^n \times \mathbb{R}^m, \qquad (3)$$

i.e. the vectorfield f is Γ-equivariant.

We also impose the condition

$$\mathcal{R}\Gamma = \Gamma\mathcal{R}, \qquad (4)$$

which is a compatibility relation between the reversibility operator $\mathcal{R}$ and the symmetry group Γ (similar to the requirement that symmetries of Hamiltonian systems should be symplectic). In fact (4) forms not really a restriction, since it can be satisfied by replacing Γ by the group generated by Γ and $\mathcal{R}\Gamma\mathcal{R}$.

We want to study Hopf bifurcation for (1) under the condition that the linearization $A := D_x f(0,0)$ has $\pm i$ as (in general non-semisimple) eigenvalues. More precisely let us denote by S the semisimple part of A; then S is Γ-equivariant. We assume the following:

(H3) $ker(S^2 + id) = \mathbb{R}^n$; hence S generates a Γ-equivariant S^1-action on $\mathbb{R}^n$ given by $(\phi, x) \in S^1 \times \mathbb{R}^n \mapsto e^{\phi S} x$;

(H4) $ker(A^2 + id)$ is irreducible for the $\Gamma \times S^1$-action generated by Γ and S.

Instead of (H3) we could assume the non-resonance condition that A has no other eigenvalues of the form $\pm il$, $l \in \mathbb{Z}$; the S^1-action is then only defined on the generalized eigenspace $ker(S^2 + id)$. Here we have imposed the stronger condition (H3) for simplicity (see [4] for the details in the more general situation). In case Γ is trivial the condition (H4) means that the eigenvalues $\pm i$ of A have geometric multiplicity equal to one.

We denote by $k \geq 1$ the smallest integer such that $ker((A^2 + id)^k) = ker(S^2 + id) = \mathbb{R}^n$; in the case of trivial symmetry k is equal to the algebraic multiplicity of the eigenvalues $\pm i$ of A. We say then that for $\lambda = 0$ the system (1) has a k-fold resonance at the equilibrium $x = 0$; we could also call this a $(1 : 1 : \cdots : 1)$-resonance.

In order to study the bifurcation of periodic solutions under the foregoing hypotheses we extend the approach elaborated in [7], [3] and [4] to the equivariant reversible case we have at hand. For that purpose we rewrite the hypotheses (H1)-(H2) as a covariance condition on f. The method then leads to a covariant bifurcation equation; for $k = 1$ we obtain from this equation an equivariant center theorem (see e.g. [6]). We describe in some detail the solutions of the bifurcation equation for the cases $k = 2$, $k = 3$ and $k = 4$. For $k = 2$ we confirm earlier results in [6]. For $k = 3$ and $k = 4$ we obtain equivariant versions of the results in [4]: periodic solutions of (1) with certain isotropies show the same bifurcation behaviour as found in [4] for the case of trivial symmetry. Since [4] contains full details of the analysis for the plain reversible case we will here mainly emphasize the problems arising from the additional symmetry.

2. Group-theoretical preliminaries

We start by reformulating the conditions (H1)-(H2) as a covariance property for the vectorfield f. Let $\mathbb{Z}_2(\mathcal{R})$ be the subgroup of $O(n)$ isomorphic to $\mathbb{Z}_2$ and generated by $\mathcal{R}$. Also let

$$\Gamma_{\mathcal{R}} := \Gamma \cup \mathcal{R}\Gamma. \tag{5}$$

Because of (4) $\Gamma_{\mathcal{R}}$ is a compact group acting linearly on $\mathbb{R}^n$ and with Γ as a normal subgroup. Also $\mathbb{Z}_2(\mathcal{R})$ is a subgroup of $\Gamma_{\mathcal{R}}$, but in general not a normal subgroup. Every element $g \in \Gamma_{\mathcal{R}}$ has a unique representation $g = h \cdot \gamma$, with $h \in \mathbb{Z}_2(\mathcal{R})$ and $\gamma \in \Gamma$ (otherwise we would have $\mathcal{R} \in \Gamma$, which by (H1)-(H2) implies $f \equiv 0$, a case excluded by our other hypotheses). We conclude that $\Gamma_{\mathcal{R}}$ is equal to the semidirect product $\Gamma \dot{+} \mathbb{Z}_2(\mathcal{R})$ of Γ and $\mathbb{Z}_2(\mathcal{R})$.

If $A = S + N$ is the unique decomposition of A in its semisimple and its nilpotent part, then $\mathcal{R}S\mathcal{R}$ is semisimple, $\mathcal{R}N\mathcal{R}$ is nilpotent, and $A = -\mathcal{R}A\mathcal{R} = -\mathcal{R}S\mathcal{R} - \mathcal{R}N\mathcal{R}$; from the uniqueness of the decomposition we conclude then that

$$S\mathcal{R} = -\mathcal{R}S \quad \text{and} \quad N\mathcal{R} = -\mathcal{R}N. \tag{6}$$

We have already introduced the $\Gamma \times S^1$-action on $\mathbb{R}^n$ given by

$$(\gamma, \phi, u) \in \Gamma \times S^1 \times \mathbb{R}^n \mapsto (\gamma, \phi) \cdot u := \gamma \, e^{\phi S} u. \tag{7}$$

By (6) the S^1-action on $\mathbb{R}^n$ combines with $\mathcal{R}$ to give us the group $S^1 \dot{+} \mathbb{Z}_2(\mathcal{R}) \cong O(2)$ acting on $\mathbb{R}^n$. Moreover, (4) and the Γ-equivariance of S imply that $g\,\Gamma = \Gamma\,g$ for all $g \in O(2)$, such that we can define the semidirect product

$$\Gamma \dot{+} O(2) = (\Gamma \times S^1) \dot{+} \mathbb{Z}_2(\mathcal{R}) = (\Gamma \times S^1) \cup \mathcal{R}(\Gamma \times S^1) \tag{8}$$

acting on $\mathbb{R}^n$. (To be precise we should denote this group as $\tilde{\Gamma} \dot{+} O(2)$, with $\tilde{\Gamma} := (\Gamma \times S^1)/S^1$; since this might create more confusion than necessary we will stick to the notation $\Gamma \dot{+} O(2)$). We can also without loss of generality assume that $\Gamma \dot{+} O(2) \subset O(n)$, which means that S is anti-symmetric and $\mathcal{R}$ is symmetric.

Next we define a *character* $\chi : \Gamma \dot{+} O(2) \to \mathbb{R}$ on $\Gamma \dot{+} O(2)$ by

$$\chi(g) := 1 \quad \text{and} \quad \chi(\mathcal{R} \cdot g) := -1, \quad \forall g \in \Gamma \times S^1. \tag{9}$$

We can then rewrite the conditions (H1)-(H2) on f as

$$f(g \cdot x, \lambda) = \chi(g) g \cdot f(x, \lambda), \quad \forall g \in \Gamma_{\mathcal{R}}. \tag{10}$$

We say that the vectorfield f is $\Gamma_{\mathcal{R}}$-covariant. The foregoing shows that then also the linearization A, its semisimple part S and its nilpotent part N are $\Gamma \dotplus O(2)$-covariant.

3. Transformation into normal form

As a first step in the bifurcation analysis we bring the system (1) into a suitable normal form. The following theorem is a covariant version of classical normal form results; for a proof in the particular context considered here compare with [3] and [4].

Theorem 1. *Let* $f : \mathbb{R}^n \times \mathbb{R}^m \to \mathbb{R}^n$ *be a smooth family of* $\Gamma_{\mathcal{R}}$-*covariant vector fields. Then there exists for each* $l \geq 1$ *a neighborhood* Ω_l *of the origin in* $\mathbb{R}^m$ *and a mapping* $\Phi \in C^\infty(\mathbb{R}^n \times \Omega_l, \mathbb{R}^n)$ *such that for each* $\lambda \in \Omega_l$ *the following holds:*

  (i) $\Phi_\lambda := \Phi(\cdot, \lambda)$ *is a* $\Gamma_{\mathcal{R}}$-*equivariant diffeomorphism on* $\mathbb{R}^n$, *with* $\Phi_\lambda(0) = 0$ *and* $D\Phi_0(0) = id$;

  (ii) *the* l-*th order Taylor polynomial* $T_l g_\lambda$ *of the pull-back* $g_\lambda(\cdot) := \Phi_\lambda^* f(\cdot, \lambda)$ *of* $f(\cdot, \lambda)$ *under* Φ_λ *is* S^1-*equivariant, i.e. we have* $T_l g_\lambda(e^{\phi S} x) = e^{\phi S} T_l g_\lambda(x)$ *for all* $\phi \in S^1$;

  (iii) *the linearization* $B_\lambda := Dg_\lambda(0)$ *is such that* $B_\lambda - A$ *commutes with the transpose* A^T *of* A *(observe that* $B_0 = A$).

Since clearly g_λ is also $\Gamma_{\mathcal{R}}$-covariant, it follows that $T_l g_\lambda$ is in fact $\Gamma \dotplus O(2)$-covariant. Assuming that such normal form reduction has been carried out we can therefore, without loss of generality, add the following hypotheses to our list:

(H5) $f(x, \lambda) = f_{NF}(x, \lambda) + O(\|x\|^{l+1})$ for some $l \geq 1$, where $f_{NF} : \mathbb{R}^n \times \mathbb{R}^m \to \mathbb{R}^n$ is a $\Gamma \dotplus O(2)$-covariant vectorfield:

$$f_{NF}(g \cdot x, \lambda) = \chi(g) g \cdot f_{NF}(x, \lambda), \quad \forall g \in \Gamma \dotplus O(2). \tag{11}$$

(H6) $A_\lambda := D_x f(0, \lambda)$ satisfies

$$(A_\lambda - A)A^T = A^T(A_\lambda - A). \tag{12}$$

We will use (H6) to obtain a convenient explicit form for A_λ.

Lemma 2. *Assume (H1)-(H4); then there exist $\Gamma \dotplus O(2)$-invariant subspaces U_j ($1 \leq j \leq k$) such that*

(i) $\mathbb{R}^n = U_1 \oplus U_2 \oplus \cdots \oplus U_k$;

(ii) $U_1 = \ker N$ and $N(U_j) = U_{j-1}$ for $2 \leq j \leq k$.

Here N is the nilpotent part of A.

Proof . The subspaces $V_j := \ker((A^2 + id)^j)$ ($0 \leq j \leq k$) are $\Gamma \dotplus O(2)$-invariant, since $A^2 + id$ is $\Gamma \dotplus O(2)$-equivariant; they form a strictly increasing sequence $(V_j)_{0 \leq j \leq k}$ of subspaces of $\mathbb{R}^n$, with $V_0 = \{0\}$ and $V_k = \mathbb{R}^n$. Moreover, $(A^2 + id)^{j-1}$ maps each complement of V_{j-1} in V_j injectively into $V_1 = \ker(A^2 + id)$; by choosing a $\Gamma \dotplus O(2)$-invariant complement it follows then from the irreducibility condition (H4) and Schur's lemma that the codimension of V_{j-1} in V_j is equal to $\dim V_1$ for $1 \leq j \leq k$. The subspaces V_j are also invariant under N, and since $A^2 + id = N(2S + N)$, with $2S + N$ an isomorphism, it follows that $V_{j-1} = (A^2 + id)(V_j) = N(V_j)$ ($1 \leq j \leq k$). Let U_k be a $\Gamma \dotplus O(2)$-invariant complement of V_{k-1} in V_k, and define $U_j := N^{k-j}(U_k)$ for $1 \leq j < k$; then $V_j = V_{j-1} \oplus U_j$, the U_j are $\Gamma \dotplus O(2)$-invariant (since N is $\Gamma \dotplus O(2)$-covariant), and the lemma follows. $\qquad\square$

Setting $U := U_1 = \ker N$ it follows from lemma 2 that we can identify $\mathbb{R}^n$ with U^k via the linear and $\Gamma \dotplus O(2)$-covariant isomorphism

$$u_1 + u_2 + \cdots + u_k \in U_1 \oplus U_2 \oplus \cdots \oplus U_k = \mathbb{R}^n \mapsto (u_1, Nu_2, \ldots, N^{k-1}u_k) \in U^k;$$

using this identification A gets the form

$$A(u_1, u_2, \ldots, u_k) = (Su_1 + u_2, Su_2 + u_3, \ldots, Su_{k-1} + u_k, Su_k), \quad \forall u_1, \ldots, u_k \in U; \tag{13}$$

in the same representation the semisimple part S of A, the reversibility operator $\mathcal{R}$ and the symmetry operators $\gamma \in \Gamma$ take the diagonal form

$$\begin{aligned} S \cdot (u_1, u_2, \ldots, u_k) &= (Su_1, Su_2, \ldots, Su_k) \\ \mathcal{R} \cdot (u_1, u_2, \ldots, u_k) &= (\mathcal{R}u_1, -\mathcal{R}u_2, \ldots, (-1)^{k-1}\mathcal{R}u_k), \\ \gamma \cdot (u_1, u_2, \ldots, u_k) &= (\gamma u_1, \gamma u_2, \ldots, \gamma u_k). \end{aligned} \tag{14}$$

¿From (12), (13) and (14) it is not difficult to verify that $A_\lambda := D_x f(0, \lambda)$ must have the form

$$
A_\lambda = \begin{pmatrix}
(1 + \alpha_1(\lambda))S & id & 0 & \cdots & 0 \\
\alpha_2(\lambda)S^2 & (1 + \alpha_1(\lambda))S & id & \ddots & 0 \\
\alpha_3(\lambda)S^3 & \alpha_2(\lambda)S^2 & \ddots & \ddots & \ddots \\
\vdots & \ddots & \ddots & \ddots & \ddots \\
\alpha_k(\lambda)S^k & \alpha_{k-1}(\lambda)S^{k-1} & \cdots & \alpha_2(\lambda)S^2 & (1 + \alpha_1(\lambda))S
\end{pmatrix},
$$

where the α_i $(1 \le i \le k)$ are smooth functions of the parameter λ such that $\alpha_i(0) = 0$. Of course in this matrix S and id actually stand for the restrictions $S|_U$ and $id|_U$, respectively; the same holds in (13) and (14). We hope no confusion will arrise from this abuse of notation. By a rescaling of the time t and the different components of $x = (u_1, \ldots, u_k)$ one can put $\alpha_1(\lambda) \equiv 0$ (see [6]). We also assume the transversality condition

(H7) $m = k - 1$ and $\dfrac{\partial(\alpha_2, \cdots, \alpha_k)}{\partial(\lambda_1, \cdots, \lambda_{k-1})}(0, \ldots, 0) \ne 0.$

Then we can transform the parameters such that $\alpha_i(\lambda) = \lambda_{i-1}$ for $2 \le i \le k$, i.e. A_λ has the form

$$
A_\lambda = \begin{pmatrix}
S & id & 0 & \cdots & 0 \\
\lambda_1 S^2 & S & id & \ddots & 0 \\
\lambda_2 S^3 & \lambda_1 S^2 & \ddots & \ddots & \ddots \\
\vdots & \ddots & \ddots & \ddots & \ddots \\
\lambda_{k-1}S^k & \lambda_{k-2}S^{k-1} & \cdots & \lambda_1 S^2 & S
\end{pmatrix}. \tag{15}
$$

We will use this linear normal form together with the representations (14) of the symmetry operators when calculating the bifurcation equations for our Hopf bifurcation problem; in the next section we first obtain a general form for these bifurcation equations.

4. Reduction of the problem

Under the hypotheses stated in the foregoing sections we want to study the small periodic solutions of (1) with period near 2π and for parameter values near zero. To obtain the corresponding bifurcation equations we proceed as in [3], [4] and [7]; for more details on the reversible case which interests us here we refer in particular to [4].

By a time rescale we have to find all small 2π-periodic solutions of

$$(1 + \sigma)\dot{x} = f(x, \lambda) \tag{16}$$

for (σ, λ) near $(0, 0)$. In the standard way we rewrite this as an operator equation

$$F(x, \sigma, \lambda) = 0, \tag{17}$$

with $F : C_{2\pi}^1 \times \mathbb{R} \times \mathbb{R}^{k-1} \to C_{2\pi}^0$ defined by

$$F(x, \sigma, \lambda)(t) := -(1 + \sigma)\dot{x}(t) + f(x(t), \lambda). \tag{18}$$

In the spaces $C_{2\pi}^0$ and $C_{2\pi}^1$ of respectively continuous and continuously differentiable 2π-periodic maps $x : \mathbb{R} \to \mathbb{R}^n$ we use the usual supremum norms; the mapping F is smooth and satisfies $F(0, \sigma, \lambda) = 0$. In $C_{2\pi}^0$ we define a $\Gamma \dot{+} O(2)$-action by

$$(\gamma, \phi) \cdot x(t) := \gamma \cdot x(t + \phi)) \text{ for } (\gamma, \phi) \in \Gamma \times S^1 \text{ and } (\mathcal{R} \cdot x)(t) := \mathcal{R}x(-t); \tag{19}$$

then F is $\Gamma \dot{+} O(2)$-covariant.

We split the linearization $L := D_x F(0, 0, 0)$ as $L = L_S + L_N$, with $(L_S\, x)(t) := -\dot{x}(t) + Sx(t)$ and $(L_N\, x)(t) := Nx(t)$; we call L_S the "semisimple part" of L. We know from classical Floquet theory that $L_S \in \mathcal{L}(C_{2\pi}^1, C_{2\pi}^0)$ is a Fredholm operator of index zero. We now perform a $\Gamma \dot{+} O(2)$-equivariant Liapunov-Schmidt reduction *with respect to* L_S, that is, instead of reducing to $ker\, L \cong ker(A^2 + id) = ker\, N = U$ we reduce only to $ker\, L_S \cong \mathbb{R}^n = U^k$; using the $\Gamma \dot{+} O(2)$-equivariant isomorphism $\zeta : U^k \to ker\, L_S$ given by $(\zeta u)(t) := e^{tS}u = (e^{tS}u_1, \cdots, e^{tS}u_n)$ we obtain the bifurcation equation

$$G(u, \sigma, \lambda) = 0, \tag{20}$$

where G is a smooth $\Gamma \dot{+} O(2)$-covariant mapping from $U^k \times \mathbb{R} \times \mathbb{R}^{k-1}$ into U^k, of the form

$$G(u, \sigma, \lambda) = -(1 + \sigma)Su + f_{NF}(u, \lambda) + \tilde{r}(u, \sigma, \lambda), \tag{21}$$

with $\tilde{r}(u, \sigma, \lambda) = O(\|u\|^{l+1})$ as $u \to 0$, uniformly for small (σ, λ).

Taking (for $u \neq 0$) the inner product of (20) with Su and dividing by $\|u\|^2$ gives a scalar equation which can be solved for $\sigma = \tilde{\sigma}(u, \lambda)$; substitution of this solution into $\tilde{r}(u, \sigma, \lambda)$ (see (21)) gives us the bifurcation equation in the form

$$(1 + \sigma)Su = \tilde{f}(u, \lambda), \quad \text{with} \quad \tilde{f}(u, \lambda) = f_{NF}(u, \lambda) + O(\|u\|^{l+1}). \tag{22}$$

The *reduced vectorfield* $\tilde{f}(u, \lambda)$ is $\Gamma \dot{+} O(2)$-covariant, i.e. it is Γ-equivariant and reversible (such as the original vectorfield f), but it has also an additional S^1-symmetry (such as the normal form part of the original vectorfield f). The reduction procedure introduces a loss of differentiability at the origin, i.e. $\tilde{f}$ is only of class C^l; it remains however of class C^∞ outside of $u = 0$.

We can summarize the reduction as follows.

Theorem 3. *Assume (H1)-(H4). Then there exist for each $l \geq 2$ a neighborhood ω_l of the origin in $\mathbb{R}^m$, a C^l-mapping $\Phi : U^k \times \omega_l \to \mathbb{R}^n$ and a C^l-vectorfield $\tilde{f} : U^k \times \omega_l \to U^k$ with the following properties:*

(i) Φ is $\Gamma_{\mathcal{R}}$-equivariant, $\Phi(0, \lambda) = 0$ and $D_u\Phi(0,0) = id$;

(ii) $\tilde{f}$ is $\Gamma \dotplus O(2)$-covariant, $\tilde{f}(0, \lambda) = 0$ and $D_u\tilde{f}(0,0) = A$;

(iii) for each $\lambda \in \omega_l$ and each T near 2π a function $\tilde{x} : \mathbb{R} \to \mathbb{R}^n$ is a sufficiently small T-periodic solution of (1) if and only if

$$\tilde{x}(t) = \Phi(\tilde{u}(t, \lambda), \lambda), \quad \forall t \in \mathbb{R}, \tag{23}$$

where $\tilde{u} : \mathbb{R} \to U^k$ is a small T-periodic solution of the reduced equation

$$\dot{u} = \tilde{f}(u, \lambda). \tag{24}$$

Moreover, these periodic solutions of (24) all have the form

$$\tilde{u}(t) = e^{(1+\sigma)St} u \tag{25}$$

with $u \in U^k$ and $\sigma = (2\pi/T) - 1$. Finally, if (H5) holds then the reduction can be done such that

$$\tilde{f}(u, \lambda) = f_{NF}(u, \lambda) + O(\|u\|^{l+1}). \tag{26}$$

It follows from Theorem 3 that the bifurcation equations for our Hopf bifurcation problem take the form

$$(1 + \sigma)Su_j = \tilde{f}_j(u_1, u_2, \ldots, u_k, \lambda), \quad j = 1, 2, \ldots, k. \tag{27}$$

Assuming (15) we can solve the first $k - 1$ equations in (27) for $u_j = \tilde{u}_j(u_1, \lambda, \sigma)$ $(j = 2, \ldots, k)$; these solutions are such that $\tilde{u}_j(g \cdot u_1, \lambda, \sigma) = \chi(g)^{j-1}g \cdot \tilde{u}_j(u_1, \lambda, \sigma)$ for all $g \in \Gamma \dotplus O(2)$. More explicitly we have

$$\tilde{u}_j(u_1, \lambda, \sigma) = h_{j-1}(\sigma, \lambda)S^{j-1}u_1 + \bar{u}_j(u_1, \lambda, \sigma), \text{ with } \bar{u}_j(u_1, \lambda, \sigma) = O(\|u_1\|^3), \tag{28}$$

and where the polynomials $h_j(\sigma, \lambda)$ $(0 \leq j \leq k)$ are defined by the following iteration scheme:

$$
\begin{aligned}
h_0(\sigma, \lambda) &= 1, \quad h_1(\sigma, \lambda) = \sigma, \\
h_j(\sigma, \lambda) &= \sigma h_{j-1}(\sigma, \lambda) - \sum_{i=1}^{j-1} \lambda_i h_{j-1-i}(\sigma, \lambda), \quad 2 \leq j \leq k.
\end{aligned}
\tag{29}
$$

(We also take $l \geq 3$ in Theorem 3). Bringing these solutions in the last equation of (27) and applying S^{-k} we obtain a bifurcation equation in the form

$$
g(u_1, \lambda, \sigma) = h_k(\sigma, \lambda) u_1 + \bar{g}(u_1, \lambda, \sigma) = 0, \tag{30}
$$

with $\bar{g}(u_1, \lambda, \sigma) = O(\|u_1\|^3)$ as $u_1 \to 0$; the mapping $g : U \times \mathbb{R}^{k-1} \times \mathbb{R} \to U$ is $\Gamma \dot{+} O(2)$-equivariant.

Using this equivariance one can apply the main idea of the Equivariant Branching Lemma to obtain some solution branches of (30). Indeed, let $\Sigma \subset \Gamma \dot{+} O(2)$ be an isotopy subgroup for the action of $\Gamma \dot{+} O(2)$ on U such that

$$
\dim U^\Sigma = 1, \tag{31}
$$

where $U^\Sigma := \{u \in U \mid g \cdot u = u, \ \forall g \in \Sigma\}$ is the fixed point subspace of Σ. A characterization of such isotropy subgroups is given in Lemma 4.1 of [6]; if Γ is trivial one can take $\Sigma = \mathbb{Z}_2(\mathcal{R})$. Let $u_\Sigma \in U$ be such that $U^\Sigma = \{\rho u_\Sigma \mid \rho \in \mathbb{R}\}$. Then for solutions $u_1 \in U^\Sigma$ the bifurcation equation (30) reduces to a *scalar* equation, of the form

$$
g_\Sigma(\rho, \lambda, \sigma) = h_k(\sigma, \lambda)\rho - \alpha_\Sigma(\rho, \lambda, \sigma)\rho^3 = 0, \tag{32}
$$

where the function $g_\Sigma(\rho, \lambda, \sigma)$ depends on the choice of Σ and is even in ρ. This equation gives us all solutions of our original problem which have an isotropy containing Σ; it follows from Lemma 4.1 of [4] that such periodic solutions are reversible.

The easiest strategy to solve (32) would be to observe that for non-zero solutions we can divide (32) by ρ, while the definition of h_k implies that

$$
\frac{\partial h_k}{\partial \lambda_{k-1}}(0,0) = -1 \neq 0,
$$

such that (32) can be solved for λ_{k-1} as a function of ρ, σ and the other components of λ. Except for the case $k = 1$ where one can solve (32) for σ as a function of ρ and which leads to an equivariant version of the Liapunov Center Theorem for reversible systems (see [6], [2] and the next section) this is however not what we want: we would like to describe for each fixed (but sufficiently small) λ the corresponding solution set of (30) or (32). At least qualitatively this can be done as follows. Assuming that

$$a_\Sigma := \alpha_\Sigma(0,0,0) \neq 0 \tag{33}$$

one can (for non-zero solutions) further transform (32) into the simple polynomial form

$$h_k(\sigma, \lambda) = a_\Sigma \rho^2. \tag{34}$$

(See [6] for details). Also, since $\rho \mapsto -\rho$ corresponds to a phase shift over half a period we can restrict to $\rho > 0$. The constant a_Σ can be calculated easily; assuming (H5) for some $l \geq 3$ and setting $x_\Sigma := (u_\Sigma, 0, \ldots, 0)$ (we identify $\mathbb{R}^n$ with U^k) we find

$$a_\Sigma u_\Sigma = \lim_{\rho \to 0} \rho^{-3} S^{-k} f_{NF,k}(\rho x_\Sigma, 0). \tag{35}$$

In the next section we briefly discuss the solution set of (34) for $k = 2$, $k = 3$ and $k = 4$; see [4] for more details and pictures.

5. The bifurcation results

Throughout this section we fix some isotropy subgroup Σ of $\Gamma \dotplus O(2)$ satisfying (31) and (33); the equation (34) describes then the nontrivial solutions of our Hopf bifurcation problem which have at least the symmetry Σ. For $k = 1$ (34) reduces to $\sigma = a_\Sigma \rho^2$, giving us a local one-parameter family of periodic orbits emanating from the equilibrium point (one can take ρ as a parameter measuring the amplitude of the periodic orbit). For $k > 1$ there will be open regions C_j ($0 \leq j \leq k/2$) in parameter space, each having the origin at their boundary, such that for $\lambda \in C_j$ and sufficiently close to the origin the linearization A_λ has $k - 2j$ different pairs of (semisimple) purely imaginary eigenvalues close to $\pm i$, and j different quadruples of complex eigenvalues off the imaginary axis. (Remember: since A_λ is reversible the fact that μ is an eigenvalue implies that also $-\mu$, $\bar{\mu}$ and $-\bar{\mu}$ are eigenvalues). The $k - 2j$ pairs of purely imaginary eigenvalues are not in resonance (at least not for λ sufficiently small) and the result for $k = 1$ can be applied separately for each of these pairs; locally near the equilibrium we see $k - 2j$ distinct families of periodic orbits emanating from that equilibrium point, each with a slightly different limiting period. The study of (34) gives a more comprehensive picture: it shows whether or not some of these families are connected to each other, and how the different families appear or disappear as one crosses the boundaries of the regions C_j.

For $k = 2$ we have $\lambda \in \mathbb{R}$ and (34) takes the form

$$\sigma^2 - \lambda = a_\Sigma \rho^2. \tag{36}$$

One can easily verify from (15) that $C_0 = \{\lambda > 0\}$ and $C_1 = \{\lambda < 0\}$. There are two cases, depending on the sign of a_Σ. If $a_\Sigma < 0$ (the *elliptic* case) then the two local families of periodic orbits (with symmetry Σ) which exist for $\lambda \in C_0$ are connected to each other and actually form just one branch of periodic orbits, with both ends emanating from the equilibrium point; since this branch is completely contained in a small neighborhood of the equilibrium we call this a *local branch*. As λ passes from C_0 into C_1 this local branch shrinks down into the equilibrium and disappears, leaving no nontrivial periodic orbits with symmetry Σ for $\lambda \in C_1$. If $a_\Sigma > 0$ (the *hyperbolic* case) then the two local families which emanate from the equilibrium for $\lambda \in C_0$ are not connected to each other (at least not within a small neighborhood of the equilibrium), and both intersect the boundary of some fixed neighborhood of this equilibrium; we say that we have two *global branches* (although our methods of course only determine these branches locally). As λ passes from C_0 into C_1 the two global branches detach from the equilibrium and merge into one smooth family of periodic orbits which no longer contains the equilibrium.

For $k = 3$ we have $\lambda = (\lambda_1, \lambda_2) \in \mathbb{R}^2$ and (34) takes the following explicit form:

$$\sigma^3 - 2\lambda_1 \sigma - \lambda_2 = a_\Sigma \rho^2. \tag{37}$$

One calculates from (15) that $C_0 = \{(\lambda_1, \lambda_2) \mid 27\lambda_2^2 < 32\lambda_1^3\}$ and $C_1 = \{(\lambda_1, \lambda_2) \mid 27\lambda_2^2 > 32\lambda_1^3\}$. For $\lambda \in C_0$ we have one local and one global branch of periodic solutions with symmetry Σ; for $\lambda \in C_1$ there is just one global branch. At boundary points $\lambda \in \partial C_0 = \partial C_1$ there is an elliptic transition if $a_\Sigma \lambda_2 > 0$: as λ moves from C_0 into C_1 the local branch shrinks down into the equilibrium, just leaving the global branch. If $a_\Sigma \lambda_2 < 0$ at such boundary point then we have a hyperbolic transition: one of the ends of the local branch and the global branch detach from the equilibrium and merge, leaving one single global branch emanating from the equilibrium point.

For $k = 4$ the equation (34) has the form

$$\sigma^4 - 3\lambda_1 \sigma^2 - 2\lambda_2 \sigma + \lambda_1^2 - \lambda_3 = a_\Sigma \rho^2. \tag{38}$$

In this case the bifurcation set $\mathcal{B} := \partial C_0 \cup \partial C_1 \cup \partial C_2$ forms a so-called "swallowtail" in $\mathbb{R}^3$, a surface well known from catastrophe theory; C_0 corresponds to the cone-like region having the cusp lines and the self-intersection line at its boundary, C_1 to the region surrounding C_0 and C_2 to the remaining region (the one which contains the "whisker"). If $a_\Sigma > 0$ then there are two global branches and one local branch for

$\lambda \in C_0$, two global branches for $\lambda \in C_1$, and one branch detached from the equilibrium for $\lambda \in C_2$. The transition from C_0 to C_1 is either elliptic (the local branch shrinks and disappears) or hyperbolic (the local branch merges with one of the two global branches into one global branch); the transition from C_1 to C_2 is hyperbolic: the two global branches detach from the equilibrium and merge. If $a_\Sigma < 0$ then there are two local branches for $\lambda \in C_0$, one local branch for $\lambda \in C_1$, and no nontrivial periodic solutions with symmetry Σ for $\lambda \in C_2$. The transition from C_0 into C_1 is again either elliptic (one of the two local branches shrinks and disappears) or hyperbolic (one end of each of the two local branches detaches from the equilibrium and the two branches merge into one local branch); the transition from C_1 into C_2 is elliptic: the one local branch shrinks and disappears.

We finish with some results which are valid for all values of k. For general k the set C_j ($0 \le j \le k/2$) consists of those $\lambda \in \mathbb{R}^{k-1}$ for which the function $\sigma \mapsto h_k(\sigma, \lambda)$ has exactly $k - 2j$ real and simple zeros. If k is even (say $k = 2k'$) and $a_\Sigma > 0$ then there are two global branches and $k' - j - 1$ local branches for $\lambda \in C_j$ ($0 \le j \le k' - 1$), and just one branch detached from the equilibrium for $\lambda \in C_{k'}$. If $k = 2k'$ and $a_\Sigma < 0$ then there are $k' - j$ local branches for $\lambda \in C_j$ ($0 \le j \le k'$). If k is odd (say $k = 2k'+1$) then we have one global branch and $k' - j$ local branches for $\lambda \in C_j$ ($0 \le j \le k'$). The bifurcation set $\mathcal{B}$ (that is, the union of the boundaries of the regions C_j, $0 \le j \le k/2$) contains those $\lambda \in \mathbb{R}^{k-1}$ for which the function $h_k(\cdot, \lambda)$ has at least one real zero of higher multiplicity, i.e.

$$\mathcal{B} = \{\lambda \in \mathbb{R}^{k-1} \mid h_k(\sigma_0, \lambda) = \frac{\partial h_k}{\partial \sigma}(\sigma_0, \lambda) = 0 \text{ for some } \sigma_0 \in \mathbb{R}\}. \tag{39}$$

This bifurcation set can be stratified; the stratum of codimension one contains those λ for which $h_k(\cdot, \lambda)$ has exactly one real zero σ_0 of multiplicity two. Crossing $\mathcal{B}$ at such point results in an elliptic transition if $a_\Sigma \frac{\partial^2 h_k}{\partial \sigma^2}(\sigma_0, \lambda) < 0$ and a hyperbolic transition if $a_\Sigma \frac{\partial^2 h_k}{\partial \sigma^2}(\sigma_0, \lambda) > 0$.

References

[1] R. Devaney. Reversible diffeomorphisms and flows. Trans. Amer. Math. Soc. 218 (1976) 89–113.

[2] M. Golubitsky, M. Krupa and C. Lim. Time-reversibility and particle sedimentation. Siam J. Appl. Math. 51 (1991) 49–72.

[3] J. Knobloch. Reduction of Hopf bifurcation problems with symmetries. To appear in the Bull. Belg. Math. Soc.

[4] J. Knobloch and A. Vanderbauwhede. Hopf bifurcation for k-fold resonances in reversible systems. In preparation.

[5] A. Vanderbauwhede. *Local bifurcation and symmetry*, Res. Notes Math. 75, Pitman, Boston, 1982.

[6] A. Vanderbauwhede. Hopf bifurcation for equivariant conservative and time-reversible systems. Proc. Royal Soc. Edinburgh 116A (1990) 103-128.

[7] A. Vanderbauwhede and J.-C. van der Meer. A general reduction method for periodic near equilibria in Hamiltonian systems. To appear in: *Normal Forms and Homoclinic Chaos*, Fields Institute Communications, A.M.S.

SYMMETRIES AND REVERSING SYMMETRIES IN KICKED SYSTEMS

JEROEN S.W. LAMB and HANNA BRANDS
Institute for Theoretical Physics, University of Amsterdam
Valckenierstraat 65, 1018 XE Amsterdam, The Netherlands

ABSTRACT. The concepts of (reversing) symmetries and (reversing) k-symmetries in dynamical systems are reviewed and their occurrence is discussed in dynamical systems of which the time one map naturally is written as the concatenation of different types of evolutions.

1 Introduction

In the theory of dynamical systems, the interest for symmetry properties is steadily growing. Yet a systematic study of symmetries involving a time reversal, so-called *reversing symmetries*, has started only recently, after some years in which the interest for reversing symmetries was mainly focussed on involutions (i.e. reversing symmetries that are their own inverse).

In physics, the conventional notion of time reversal symmetry is related to the observation that the motion picture of some dynamical phenomenon may be reversed without becoming unphysical, i.e. the trajectory in configuration space in the forward running movie is ruled by the same 'laws of motion' as the trajectory in the backward running movie. For instance, this phenomenon occurs in mechanics if there is no friction. The reversed motion picture of a ball moving on a surface without friction still looks fine. However, if there is some friction the ball will slow down and eventually stop. The reversed motion picture then shows the ball accellerating by itself. The latter phenomenon clearly does not obey the same rules of motion as the first one. In an analogous way, in spin systems magnetic fields tend to break conventional time reversal symmetry.

After this introduction let us define conventional time reversal symmetry in a more mathematical way. In mechanics, the equations of motion are derived from a Hamiltonian, $H(\mathbf{p}, \mathbf{q})$, that is the energy functional defined on the phase space consisting of momenta $\mathbf{p}$ and positions $\mathbf{q}$. Such a system is reversible in the conventional sense if and only if

$$H(\mathbf{p}, \mathbf{q}) = H(-\mathbf{p}, \mathbf{q}). \tag{1.1}$$

This property (1.1) is called reversibility since the equations of motion

$$\dot{\mathbf{p}} = -\partial H(\mathbf{p}, \mathbf{q})/\partial \mathbf{q}, \quad \dot{\mathbf{q}} = \partial H(\mathbf{p}, \mathbf{q})/\partial \mathbf{p} \tag{1.2}$$

P. Chossat (ed.), Dynamics, Bifurcation and Symmetry, 181–196.
© 1994 *Kluwer Academic Publishers.*

are invariant under the transformation in phase space $(\mathbf{p}, \mathbf{q}) \mapsto (-\mathbf{p}, \mathbf{q})$ together with a time reversal $t \mapsto -t$. This type of time reversal symmetry causes the backward running motion picture of the dynamics (in configuration space) of (1.2) to be ruled by (1.2) too (only with different initial conditions).

Conventional time reversal symmetry is widely discussed in physics literature. In statistical physics there has been extensive discussions about the paradox that arises from the observation a macroscopic number of particles that individually obey reversible equations of motion, collectively display irreversible behaviour (entropy always increases). This problem is known as Loschmidt's paradox [1] and - although apparently solved by Gibbs [2] in 1902 - still actively discussed today.

In the early thirties, Wigner [3] succesfully introduced the quantum mechanical equivalent of the classical conventional time reversal operator. With it, he managed to explain the so-called Kramers degeneracy observed in atomic and nuclear energy spectra with time reversal symmetry (in the absense of a magnetic field) [4]. If the time reversal symmetry is broken (in the presence of a magnetic field) the degeneracy disappears yielding a band splitting. The notion of conventional time reversal symmetry is also important in relativistic quantum field theories of elementary particles in high energy physics. For a review of the applications of conventional time reversal symmetry in physics, the reader is refered to Sachs [5].

As such, time reversal symmetry has been discussed over the years as a special type of symmetry property a dynamical system may possess. However, hardly anyone has discussed time reversal symmetry within the context of the other 'normal' symmetries that leave the equations of motion invariant without an additional time reversal. An exception may be found in the work of Wigner who distinguishes symmetries as unitary operators commuting with the Hamiltonian from reversing symmetries that are anti-unitary operators commuting with the Hamiltonian. In this context it is interesting to note that the parallel between unitary and anti-unitary operators versus symmetries and reversing symmetries only holds within the class of operators commuting with the Hamiltonian. In fact, for operators anti-commuting with the Hamiltonian the unitary operators are reversing symmetries whereas anti-unitary operators are symmetries.

In the theory of dynamical systems, the importance of the notion of time reversal symmetry was already appreciated by Birkhoff, who used reversibility in his studies into the restricted three body problem [6]. In the work of Birkhoff [6, 7], and later Devaney [8], it became clear that the consequences of conventional time reversal symmetry are shared by dynamical systems which have a different but equivalent reversing symmetry.

One was led to give definitions of reversibility in which the involutory character of the symmetry transformation was incorporated, rather than the fact that it should be precisely the mirror $(\mathbf{p}, \mathbf{q}) \mapsto (-\mathbf{p}, \mathbf{q})$. For instance, Sevryuk [9] defines *reversible* systems as dynamical systems possessing an involution that reverses the motion. In case the reversing symmetry is not an involution he calls the system *weakly reversible*. We would rather not use this this terminology here, because it suggests that involutory reversing symmetries are in some sense 'stronger' than noninvolutory ones. However, most prominent properties due to reversing symmetries can be formulated without using the order[1] of the reversing

---

[1] The order of a (reversing) symmetry U is defined to be the smallest positive integer n such that $U^n = Id$.

symmetry. Most typical properties for systems possessing an involutory time reversing symmetry directly extend (or are easily generalized) to systems possessing noninvolutory reversing symmetries [10]. In practice, the representation of the reversing symmetry turns out to be at least as important for the dynamics as the order of the reversing symmetry, see e.g. [11].

The occurrence of a reversing symmetry does not imply a system to be Hamiltonian or even measure preserving. In a recent review of Roberts and Quispel [12], various consequences of reversibility for dynamical systems that are not measure preserving were discussed, with an emphasis on reversible mappings of the plane displaying pairs of attractors and repellors. In these type of systems a main difference between the symmetric and asymmetric phenomena was found ('symmetric' with respect to the reversing symmetry). The symmetric features are similar to those in measure preserving maps, including structurally stable elliptic fixed points and a KAM theorem [9], whereas the asymmetric behaviour is as in dissipative maps with pairs of (strange) attractors and repellors. Although most reversible systems discussed in literature are Hamiltonian, there are several examples in which reversibility was recognized in physically motivated dynamical systems that are not measure preserving [13, 14].

In this paper we will discuss the occurrence of reversing symmetries in dynamical systems within the context of symmetries. In particular we will show how nonconventional reversing symmetries can occur in kicked systems. Thereafter we will discuss a more general concept of (reversing) symmetries occurring only at specific time scales. The analysis of these so-called (reversing) k-symmetries will also be treated specifically in the context of kicked systems.

We will illustrate our results with two examples: the kicked rotator and the web map.

2 Reversing Symmetry Group

2.1 GENERAL CONCEPTS

Consider a dynamical system on a state space (or phase space) Ω. Throughout this paper we will confine ourselves to invertible dynamical systems and invertible (reversing) symmetries. In the analysis of symmetry properties of a dynamical system we distinguish between symmetries and reversing symmetries. An autonomous[2] dynamical system with continuous time, governed by the ordinary differential equation

$$\frac{d}{dt}\mathbf{x} = F(\mathbf{x}), \quad (\mathbf{x} \in \Omega) \tag{2.1}$$

the map $M : \Omega \mapsto \Omega$ is a symmetry of (2.1) if

$$\frac{d}{dt}(M\mathbf{x}) = F(M\mathbf{x}). \tag{2.2}$$

The map $S : \Omega \mapsto \Omega$ is a reversing symmetry of (2.1) if

$$\frac{d}{dt}(S\mathbf{x}) = -F(S\mathbf{x}). \tag{2.3}$$

---

[2]Note that a nonautonomous dynamical system can always be written as an autonomous one by extending the state space.

In the case of discrete time we consider an invertible map $L : \Omega \mapsto \Omega$. Symmetries M and reversing symmetries S of this map satisfy

$$M \circ L \circ M^{-1} = L, \tag{2.4}$$

$$S \circ L \circ S^{-1} = L^{-1}. \tag{2.5}$$

If the order of the reversing symmetry S is odd then $L^2 = Id$ and the dynamics of L is trivial. So any nontrivial dynamical system may only possess reversing symmetries of even or infinite order.

The reader should note that in the above definitions we do not require the action of the (reversing) symmetries to be linear.

In this paper the emphasis will be on discrete time maps rather than continuous time flows. Most discrete time results easily extend to continuous time flows, the map L denoting the time one map of the continuous time flow.

If we restrict ourselves to invertible maps and invertible (reversing) symmetries, then the set $\mathcal{G}(L)$ of symmetries of the map L is a group (under composition). The set $\mathcal{R}(L)$ of reversing symmetries, however, is not a group under composition since the composition of two reversing symmetries yields a symmetry. Yet, since the composition of a symmetry and a reversing symmetry gives a reversing symmetry, the union of all symmetries and reversing symmetries of L is again a group. This group is denoted $\mathcal{E}(L)$ and called the *reversing symmetry group* of the map L [10].

The reversing symmetry group has a very particular structure. Namely, the symmetry group $\mathcal{G}(L)$ is a normal subgroup of $\mathcal{E}(L)$. Moreover, if L possesses a reversing symmetry S then

$$\mathcal{R}(L) = S\mathcal{G}(L). \tag{2.6}$$

Therefore,

$$\mathcal{E}(L)/\mathcal{G}(L) \simeq \mathbb{Z}_2, \tag{2.7}$$

implying that $\mathcal{G}(L)$ is a normal subgroup of index 2. Note that $\mathcal{E}(L)$ can be written as the semi-direct product of $\mathcal{G}(L)$ and $\langle S \rangle$ [3] if and only if S is an involution.

It is interesting to note that the group structure above has been encountered before in symmetry studies with a similar nature. For example, in theory of magnetic crystals the symmetry groups of spin patterns have an analogous structure. There are 'normal' symmetries that transform the spin pattern into itself, and reversing symmetries that transform the spin pattern into itself after an additional spin reversal (see e.g. [15]). Also in the theory of dichromatic (black and white) patterns, the notion of symmetries versus reversing symmetries has been discussed extensively. In this context, a classification of magnetic crystallographic groups on the plane is available [16]. Recently these groups were shown to be relevant in a dynamical systems application: the study of so-called stochastic webs [17, 18].

From the group structure of reversing symmetry groups a very convenient decomposition property follows [10]: a map L possesses a reversing symmetry K_0 of even or infinite order if and only if L can be written as

$$L = K_0 \circ K_1, \tag{2.8}$$

---

[3] $\langle a_1 \ldots a_n \rangle$ denotes the group generated by $a_1 \ldots a_n$ (and their inverses).

where K_0 and K_1 satisfy

$$K_0^2 \circ K_1^2 = Id. \tag{2.9}$$

From this result the well known observation of Birkhoff [6] that every reversible map (i.e. a map possessing an involutary reversing symmetry) is the composition of two involutions follows as a special (and most trivial) case.

As an example of this decomposition property let us consider the Anosov map (also known as Arnold's 'cat map' [19]), i.e. a linear map on a two-torus

$$\begin{pmatrix} x' \\ y' \end{pmatrix} = \begin{pmatrix} 1 & 1 \\ 1 & 2 \end{pmatrix} \cdot \begin{pmatrix} x \\ y \end{pmatrix} \bmod 1. \tag{2.10}$$

This map possesses a the rotation around $(0,0)$ over an angle $\pi/2$ as a reversing symmetry, inducing the following decomposition

$$\begin{pmatrix} 1 & 1 \\ 1 & 2 \end{pmatrix} = \begin{pmatrix} 0 & 1 \\ -1 & 0 \end{pmatrix} \cdot \begin{pmatrix} -1 & -2 \\ 1 & 1 \end{pmatrix}. \tag{2.11}$$

It is easily verified that the above decomposition satisfies (2.8) and (2.9). Fourfold reversing rotocenters are particularly interesting since they are almost always fixed saddle points [17], like in the case of the Anosov map.

The above approach includes in a natural way reversing symmetries within the scope of symmetry properties of a dynamical system. Note that the fact whether some reversing symmetry is an involution is not of crucial importance to the structure.

Symmetries and reversing symmetries both appear as symmetries in the phase portrait of a dynamical system since for instance every periodic orbit of the system is mapped onto another periodic orbit of the system, preserving the period of the orbit. In section 3 we will see however that there are even more properties that determine the symmetries of the phase portrait.

For a more detailed discussion on the structure of reversing symmetry groups, see Lamb [10].

2.2 KICKED SYSTEMS

In many applications, maps naturally are written as the composition of two distinct maps,

$$L = A \circ B, \quad \text{or} \quad L = A \circ B \circ A, \tag{2.12}$$

where A and B are invertible maps on a state space Ω. Important examples of such maps are stroboscopic maps of kicked systems like the Chirikov-Taylor standard map [20], the kicked rotator [21] and the Zaslavsky web maps [22].

Part of the reversing symmetry group of the map L follows from the symmetry properties of the constituting parts A and B.

Proposition 2.1 *If $\mathcal{G}(A_i)$ is the symmetry group of A_i, $i = 1 \ldots n$, then*

$$\mathcal{G}(A_1) \cap \ldots \cap \mathcal{G}(A_n) \leq \mathcal{G}(A_1 \circ \ldots \circ A_n). \tag{2.13}$$

Proposition 2.2 *If $\mathcal{R}(A)$ is the set of reversing symmetries of A and $\mathcal{R}(B)$ is the set of reversing symmetries of B, then*

$$\mathcal{R}(A) \cap \mathcal{R}(B) \subset \mathcal{R}(A \circ B \circ A), \tag{2.14}$$

and

$$[\mathcal{R}(A) \cap \mathcal{R}(B)] \circ B \subset \mathcal{R}(A \circ B), \tag{2.15}$$

$$A \circ [\mathcal{R}(A) \cap \mathcal{R}(B)] \subset \mathcal{R}(A \circ B). \tag{2.16}$$

Proof

If $S \in \mathcal{R}(A) \cap \mathcal{R}(B)$, we have

$$S \circ A \circ B \circ A = A^{-1} \circ B^{-1} \circ A^{-1} \circ S, \tag{2.17}$$

and

$$S \circ B \circ A \circ B = B^{-1} \circ A^{-1} \circ S \circ B. \tag{2.18}$$

Because $A \circ B$ is a symmetry of $A \circ B$, and $S \circ B$ is a reversing symmetry of $A \circ B$, their product

$$A \circ B \circ S \circ B = A \circ S, \tag{2.19}$$

is also a reversing symmetry of $A \circ B$. □

The result of proposition 2.1 is not very surprising, but the result of proposition 2.2 is more interesting. It mainly tells us that if two maps share a reversing symmetry S, their composition possesses a reversing symmetry that is not necessarily S itself, but is strongly related to S. As such, nonconventional (even nonlinear) time-reversal symmetry may be found to be caused by conventional time-reversal symmetry in case the dynamics is a concatenation of two types of time-reversible evolutions.

In [12, 23] it was already noted that a map is reversible if it is the composition of two maps that are reversible with respect to the same involution. Haake [24] moreover remarks that if the composition of two maps can be taken symmetrically (as in (2.14)), the shared reversing involution is preserved. Proposition 2.2 puts above findings in their natural context: that of the reversing symmetry group.

If we combine proposition 2.1 and proposition 2.2, it seems natural to regard the group

$$[\mathcal{G}(A) \cap \mathcal{G}(B)] \cup [\mathcal{R}(A) \cap \mathcal{R}(B)] \le \mathcal{E}(A) \cap \mathcal{E}(B), \tag{2.20}$$

This group is a reversing symmetry group of the symmetric map $A \circ B \circ A$, but not of the map $A \circ B$. However, although the reversing symmetries of $A \circ B$ differ from those of A and B, we find that

$$[\mathcal{G}(A) \cap \mathcal{G}(B)] \cup [\mathcal{R}(A) \cap \mathcal{R}(B)] \circ B, \tag{2.21}$$

$$[\mathcal{G}(A) \cap \mathcal{G}(B)] \cup A \circ [\mathcal{R}(A) \cap \mathcal{R}(B)], \tag{2.22}$$

are reversing symmetry groups of $A \circ B$, because both A and B commute with $[\mathcal{G}(A) \cap \mathcal{G}(B)]$ and $S^2 = (A \circ S)^2 = (S \circ B)^2$. Moreover, the groups (2.21) and (2.22) are isomorphic to (2.20). If $A = A' \circ A'$ such that A and A' have the same symmetry properties, we may regard $A' \circ B \circ A'$ instead of $A \circ B$, yielding equivalent dynamics. It can be easily seen that equations (2.20), (2.21) and (2.22) lead to the same reversing symmetry groups of L if we add the trivial symmetry L as a generator.

3   Reversing k-Symmetry Group

Recently it was found in various case studies of dynamical systems [21, 23, 25, 17] that symmetries and reversing symmetries might occur only if the system is considered on some specific time scale, i.e. a map L may possess less symmetries than L^k, for some integer value of k. In a recent work these type of generalized symmetries were discussed extensively by Lamb and Quispel [26].

Particular examples of this phenomenon are found in kicked systems satisfying some resonance condition. Before we enter details about that, let us first discuss the conceptional aspects more closely.

3.1   GENERAL CONCEPTS

Consider an invertible dynamical system $L : \Omega \mapsto \Omega$. Then the invertible map M is called a *k-symmetry* of L if k is the smallest integer such that

$$M \circ L^k \circ M^{-1} = L^k. \tag{3.1}$$

This is denoted as $\#_L(M) = k$. Analogously S is called a *revering k-symmetry* of L if k is the smallest positive integer such that

$$S \circ L^k \circ S^{-1} = L^{-k}, \tag{3.2}$$

and denoted $\#_L(S) = k$. Note that a (reversing) 1-symmetry of L is precisely a (reversing) symmetry of L.

These definitions imply that a (reversing) k-symmetry of L automatically is a (reversing) symmetry of L^k, but not vice-versa. We will denote the symmetry group of L^k as $\mathcal{G}_k$ and the reversing symmetry group of L^k as $\mathcal{E}_k$. The set of reversing symmetries of L^k will be denoted $\mathcal{R}_k$.

Any subgroup $\mathcal{D}$ of $\mathcal{E}_k$ can be given a number $\#_L(\mathcal{D})$ indicating the smallest integer $\tilde{k}$ (that divides k) for which all $U \in \mathcal{D}$ are elements of $\mathcal{E}_{\tilde{k}}$.

The notion of k-symmetries induces a map ϕ_L on $\mathcal{G}_k$: let $M \in \mathcal{G}_k$, then

$$\phi_L(M) = L \circ M \circ L^{-1}. \tag{3.3}$$

A k-symmetry of L corresponds to a k-cycle of ϕ_L in $\mathcal{G}_k$. Analogously we can define ϕ_L on $\mathcal{R}_k$ to act on a reversing symmetry S of L^k in the following way

$$\phi_L(S) = L \circ S \circ L. \tag{3.4}$$

Again, a reversing k-symmetry of L corresponds to a k-cycle of ϕ_L in $\mathcal{R}_k$.

We defined ϕ_L both on the reversing and non-reversing part of $\mathcal{E}_k$, so

$$\phi_L \; : \; \mathcal{E}_k \mapsto \mathcal{E}_k. \tag{3.5}$$

Note that ϕ_L is a group automorphism of $\mathcal{G}_k$, but not of $\mathcal{E}_k$.

The operator ϕ_L generates orbits on $\mathcal{G}_k$ and $\mathcal{R}_k$:

$$\Phi_L(U) = \{U_i \in \mathrm{Inv}(\Omega) \mid U_i = \phi_L^i(U), \; i \in \mathbb{Z}\}. \tag{3.6}$$

The natural embedding of such an orbit is its orbit group $\mathcal{O}_L(U)$, that is the group generated by all elements of the orbit of U under ϕ_L. Orbit groups are useful since ϕ_L works within them

$$\phi_L : \mathcal{O}_L(U) \mapsto \mathcal{O}_L(U). \tag{3.7}$$

Moreover, we have $\#_L(\mathcal{O}_L(U)) = k$ if and only if U is a (reversing) k-symmetry of L.

In general, $\mathcal{E}_k$ is called <u>the</u> *reversing k-symmetry group* of L if and only if $\mathcal{E}_k$ possesses a genuine (reversing) k-symmetry, i.e. $\#_L(\mathcal{E}_k) = k$. A subgroup $\mathcal{D}$ of $\mathcal{E}_k$ is called <u>a</u> reversing k-symmetry group if and only if $\mathcal{D}$ is closed under ϕ_L and $\#_L(\mathcal{D}) = k$.

In the phase portrait of a dynamical system all k-symmetries and reversing k-symmetries manifest themselves in the following way: let $\mathcal{P}$ be the set of all periodic orbits and U a (reversing) k-symmetry (for some $k \in \mathbb{N}$), then $U\mathcal{P} = \mathcal{P}$. U does not necessarily preserve the period of the orbits (but this is not visible in the phase portrait). Thus, (reversing) k-symmetries cause symmetries of phase portraits, see for an example figure 1. In this respect it is useful to define

$$\mathcal{G}_{\mathbb{N}} := \cup_{k \in \mathbb{N}} \mathcal{G}_k, \tag{3.8}$$

$$\mathcal{R}_{\mathbb{N}} := \cup_{k \in \mathbb{N}} \mathcal{R}_k, \tag{3.9}$$

$$\mathcal{E}_{\mathbb{N}} := \cup_{k \in \mathbb{N}} \mathcal{E}_k = \mathcal{G}_{\mathbb{N}} \cup \mathcal{R}_{\mathbb{N}}. \tag{3.10}$$

Ultimately, the phase portrait is symmetric with respect to $\mathcal{E}_{\mathbb{N}}$.

For a more detailed discussion on (reversing) k-symmetries, including a discussion on so-called k-symmetric orbits, we refer the reader to [26]. In the remaining part of this paper, we will focus on an explanation for the occurrence of (reversing) k-symmetries.

3.2 KICKED SYSTEMS

Now we want to extend the results of section 2.2 to the case of (reversing) k-symmetries. Therefore let us regard the action of ϕ_L if $L = A \circ B$ or $L = A \circ B \circ A$. The action on a symmetry of $\phi_{A \circ B}$ decomposes into

$$\phi_{A \circ B}(M) = [\phi_A \circ \phi_B](M). \tag{3.11}$$

For the action on a reversing symmetry S we derive

$$\phi_{A \circ B}(S \circ B) = [\phi_A \circ \phi_B](S) \circ B, \tag{3.12}$$

$$\phi_{A \circ B}(A \circ S) = A \circ [\phi_B \circ \phi_A](S), \tag{3.13}$$

$$\phi_{A \circ B \circ A}(S) = [\phi_A \circ \phi_B \circ \phi_A](S). \tag{3.14}$$

In general, finding periodic orbits of ϕ_L is not easy, but the above relations facilitate us with a means of tracking them down. Therefore we need to find a group of which all elements have finite orbits under ϕ_A and ϕ_B. Such a group has to be a subgroup of

$$[\mathcal{G}_{\mathbb{N}}(A) \cap \mathcal{G}_{\mathbb{N}}(B)] \cup [\mathcal{R}_{\mathbb{N}}(A) \cap \mathcal{R}_{\mathbb{N}}(B)]. \tag{3.15}$$

The orbits under ϕ_A and ϕ_B have to be finite within this invariant subgroup. However, this does not imply that orbits under $\phi_A \circ \phi_B$ or $\phi_A \circ \phi_B \circ \phi_A$ are finite. Other conditions are needed to ensure this. The following propositions deal with sufficient conditions for which we find (reversing) k-symmetries of $A \circ B$ or $A \circ B \circ A$ from (3.15).

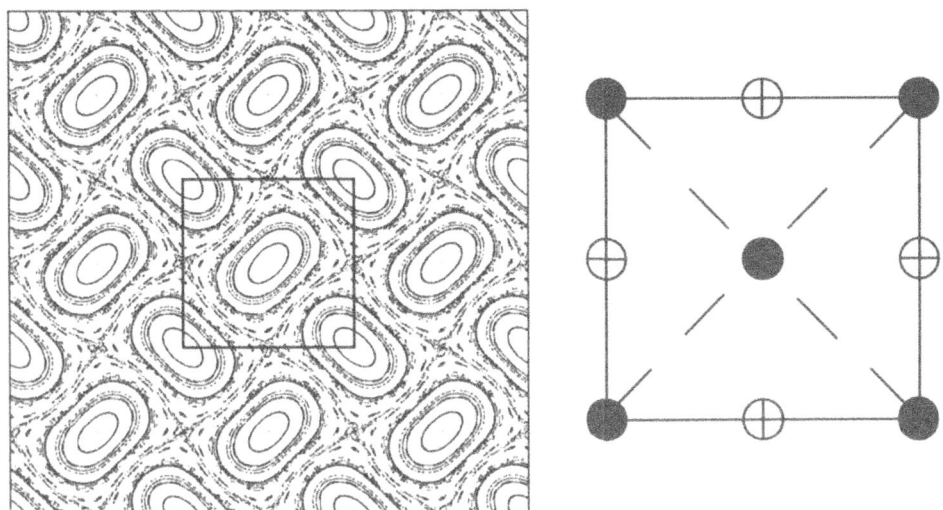

Figure 1: *Part of the phase portrait of the web map* (4.17) *with* $f(y) = \sin(2\pi y)$ *on* $[-1.5, 1.5]$ $\times$ $[-1.5, 1.5]$, *for* $\alpha = \pi/2$. *Also a square unit cell is drawn. The symmetry decoration of the square unit cell is indicated in the accompanying picture. Dashed lines denote reversing mirrors,* $\oplus$*'s fourfold reversing rotocenters and* $\bullet$*'s twofold rotocenters. The translation symmetries in the phase portrait are not (reversing) symmetries of the map but 4-symmetries.*

Proposition 3.1 *Let* Σ *be a group*

$$\Sigma \leq \mathcal{G}_{\mathbb{N}}(A) \cap \mathcal{G}(B), \tag{3.16}$$

satisfying

$$\phi_A(\Sigma) = \Sigma. \tag{3.17}$$

Then

$$\Sigma \leq \mathcal{G}_{\mathbb{N}}(L), \tag{3.18}$$

if L *is a composition of* A*'s and* B*'s. If* $L = A \circ B$, *and* $M \in \Sigma$, *then* $\#_L(M) = \#_A(M)$.

Proof
Let $M \in \Sigma$ and $\phi_A(\Sigma) = \Sigma$, then for all $i \in \mathbb{Z}$

$$[\phi_B \circ \phi_A^i](M) = \phi_A^i(M). \tag{3.19}$$

Hence, we find for all L consisting of A's and B's

$$L^k \circ M = \phi_L^k(M) \circ L^k = M \circ L^k. \square \tag{3.20}$$

Proposition 3.2 *Let* Θ *be a set*

$$\Theta \subset \mathcal{R}_{\mathbb{N}}(A) \cap \mathcal{R}(B), \tag{3.21}$$

such that

$$\phi_A(\Theta) = \Theta. \tag{3.22}$$

Then

$$\Theta \circ B \subset \mathcal{R}_{\mathbb{N}}(A \circ B), \tag{3.23}$$

$$A \circ \Theta \subset \mathcal{R}_{\mathbb{N}}(A \circ B), \tag{3.24}$$

and

$$\Theta \subset \mathcal{R}_{\mathbb{N}}(A \circ B \circ A). \tag{3.25}$$

If $L = A \circ B$ *and* $S \in \Theta$, *then* $\#_L(S) = \#_A(S)$.

Proof
Let $S \in \Theta$ and $\phi_A(\Theta) = \Theta$, then for all $i \in \mathbb{Z}$

$$[\phi_B \circ \phi_A^i](S) = \phi_A^i(S). \tag{3.26}$$

So we find for $L = A \circ B$

$$L^k \circ S \circ B = \phi_L^k(S \circ B) \circ L^k = (\phi_A \circ \phi_B)^k(S) \circ B \circ L^k = S \circ B \circ L^k. \tag{3.27}$$

In a similar way one proofs (3.24) and (3.25).$\square$
Note that a similar result applies in case $\Theta \subset \mathcal{R}(A) \cap \mathcal{R}_{\mathbb{N}}(B)$ and $\phi_B(\Theta) = \Theta$.

Proposition 3.3 *Let* Σ *be a group*

$$\Sigma \leq \mathcal{G}_{\mathbb{N}}(A) \cap \mathcal{G}_{\mathbb{N}}(B), \tag{3.28}$$

satisfying

$$\phi_A(\Sigma) = \phi_B(\Sigma) = \Sigma, \tag{3.29}$$

and Σ *is finite. Moreover, let* L *be composed of* A*'s and* B*'s, then*

$$\Sigma \leq \mathcal{G}_{\mathbb{N}}(L), \tag{3.30}$$

For $M \in \Sigma$, $\#_L(M)$ *can be at most the number of elements in* Σ *minus one.*

Proof
If Σ consists of a finite number of elements and ϕ_L maps Σ onto itself then it directly follows that every orbit of ϕ_L within Σ has finite length. $\square$

Proposition 3.4 *Let Θ be a set*

$$\Theta \subset \mathcal{R}_{\mathbb{N}}(A) \cap \mathcal{R}_{\mathbb{N}}(B), \tag{3.31}$$

such that

$$\phi_A(\Theta) = \phi_B(\Theta) = \Theta, \tag{3.32}$$

and Θ is finite. Then

$$\Theta \circ B \subset \mathcal{R}_{\mathbb{N}}(A \circ B), \tag{3.33}$$

$$A \circ \Theta \subset \mathcal{R}_{\mathbb{N}}(A \circ B), \tag{3.34}$$

and

$$\Theta \subset \mathcal{R}_{\mathbb{N}}(A \circ B \circ A). \tag{3.35}$$

For $S \in \Theta$, $\#_L(S)$ can be at most the number of elements in Θ.

Proof
The proof is analogous to the proof of proposition 3.3, using relations (3.12)-(3.14). $\square$

Above results can always be combined such that, if Θ is not empty, $\Theta = S \circ \Sigma$, for some $S \in \Theta$, yielding the group

$$[\Sigma \cup \Theta] \leq \mathcal{E}_{\mathbb{N}}(A) \cap \mathcal{E}_{\mathbb{N}}(B). \tag{3.36}$$

And we find

$$\Sigma \cup (\Theta \circ B) \ \leq \ \mathcal{E}_{\mathbb{N}}(A \circ B), \tag{3.37}$$

$$\Sigma \cup (A \circ \Theta) \ \leq \ \mathcal{E}_{\mathbb{N}}(A \circ B), \tag{3.38}$$

$$\Sigma \cup \Theta \ \leq \ \mathcal{E}_{\mathbb{N}}(A \circ B \circ A). \tag{3.39}$$

these groups are all isomorphic like in the case of (reversing) symmetries as found in section 2.1.

4 Some examples

4.1 THE KICKED ROTATOR

The equations of motion (of the components of the angular momentum) of a kicked rotator can be written (in a classical description) as [21]

$$L = A \circ B = R^z_{\alpha z} \circ R^y_\beta, \tag{4.1}$$

where

$$R^z_{\alpha z} : \begin{cases} x' = x \cos(\alpha z) - y \sin(\alpha z) \\ y' = y \cos(\alpha z) + x \sin(\alpha z) \\ z' = z \end{cases}, \qquad R^y_\beta : \begin{cases} x' = x \cos(\beta) + z \sin(\beta) \\ y' = y \\ z' = z \cos(\beta) - x \sin(\beta) \end{cases}. \tag{4.2}$$

This map is a result of two parts of the motion: a constant (free) precession around the y-axis and a periodic strain of impulsive non-linear kicks around the z-axis.

First we look at the symmetry properties of the two rotations separately. Symmetries of $R_{\alpha z}^z$ are all rotations around the z-axis, denoted by $\mathbf{R}^z$, and I_{xz}

$$I_{xz} : \begin{cases} x' = -x \\ y' = y \\ z' = -z \end{cases} . \tag{4.3}$$

Taking these symmetries together, we find that

$$\mathbf{R}^z \otimes_s \langle I_{xz} \rangle \leq \mathcal{G}(R_{\alpha z}^z), \tag{4.4}$$

where $\otimes_s$ denotes a semi-direct product.

The mirror in the x-z-plane, M_{xz}, is a reversing symmetry of $R_{\alpha z}^z$, so

$$\mathbf{R}^z \otimes_s \langle I_{xz}, M_{xz} \rangle \leq \mathcal{E}(R_{\alpha z}^z) \tag{4.5}$$

For R_β^y we find

$$\mathbf{R}^y \otimes_s \langle M_{xz} \rangle \quad \leq \quad \mathcal{G}(R_\beta^y) \tag{4.6}$$
$$\mathbf{R}^y \otimes_s \langle M_{xz}, I_{xy} \rangle \quad \leq \quad \mathcal{E}(R_\beta^y). \tag{4.7}$$

Using proposition 2.1 we obtain (noting that $I_{xz} = R_\pi^y$)

$$\langle I_{xz} \rangle \leq \mathcal{G}(R_{\alpha z}^z) \cap \mathcal{G}(R_\beta^y) \leq \mathcal{G}(L). \tag{4.8}$$

$I_{yz} \circ M_{xz} = M_{xy}$ is a reversing symmetry of both $R_{\alpha z}^z$ and R_β^y, so proposition 2.2 gives

$$\langle I_{xz}, M_{xy} \circ R_\beta^y \rangle \leq \mathcal{E}(L). \tag{4.9}$$

Using the results of the previous propositions, we might be able to find (reversing) k-symmetries of L if $R_{\alpha z}^z$ or R_β^y would have a (reversing) k-symmetry group, for some value of k. It is obvious that R_β^y possesses such a group if $\beta = 2\pi \frac{p}{q}$, where p and q are coprime integers. In fact, if q is even, $(R_\beta^y)^{\frac{q}{2}} = R_\pi^y = (R_\pi^y)^{-1}$, so

$$\mathbf{R}^y \otimes_s \langle M_{xz}, I_{xy} \rangle \leq \mathcal{G}_{\frac{q}{2}}(R_\beta^y) = \mathcal{E}_{\frac{q}{2}}(R_\beta^y). \tag{4.10}$$

Furthermore, $(R_\beta^y)^q = \mathbb{1}$, so all invertible maps on the state space are symmetries and reversing symmetries of $(R_\beta^y)^q$

$$\mathrm{Inv}(\mathbb{R}^3) = \mathcal{G}_q(R_\beta^y) = \mathcal{E}_q(R_\beta^y). \tag{4.11}$$

If we want to know the implications of these k-symmetries of R_β^y for the map (4.1), we have to find the orbits under $\phi_{R_\beta^y}$ that are in the symmetry group of $R_{\alpha z}^z$, to be able to make use of proposition 3.1. There are only two cases for which we can find such an orbit.

In case $\beta = \pi$ the orbit is trivial as I_{xy} is an ordinary symmetry of R_β^y. As I_{xy} is also a symmetry of $R_{\alpha z}^z$, it is a symmetry of L_π too (L_π denoting L from equation (4.1) with $\beta = \pi$).

$$\langle I_{xz}, I_{xy} \rangle \leq \mathcal{G}(L_\pi) \tag{4.12}$$

$$\langle I_{xz}, I_{xy}, M_{yz} \rangle \leq \mathcal{E}(L_\pi), \tag{4.13}$$

where $M_{yz} = M_{xy} \circ R_\pi^y$.

For $\beta = \frac{\pi}{2}$ we find that I_{xy} and I_{yz} form an orbit under $\phi_{R_\beta^y}$

$$\phi_{R_\beta^y}^2(I_{xy}) = \phi_{R_\beta^y}(I_{yz}) = I_{xy}. \tag{4.14}$$

As I_{xy} and I_{yz} are symmetries of $R_{\alpha z}^z$, they are 2-symmetries of $L_{\frac{\pi}{2}}$, i.e. L with $\beta = \frac{\pi}{2}$.

$$\langle I_{xz}, I_{xy} \rangle \leq \mathcal{G}_2(L_{\frac{\pi}{2}}) \tag{4.15}$$

$$\langle I_{xz}, I_{xy}, P_{xz} \rangle \leq \mathcal{E}_2(L_{\frac{\pi}{2}}), \tag{4.16}$$

where $P_{xz} = M_{xy} \circ R_{\frac{\pi}{2}}^y$.

4.2 WEB MAPS

The dynamics of a particle in a homogeneous magnetic field and a kicking inhomogeneous electric field is modelled by the so-called web map [22]

$$L_\alpha = R_\alpha \circ T, \tag{4.17}$$

where

$$R_\alpha : \begin{cases} x' = x\cos(\alpha) + y\sin(\alpha) \\ y' = y\cos(\alpha) - x\sin(\alpha) \end{cases}, \qquad T : \begin{cases} x' = x + f(y) \\ y' = y \end{cases}, \tag{4.18}$$

where $f(y)$ is unit periodic, $f(y+1) = f(y)$. Again we look at the symmetry properties of R_α and T separately.

For R_α we find that all rotations around the origin, R, are symmetries and the mirror in the x-axis, M_x, is a reversing symmetry,

$$\mathsf{R} \otimes_s \langle M_x \rangle \leq \mathcal{E}(R_\alpha). \tag{4.19}$$

Furthermore, if $\alpha = 2\pi \frac{p}{q}$, $p, q \in \mathbb{Z}$, then $(R_\alpha)^q = \mathbb{1}$, so all invertible maps on the state space are symmetries and reversing symmetries of $(R_\alpha)^q$,

$$\mathrm{Inv}(\mathbb{R}^2) = \mathcal{E}_q(R_\alpha). \tag{4.20}$$

In table 1 we made a list of symmetries and reversing symmetries of T, for different properties of $f(y)$ (in addition to $f(y+1) = f(y)$). In this table $\mathcal{U}_{r,n}$ denotes all translations

$$U_{r,n} : \begin{cases} x' = x + r \\ y' = y + n \end{cases}, \tag{4.21}$$

property of $f(y)$	sym. of T	rev. sym. of T	sym. of L	rev. sym. of L
-	$\mathcal{U}_{r,n}$	M_y	-	$R_\alpha \circ M_y$
$f(-y) = -f(y)$	$-1\!1, \mathcal{U}_{r,n}$	M_x, M_y	$-1\!1$	$R_\alpha \circ M_x, R_\alpha \circ M_y$
$f(y+\frac{1}{2}) = -f(y) = f(-y)$	$-1\!1, \mathcal{U}_{r,n}$	$M_x, M_y, \mathcal{U}_{r,n+\frac{1}{2}}$	$-1\!1$	$R_\alpha \circ M_x, R_\alpha \circ M_y$

Table 1: *Symmetries and reversing symmetries of T and $L = R_\alpha \circ T$, for different properties of $f(y)$, in addition to $f(y+1) = f(y)$.*

with $r \in \mathbb{R}$ and $n \in \mathbb{Z}$. The symmetries and reversing symmetries of L_α, also listed in this table, are again found by using proposition 2.1 and 2.2.

Looking for reversing k-symmetries makes sense for $\alpha = 2\pi \frac{p}{q}$. $\mathcal{U}_{r,n}$ are the candidates for symmetries of L_α^k. We look at the orbit of these under ϕ_{R_α}

$$\phi_{R_\alpha}(\mathcal{U}_{r,n}) = R_\alpha \circ \mathcal{U}_{r,n} \circ R_\alpha^{-1}. \tag{4.22}$$

In [17] it was shown that for all $i \in \mathbb{N}$, $\phi_{R_\alpha}^i(\mathcal{U}_{r,n}) = \mathcal{U}_{s,m}$ with $r, s \in \mathbb{R}$ and $n, m \in \mathbb{Z}$, if and only if $\alpha = 2\pi \frac{p}{q}$, $q \in \{1, 2, 3, 4, 6\}$. In fact, this result is strongly related to the well-known crystallographic restriction in $\mathbb{R}^2$, cf. [16]. For L_α we thus find 2-,3-,4- and 6-symmetries $\mathcal{U}_{r,n}$, depending on α. For example, for $\alpha = \frac{\pi}{2}$, we find that $\mathcal{U}_{1,0}$ and $\mathcal{U}_{0,1}$ are 4-symmetries.

For reversing symmetries of L_α^k we have the candidates $\mathcal{U}_{r,n+\frac{1}{2}}$, if $f(y+1/2) = -f(y)$. Again we look at the orbit under ϕ_{R_α}

$$\phi_{R_\alpha}(\mathcal{U}_{r,n+\frac{1}{2}}) = R_\alpha \circ \mathcal{U}_{r,n+\frac{1}{2}} \circ R_\alpha. \tag{4.23}$$

For $\alpha = \pi/2$ we find

$$\begin{aligned}
\phi_{R_{\frac{\pi}{2}}}^4(\mathcal{U}_{r,n+\frac{1}{2}}) &= \phi_{R_{\frac{\pi}{2}}}^3(-1\!1 \circ \mathcal{U}_{n+\frac{1}{2},-r}) = \phi_{R_{\frac{\pi}{2}}}^2(\mathcal{U}_{-r,-(n+\frac{1}{2})}) \\
&= \phi_{R_{\frac{\pi}{2}}}(-1\!1 \circ \mathcal{U}_{-(n+\frac{1}{2}),r}) = \mathcal{U}_{r,n+\frac{1}{2}}.
\end{aligned} \tag{4.24}$$

If $f(y)$ is odd, $-1\!1$ is a symmetry of T and all these orbit elements are reversing symmetries of T if $r = n+\frac{1}{2}$, $n \in \mathbb{Z}$. So we find that $R_{\frac{\pi}{2}} \circ \mathcal{U}_{n+\frac{1}{2},m+\frac{1}{2}}$, with $n, m \in \mathbb{Z}$, is a (noninvolutory) reversing 4-symmetry of $L_{\frac{\pi}{2}}$. Note that this can be written as a rotation over an angle $\pi/2$ around a point $(\frac{k}{2}, \frac{l}{2})$, with the condition that $k+l$ is odd.

We summarize these results in table 2. The (reversing) symmetry groups in this table are well-known crystallographic groups, see e.g. [17].

property of $f(y)$	4-symmetry group of $L_{\frac{\pi}{2}}$	reversing 4-symmetry group of $L_{\frac{\pi}{2}}$
-	$\langle \mathcal{U}_{1,0}, \mathcal{U}_{0,1} \rangle$	$\langle \mathcal{U}_{1,0}, \mathcal{U}_{0,1}, R_{\frac{\pi}{2}} \circ M_y \rangle$
$f(-y) = -f(y)$	$\langle -1\!1, \mathcal{U}_{1,0}, \mathcal{U}_{0,1} \rangle$	$\langle -1\!1, \mathcal{U}_{1,0}, \mathcal{U}_{0,1}, R_{\frac{\pi}{2}} \circ M_x \rangle$
$f(y+\frac{1}{2}) = -f(y) = f(-y)$	$\langle -1\!1, \mathcal{U}_{1,0}, \mathcal{U}_{0,1} \rangle$	$\langle \mathcal{U}_{1,0}, \mathcal{U}_{0,1}, R_{\frac{\pi}{2}} \circ \mathcal{U}_{\frac{1}{2},\frac{1}{2}}, R_{\frac{\pi}{2}} \circ M_y \rangle$

Table 2: *(Reversing) 4-symmetry groups of $L_{\frac{\pi}{2}}$ for different properties of $f(y)$, in addition to $f(y+1) = f(y)$.*

Figure 1 shows a part of the phase portrait of the web map with $f(y) = \sin(2\pi y)$ for $\alpha = \pi/2$. The reversing mirrors are easily recognized, but also the picture clearly shows translation symmetries, which are (reversing) 4-symmetries of the map.

5 Concluding remarks

In this paper we have shown how nonconventional reversing symmetries and (reversing) k-symmetries naturally arise in systems that can be written as the concatenation of different types of evolutions, such as kicked systems.

Our analysis shows that the occurrence of these nonconventional reversing symmetries and/or (reversing) k-symmetries in kicked systems is not due to some model artifact, but is caused by (physically motivated) model assumptions. In our examples of section 4 the constituting maps could all be written out explicitly. But even if this were not possible, in case A and B represent formal time-one maps of flows of (nonintegrable) differential equations, this would not lead to different answers.

The present analysis shows that (reversing) k-symmetries with $k > 1$ typically tend to occur in kicked systems of which the constituting evolutions are in resonance.

Acknowledgement

We are grateful to Marty Golubitsky for useful discussions and encouragement.

References

[1] Loschmidt J 1877, *Wien Ber.* **75**, 67

[2] Gibbs J W 1931, in : *The collected works of J. Willard Gibbs*, Longmans Green, New York (originally published 1902)

[3] Wigner E P 1932,Nachr.Ges.Wiss.Göttingen Math- physik K1 *32*, 546
ibid. 1959, *Group theory and applications to the quantum mechanics of atomic spectra*, Pure and Applied Physics vol.5, Academic Press, New York

[4] Kramers H A 1930, *Proc.Kon.Ned.Akad.Wetenschap.* **33**, 959

[5] Sachs R G 1987, *The physics of time reversal*, The University of Chicago Press, Chicago

[6] Birkhoff G D 1915, *Rend.Circ.Mat.Palermo* **39**, 265

[7] Birkhoff G D 1927, *Dynamical systems*, Am.Math.Soc., New York

[8] Devaney R L 1976, *Trans.Am.Math.Soc.* **218**, 89

[9] Sevryuk M B 1986, *Reversible systems*, Lecture Notes in Mathematics Vol. 1211, Springer, Berlin.

[10] Lamb J S W 1992, *J.Phys.A* **25**, 925

[11] Lamb J S W and Capel H W 1993, *Local bifurcations on the plane with reversing point group symmetry*, University of Amsterdam preprint ITFA 93-17, to appear in *Chaos,Solitons and Fractals*

[12] Roberts J A G and Quispel G R W 1992, *Phys.Rep.* **216**, 63

[13] Hoover W G 1985, *Phys.Rev.A* **31**, 1695

[14] Politi A, Oppo G L and Badii R 1986, *Phys.Rev.A* **33**, 4055

[15] Ludwig W and Falter C 1988, *Symmetries in physics*, Springer Series in Solid-State Sciences Vol. 64, Springer, Berlin.

[16] Loeb A L 1971, *Color and symmetry*, Wiley & Sons, New York.

[17] Lamb J S W 1993, *J.Phys.A* **26**, 2921

[18] Lamb J S W 1993, *Stochastic webs with fourfold rotation symmetry*, to appear in *Hamiltonian systems: integrability and chaotic behaviour*, ed. J. Seimenis, NATO ASI series B: Physics, Plenum Press, New York.

[19] Arnold V I and Avez A 1967, *Problèmes Ergodiques de la Mécanique Classique*, Gauthier-Villars, Paris

[20] Chirikov B V 1979, *Phys.Rep.* **216**, 63

[21] Haake F, Kús M and Scharf R 1987, *Z.Phys.B* **65**, 381

[22] Zaslavsky G M, Zakharov M Yu, Sagdeev R Z, Usikov D A and Chernikov A A 1986, *JETP Lett.* **44**, 451 [*Pis'ma Zh.Eksp.Teor.Fiz.* **44**, 349], ibid. 1986, *Sov.Phys. JETP* **64**, 294 [*Zh.Eksp.Teor.Fiz.* **91**, 500]

[23] Roberts J A G and Baake M 1993, *Trace maps as 3D reversible dynamical systems with an invariant*, *J.Stat.Phys.*, to appear in *J.Stat.Phys.*

[24] Haake F 1991, *Quantum signatures of chaos*, Springer, Berlin

[25] Hoveijn I 1992, *Chaos, Solitons and Fractals* **2**, 81

[26] Lamb J S W and Quispel G R W 1993, *Reversing k-Symmetries in Dynamical Systems*, to appear in *Physica D*

EXCLUSION OF RELATIVE EQUILIBRIA

REINER LAUTERBACH
Institut fur Angewandte Analysis
und Stochastik
Berlin GERMANY

PASCAL CHOSSAT
Institut Non Linéaire de Nice
CNRS-Université de Nice
Sophia-Antipolis FRANCE

ABSTRACT. We study the question of existence of relative equilibria in a **SO(3)**- equivariant dynamical system when the dynamics is governd by the $\ell = 4$ representation of this group. This answers a question which remained open in CHOSSAT ET AL. [1].

1   Group Orbits and Sections

In this section we give a proof for a certainly well known result. However we could not find a reference for it. The proof which is presented here is due to Mike Field.

Theorem 1.1 *Let G be a compact Lie group acting smoothly and orthogonally on a Hilbert space H. Denote the orbit through the point $x \in H$ by $\mathcal{O}_x$ and let $T_y\mathcal{O}_x$ be the tangent space to the manifold $\mathcal{O}_x$ at $y \in \mathcal{O}_x$. Let $W = T_y(\mathcal{O}_x)^{\perp}$ be the orthogonal complement to $T_y\mathcal{O}_x$ for some group orbit $\mathcal{O}_x$ and some point $y \in \mathcal{O}_x$. Then every group orbit intersects W.*

Proof: Let $w \in H$ be any point. Then $\mathcal{O}_w$ and $\mathcal{O}_x$ are both compact. If $w \notin \mathcal{O}_x$ then the two orbits are disjoint. Therefore there exist points $y' \in \mathcal{O}_x$ and $w' \in \mathcal{O}_w$, such that

$$\text{dist}(\mathcal{O}_x, \mathcal{O}_w) = \text{dist}(y', w').$$

We find a group element $g \in G$ such that $gy' = y$. By the orthogonality of the action we have

$$\text{dist}(y, gw') = \text{dist}(x', w').$$

P. Chossat (ed.), Dynamics, Bifurcation and Symmetry, 197–203.

Therefore $w = gy' - y$ is an element of W. Again, by the orthogonality of the group action $y \perp T_y \mathcal{O}_x$ and hence $gw' = y + w \in W$. $\qquad\qquad\qquad\qquad\qquad\square$

Definition 1.2 *We call the space W a* **global section** *to the group action.*

Remark 1.3 If the orbit $\mathcal{O}_x$ has the dimension of G one gets a maximal reduction of dimension. This reduction is similar to the reduction one obtains using the orbit space. Although the orbit space is the geometrically more natural object to study, in this particular case it seems to be easier to use the concept of a global section. It is not very wll suited to study more deep dynamical questions because group orbits intersect such a global section several times and the number of intersections depends on the group orbit.

2 Special Coordinates for Representations of O(3)

In order to understand the representations of $O(3)$ it suffices to look at the representatins of $SO(3)$ and to extend them in a straightforward manner, compare IHRIG & GOLUBITSKY [2]. One method to construct the lowest order equivariant mappings is proposen in SATTINGER [4] and used in various papers. It consists of solving the linear eequations for the coefficients coming from the action of the raising operator. To be more precise one determines the functions in the kernel of $\mathcal{J}_0$ and $\mathcal{J}_+$ and applies the lowering operator. In physics and in several mathematical papers (for example CHOSSAT et al. [1]) a normalization is used which is practical for certain purposes but leads to rather cumbersome calculations if one tries to study the zeros of these mappings. For this reason we want to use a different normalization, which leads to rational coefficients in the mappings. The main point is an appropriate choice of the basis such that these operators get a specific form of the lowering or raising operators, see MILLER [3], SATTINGER [4] for a definition of these operators and their relation to representation theory of $O(3)$. There are various ways of constructing the bases for the representation spaces. In MILLER [3], p. 233/234, one finds a construction of irreducible representations of $s\ell\,(2,\mathbb{R})$ and normalizations of the bases, which is solely based on the commutation relations in the Lie algebra. Since $s\ell\,(2,\mathbb{R})$ is isomorphic (as a Lie algebra) to the Lie algebra $so(3)$ of $SO(3)$ this construction yields a basis $\{f_{-\ell}, \ldots, f_0, \ldots, f_\ell\}$ of V_ℓ and operators $\mathcal{J}_0$, $\mathcal{J}_\pm$ such that $\mathcal{J}_0 f_m = m f_m$ $\mathcal{J}_- f_m = f_{m-1}$, and $J_+ f_m = (\ell - m)(\ell + m + 1)f_m$ for $m = -\ell, \ldots, \ell$. With the normalization $f'_m = \gamma_m f_m$, where

$$\frac{\gamma_m}{\gamma_{m+1}} = \frac{1}{\ell + m + 1}$$

the operators $\mathcal{J}_0$, $\mathcal{J}_\pm$ have the form (write f_m for f'_m again)

$$\mathcal{J}_0 f_m = m f_m, \quad \mathcal{J}_+ f_m = (\ell - m)f_{m+1}, \quad \mathcal{J}_- f_m = (\ell + m)f_{m-1}.$$

Obviously the operators $\mathcal{J}_\pm$ are nilpotent. The following commutation relations hold for these operators

$$[\mathcal{J}_0, \mathcal{J}_\pm] = \mathcal{J}_\pm$$

and

$$[\mathcal{J}_+, \mathcal{J}_-] = 2\mathcal{J}_0.$$

The functions f_m satisfy the reality condition

$$f_{-m} = (-1)^m f_m.$$

Therefore we have constructed an irreducible representation of the Lie algebra $s\,o(3)$. This yields an irreducible representation of $\mathbf{SO}(3)$ on this space. It is well known to be equivalent to the absolutely irreducible represention on the space of spherical harmonics of order ℓ. One other ingredient is the construction of an invariant inner product. Let $u = \sum_{m=-\ell}^{\ell} \alpha_m f_m$ and $v = \sum_{m=-\ell}^{\ell} \beta_m f_m$. Then

$$(u, v) = \sum_{m=-\ell}^{\ell} (-1)^m \sigma_m \alpha_m \overline{\beta_{-m}}$$

is an invariant inner product, if

$$\frac{\sigma_{m+1}}{\sigma_m} = \frac{\ell + m + 1}{\ell - m}.$$

That this inner product is invariant is easily seen, by proving

$$(\mathcal{J}u, v) + (u, \mathcal{J}v) = 0$$

for $\mathcal{J} \in \{\mathcal{J}_0, \mathcal{J}_\pm\}$.

3 A Global Section for $\ell = 4$

According to Theorem 1.1 we have to compute the orthogonal complement to an orbit at one point to find a global section. Let $v_0 \in V_4$ be such a point, with $z_{-4} = z_4 = 1$ and all the other components being zero. Then the ortogonal complement to the orbit is characterized by

$$\mathcal{O}(v_0)^\perp = \{v \in V \mid z_3 = z_{-3} = 0, \ z_4 + z_{-4} = 0\}.$$

Call this space W. It is a global section. Let us check this. Obviously it has codimension three, so if it is perpendicular to the group orbit it is a global section if the group orbit through such a point has dimension three. To see this we have to show that no linear combination of the the operators $\mathcal{J}_0$, $\mathcal{J}_\pm$ annihilates that point. So, assume that

$$\alpha \mathcal{J}_0 z + \beta_+ \mathcal{J}_+ z + \beta_- \mathcal{J}_- z = 0,$$

with $z_4 = z_{-4} = 1$ and all other components being zero. Since $\{\mathcal{J}_{tz}, \mathcal{J}_{\pm}z\}$ is linearly independent it follows that $\alpha = \beta_\pm = 0$.

Note, that W is not orthogonal to all orbits intersecting it.

4 The Exclusion of Solutions with Low Isotropy

The equations for $\ell = 4$ are

$$
\begin{aligned}
f_0(z, \lambda) &= \lambda z_0 - 45z_0^2 + 36z_{-1}z_1 + 22z_{-2}z_2 - 12z_{-3}z_3 - z_4z_{-4} \\
f_1(z, \lambda) &= \lambda z_1 - 45z_0z_1 + 60z_{-1}z_2 - 10z_{-2}z_3 - 5z_{-3}z_4 \\
f_2(z, \lambda) &= \lambda z_2 - 60z_1^2 + 55z_0z_2 + 20z_{-1}z_3 - 15z_{-2}z_4 \\
f_3(z, \lambda) &= \lambda z_3 - 70z_1z_2 + 105z_0z_3 - 35z_{-1}z_4 \\
f_4(z, \lambda) &= \lambda z_4 - 210z_2^2 + 280z_1z_3 - 70z_0z_4.
\end{aligned}
$$

Here we use the standard notation $z_m = x_m + iy_m$. With $f_{-m}(z, \lambda) = (-1)^m \overline{f_m(z, \lambda)}$ and $f = (f_{-\ell}, \ldots, f_0, \ldots, f_\ell)$ we have to solve $f(z, \lambda) = 0$. By construction it suffices to solve the set of equations $f_0 = 0, \ldots, f_\ell = 0$. We have $z_{-m} = (-1)^m \overline{z_m}$ for the real subspace. Therefore z_0 is a real variable and in addition λ is considered to be real. Restricting to the space W, orthogonal to a point with coordinates $z_4 = 1$, $z_{-4} = 1$ we get z_4 is real and $z_3 = z_{-3} = 0$ and hence the equations reduce to

$$\lambda z_0 - 45z_0^2 + 36z_{-1}z_1 + 22z_{-2}z_2 - x_4^2 = 0 \tag{1}$$

$$\lambda z_1 - 45z_0z_1 + 60z_{-1}z_2 = 0 \tag{2}$$

$$\lambda z_2 - 60z_1^2 + 55z_0z_2 - 15z_{-2}x_4 = 0 \tag{3}$$

$$2z_1z_2 + z_{-1}x_4 = 0 \tag{4}$$

$$\lambda x_4 - 210z_2^2 - 70z_0x_4 = 0 \tag{5}$$

Lemma 4.1 *All solutions of the equations (1) to (5) are in the subspace with $z_1 = 0$.*

Proof: If $z_2 = 0$ from (3) we get $z_1 = 0$. Therefore we assume $z_2 \neq 0$. Set $\alpha = \frac{z_1}{z_2}$, i.e. $z_1 = \alpha z_2$ and $z_{-1} = -\overline{z_1} = -\overline{\alpha z_2} = -\overline{\alpha} z_{-2}$. Then the second equation implies

$$\lambda \alpha z_2 - 45\alpha z_0 z_2 - 60\overline{\alpha} z_{-2} z_2 = 0 \tag{6}$$

or (since $z_2 \neq 0$)

$$\lambda \alpha - 45\alpha z_0 - 60\overline{\alpha} z_{-2} = 0. \tag{7}$$

Since $\alpha \neq 0$ we get

$$\lambda = 45z_0 + 60\frac{\overline{\alpha}}{\alpha}z_{-2}. \tag{8}$$

Since $\lambda,\ z_0$ are real numbers we find

$$\frac{\overline{\alpha}}{\alpha}z_{-2} \in \mathbb{R}. \tag{9}$$

Equation (4) yields

$$2\alpha z_z^2 - \overline{\alpha}z_{-2}x_4 = 0$$

and therefore

$$x_4 = 2\frac{\alpha}{\overline{\alpha}z_{-2}}z_2^2. \tag{10}$$

Due to (9) z_2^2 is real and $z_2 = iy_2$ and $z_{-2} = -iy_2$. We write $\alpha = \rho e^{i\beta}$ and from (9) we conclude

$$\frac{\overline{\alpha}}{\alpha}i \in \mathbb{R} \tag{11}$$

and hence $e^{-2i\beta} \in i\mathbb{R}$. Therefore $\beta = \varepsilon\frac{\pi}{4}$ with $\varepsilon = \pm 1$. Observe that $\frac{e^{2i\beta}}{i} = \varepsilon$. Altogether x_4 can be expressed as

$$x_4 = 2\frac{e^{2i\beta}}{i} = 2\varepsilon y_2. \tag{12}$$

Now we turn our attention to equation (5). We get with

$$\lambda = 45z_0 - 60e^{-2i\beta}iy_2$$

the equation

$$2(45z_0 - 60e^{-2i\beta}iy_2)\frac{e^{2i\beta}}{i}y_2 + 210y_2^2 - 140z_0\frac{e^{2i\beta}}{i}y_2 = 0. \tag{13}$$

From here we get

$$-50\varepsilon z_0y_2 + 90y_2^2 = 0. \tag{14}$$

This gives

$$y_2 = \varepsilon\frac{5}{9}z_0. \tag{15}$$

Now we express all terms in z_0. Especially

$$\lambda = 45z_0 - 60e^{-2i\beta}iy_2 = (45 - 60(-\varepsilon)\varepsilon\frac{5}{9})z_0 = (45 + \frac{100}{3}\varepsilon^2)z_0 = \frac{235}{3}z_0$$

Now, equation (1) is

$$\frac{100}{3}z_0^2 - 36\rho^2z_2z_{-2} + 22z_2z_{-2} - x_4^2 = 0. \tag{16}$$

We have

$$\frac{100}{3}z_0^2 - (36\rho^2 - 22)y_2^2 - 4y_2^2 = 0 \qquad (17)$$

and therefore

$$(100 + (22 - 4)\frac{25}{27})z_0^2 = 36\frac{25}{27}\rho^2 z_0^2 \qquad (18)$$

i.e. we find $z_0 = 0$ or

$$\rho^2 = \frac{7}{2}.$$

The first case is excluded because $z_0 = 0$ implies $y_2 = 0$ contradicting the fact $z_2 \neq 0$. Finally we have to consider (3). Note $\lambda = \frac{235}{3}z_0$, $z_2 = i\varepsilon\frac{5}{9}z_0$, $x_4 = \frac{10}{9}z_0$ and $z_1 = \rho\frac{5}{9}i\varepsilon e^{i\beta}z_0$ and finally $z_1^2 = -\rho^2\frac{25}{81}e^{2i\beta}z_0^2$. This last expression is equal to $z_1^2 = -\rho^2\frac{25}{81}\varepsilon i z_0^2$. Then equation (3) gives

$$
\begin{aligned}
0 &= \lambda z_2 - 60z_1^2 + 55z_0 z_2 - 15z_{-2}x_4 \\
&= \frac{235}{3}\frac{5}{9}\varepsilon i z_0^2 - 60(-\rho^2\varepsilon i)\frac{25}{81}z_0^2 + 55z_0(\frac{5}{9}\varepsilon i z_0) - 15(-\frac{5}{9}i\varepsilon z_0)\frac{10}{9}z_0 \\
&= \varepsilon i z_0^2(-\frac{1175}{27} + \rho^2\frac{500}{27} + \frac{825}{27} + \frac{250}{27}) \\
&= \frac{1}{27}i\varepsilon z_0^2(500\rho^2 - 100).
\end{aligned}
$$

It follows $\rho^2 = \frac{1}{5}$ contradicting the previous result. $\qquad\square$

Theorem 4.2 *All solutions of the set of equations (1) to (5) have maximal isotropy.*

Proof: If $z_1 = 0$ all solutions are found in the z_0, z_2, z_4–subspace. All solutions in this space have a maximal isotropy subgroup, compare CHOSSAT ET AL. [1] p. 341. Observe that this subspace is the fixed point space for some copy of a $\mathbb{Z}_2$ subgroup. $\qquad\square$

Remark 4.3 *This computation hints that a similar result should hold in higher representations with even ℓ as well. Using the global section one has more equations than variables. However it is not possible to use genericity arguments since one has only one quadratic equivariant mapping. This mapping either allow bifurcating relative equilibria or not.*

References

[1] P. CHOSSAT, R. LAUTERBACH & I. MELBOURNE. Steady-state bifurcation with O(3)-symmetry. *Arch. Rat. Mech. Anal.*, 113(4), 313-376, 1991.

[2] E. IHRIG & M. GOLUBITSKY. Pattern selection with O(3)-symmetry. *Physica 13D*, pages 1-13, 1984.

[3] W. MILLER. *Symmetry Groups and their Applications.* Academic Press, 1972.

[4] D. H. SATTINGER. *Group Theoretic Methods in Bifurcation Theory*, volume 762 of *Lecture Notes in Mathematics.* Springer Verlag, 1978.

BIFURCATION OF PERIODIC ORBITS IN 1:2 RESONANCE: A SINGULARITY THEORY APPROACH

VICTOR G. LEBLANC
Dept. of Applied Mathematics
University of Waterloo
CANADA

WILLIAM F. LANGFORD
Dept. of Mathematics and Statistics
University of Guelph
CANADA

ABSTRACT. The Liapunov-Schmidt reduction procedure is used to study the existence of periodic orbits in a parametrized family of autonomous differential equations near a 1:2 resonant equilibrium point. This corresponds to a Hopf-Hopf mode interaction where the imaginary eigenvalues are in 1 to 2 ratio. We assume the existence of a distinguished bifurcation parameter, and then use singularity theory in order to classify the generic perfect and perturbed (unfolded) bifurcation diagrams for periodic orbits.

1 Liapunov-Schmidt Reduction

We consider an ordinary differential equation on $\mathbb{R}^n$, $n \geq 4$

$$\dot{x} = f(x, \alpha) \tag{1.1}$$

where $x \in \mathbb{R}^n$ the state space, $\alpha \in \mathbb{R}^p$ the parameter space, and $f : \mathbb{R}^{n+p} \longrightarrow \mathbb{R}^n$ is smooth. We assume that (1.1) has an equilibrium point at the origin for $\alpha = 0$, i.e. $f(0,0) = 0$. Let $D_x f(0,0)$ denote the Jacobian matrix of f at this equilibrium point.

Hypothesis 1.1 $D_x f(0,0)$ *has simple eigenvalues* $\pm \omega i$, $\pm 2\omega i$, $\omega > 0$, *and no other eigenvalues on the imaginary axis.*

We can then perform a time rescaling in order for these eigenvalues to be respectively $\pm i$ and $\pm 2i$. Hypothesis 1.1, along with the implicit function theorem implies that there exists a unique equilibrium point $x = x(\alpha)$ of (1.1) for α near the origin, and $x(0) = 0$. Without loss of generality, we will translate this equilibrium point to the

P. Chossat (ed.), Dynamics, Bifurcation and Symmetry, 205–219.
© 1994 Kluwer Academic Publishers.

origin in order to get $f(0, \alpha) = 0$ for α close to 0 in $\mathbb{R}^p$. Finally, we will assume that we have made an appropriate change of basis in $\mathbb{R}^n$ so that

$$D_x f(0,0) = \left(\begin{array}{c|c} J & 0_{4 \times (n-4)} \\ \hline 0_{(n-4) \times 4} & B_{(n-4) \times (n-4)} \end{array} \right),$$

where

$$J = \begin{pmatrix} 0 & -1 & 0 & 0 \\ 1 & 0 & 0 & 0 \\ 0 & 0 & 0 & -2 \\ 0 & 0 & 2 & 0 \end{pmatrix},$$

$0_{p \times q}$ denotes the p by q matrix of zeroes and $B_{(n-4) \times (n-4)}$ is an $n - 4$ by $n - 4$ matrix with no eigenvalues on the imaginary axis.

Using Arnold's theory on unfoldings of matrices, (see [A 81], pp. 46-60) a versal unfolding of the matrix J is given by the 4-parameter family

$$\mathcal{J}_\mu = \begin{pmatrix} \mu_1 & -(1+\mu_2) & 0 & 0 \\ 1+\mu_2 & \mu_1 & 0 & 0 \\ 0 & 0 & \mu_3 & -2(1+\mu_4) \\ 0 & 0 & 2(1+\mu_4) & \mu_3 \end{pmatrix}. \tag{1.2}$$

We will assume that an appropriate change of basis in $\mathbb{R}^n$ (depending smoothly on α, [A 81]) has been made so that

$$D_x f(0, \alpha) = \left(\begin{array}{c|c} \mathcal{J}_\alpha & 0_{4 \times (n-4)} \\ \hline 0_{(n-4) \times 4} & B_{(n-4) \times (n-4)} \end{array} \right), \tag{1.3}$$

where $\mathcal{J}_\alpha$ is the same as in (1.2) with $\mu_j = \alpha_j$, $j = 1, 2, 3, 4$. We can therefore write (1.1) as

$$\dot{x} = D_x f(0, \alpha) \cdot x + h(x, \alpha),$$

where $D_x f(0, \alpha)$ is as in (1.3) and h consists of terms which are $O(\| x \|^2)$ uniformly in α. By rescaling time and relabeling parameters, we can achieve $\alpha_4 = 0$ and obtain

$$\mathcal{J}_\alpha = \begin{pmatrix} \alpha_1 & -(1+\alpha_2) & 0 & 0 \\ 1+\alpha_2 & \alpha_1 & 0 & 0 \\ 0 & 0 & \alpha_3 & -2 \\ 0 & 0 & 2 & \alpha_3 \end{pmatrix}. \tag{1.4}$$

Let Y denote the Banach space of continuous 2π-periodic functions from $\mathbb{R}$ to $\mathbb{R}^n$, and let $X \subset Y$ be the subspace of continuously differentiable functions. Define a mapping

$$\Phi : X \times \mathbb{R}^p \times \mathbb{R} \longrightarrow Y$$

by

$$\Phi(x(s), \alpha, \tau) = -(1 + \tau)\frac{dx}{ds}(s) + f(x(s), \alpha).$$

Obviously, $\Phi(0, 0, 0) = 0$, and we will use the standard Liapunov-Schmidt reduction method (see [GS 85], Chapter VII for a nice treatment) in order to solve $\Phi = 0$ in a neighborhood of $(0, 0, 0)$ in $X \times \mathbb{R}^p \times \mathbb{R}$. These solutions to $\Phi = 0$ correspond to small-amplitude periodic solutions of (1.1), with period equal to $2\pi/(1 + \tau)$.

Hypothesis 1.1 implies that the kernel K of the linearization of Φ at $(0, 0, 0)$ is 4-dimensional, and we thus reduce the search for solutions of $\Phi = 0$ to a search for solutions of a reduced bifurcation equation $g(z_1, z_2, \alpha, \tau) = 0$, where

$$g : \mathbb{C}^2 \times \mathbb{R}^p \times \mathbb{R} \longrightarrow \mathbb{C}^2.$$

The form of g is restricted by the fact that Φ commutes with the S^1 symmetry corresponding to phase shifts on elements of Y,

$$\theta(x(s)) = x(s - \theta), \quad \theta \in S^1, \ x \in Y. \tag{1.5}$$

The mapping g must commute with the action of S^1 on K which is induced by (1.5):

$$\theta(z_1, z_2) = (e^{-i\theta}z_1, e^{-2i\theta}z_2), \quad \theta \in S^1, \ z_1, z_2 \in \mathbb{C}.$$

A straightforward computation then shows that g must have the following form:

$$g(z_1, z_2) = (Az_1 + B\bar{z}_1 z_2, Cz_2 + Dz_1^2), \tag{1.6}$$

where A, B, C, D are complex-valued functions of $z_1\bar{z}_1$, $z_2\bar{z}_2$, $\mathrm{Re}(\bar{z}_1^2 z_2)$, $\mathrm{Im}(\bar{z}_1^2 z_2)$, α and τ. Moreover, we have

$$A(0, \ldots, 0) = C(0, \ldots, 0) = 0,$$

$$A_\tau(0, \ldots, 0) = -i, \quad C_\tau(0, \ldots, 0) = -2i. \tag{1.7}$$

2 The Bifurcation Equations

Periodic solutions of (1.1) with period close to 2π are in 1:1 correspondence with solutions in $\mathbb{C}^2$ of the *bifurcation equation* $g(z_1, z_2) = 0$, which by (1.6) can be written in the form

$$Az_1 + B\bar{z}_1 z_2 = 0 \tag{2.1}$$
$$Cz_2 + Dz_1^2 = 0 \tag{2.2}$$

We notice that if $z_1 = 0$, equation (2.1) is identically satisfied and (2.2) reduces to

$$P(z_2\bar{z}_2, \alpha, \tau)z_2 = 0, \tag{2.3}$$

where $P(z_2\bar{z}_2, \alpha, \tau) = C(0, z_2\bar{z}_2, 0, 0, \alpha, \tau)$. This is the bifurcation equation for the *classical* π-periodic solutions (i.e. corresponding to the $2i$ eigenvalue for which the classical Hopf bifurcation theorem holds), and has been studied in [GL 81]. We will thus concentrate on solutions of (2.1)-(2.2) such that $z_1 \neq 0$. Using the S^1 symmetry of these equations, we can rotate z_1 to the real number $r > 0$. Equations (2.1)-(2.2) then reduce to

$$A + Bz_2 = 0 \tag{2.4}$$
$$Cz_2 + Dr^2 = 0, \tag{2.5}$$

where now A, B, C, D are complex-valued functions of r^2, $z_2\bar{z}_2$, $r^2\mathrm{Re}(z_2)$, $r^2\mathrm{Im}(z_2)$, α and τ.

We assume that the following generic condition is satisfied:

Hypothesis 2.1 $B(0, \dots, 0) \equiv b \neq 0$.

If we rescale z_2 by b, we can use the implicit function theorem to solve (2.4) for z_2 as a function of τ, r^2 and α, and substitute into (2.5) to get

$$\tilde{C}(r^2, \tau, \alpha)w(r^2, \tau, \alpha) + \tilde{D}(r^2, \tau, \alpha)r^2 = 0, \tag{2.6}$$

where

$$\tilde{C}(r^2, \tau, \alpha) = C(r^2, z_2\bar{z}_2, r^2\,\mathrm{Re}\,(z_2), r^2\,\mathrm{Im}\,(z_2), \alpha, \tau),$$
$$\tilde{D}(r^2, \tau, \alpha) = b\,D(r^2, z_2\bar{z}_2, r^2\,\mathrm{Re}\,(z_2), r^2\,\mathrm{Im}\,(z_2), \alpha, \tau),$$

$z_2 = z_2(r^2, \tau, \alpha)$, $z_2(0, 0, 0) = 0$.

In a controlled physical experiment, we usually choose to vary only one auxiliary parameter at a time. For this reason, we will assume that we have a distinguished parameter λ, and then study the resulting bifurcation diagrams in (2.6). The parameters α_1, α_2, and α_3 in $\mathcal{J}_\alpha$ (see (1.4)) are assumed to be smooth functions of λ in a neighborhood of $\lambda = 0$:

$$\alpha_1(\lambda) = \varepsilon_1 + \alpha_1'(0)\lambda + O(\lambda^2)$$

$$\alpha_2(\lambda) = \varepsilon_2 + \alpha_2'(0)\lambda + O(\lambda^2) \tag{2.7}$$

$$\alpha_3(\lambda) = \varepsilon_3 + \alpha_3'(0)\lambda + O(\lambda^2).$$

When $\varepsilon_1 = \varepsilon_2 = \varepsilon_3 = 0$, we have a "perfect" 1:2 resonance bifurcation. In analogy to the classical Hopf bifurcation theorem, we assume the following generic hypothesis:

Hypothesis 2.2

$$\alpha_1'(0) \neq 0, \quad \alpha_3'(0) \neq 0.$$

Using Hypothesis 2.2, we can rescale λ by a positive factor in order to achieve

$$\alpha_1'(0) = \beta_1 \neq 0, \quad \alpha_3'(0) = \pm 1.$$

If we denote $\beta_2 \equiv \alpha_2'(0)$, the Liapunov-Schmidt reduction yields

$$\mathrm{Re}\,(A_\lambda(0,\dots,0)) = \beta_1, \quad \mathrm{Re}\,(C_\lambda(0,\dots,0)) = \pm 1,$$
$$\mathrm{Im}\,(A_\lambda(0,\dots,0)) = \beta_2, \quad \mathrm{Im}\,(C_\lambda(0,\dots,0)) = 0.$$

Using (1.7), equation (2.6) becomes

$$(-2i\tau \pm \lambda + \varepsilon_3 + p_1)(i\tau - \beta_1\lambda - i\beta_2\lambda - \varepsilon_1 - i\varepsilon_2 + p_2) + b\tilde{D}r^2 = 0, \qquad (2.8)$$

where p_1 and p_2 are at least quadratic in $\tau, \lambda, r, \varepsilon_1, \varepsilon_2, \varepsilon_3, \alpha_4, \dots, \alpha_p$. Let us denote the left-hand side of equation (2.8) by $\xi(\tau, r^2, \lambda, \varepsilon_1, \varepsilon_2, \varepsilon_3, \alpha_4, \dots, \alpha_p)$, and denote

$$\eta_1(\tau, r^2, \lambda, \varepsilon_1, \varepsilon_2, \varepsilon_3, \alpha_4, \dots, \alpha_p) = -2i\tau \pm \lambda + \varepsilon_3 + p_1$$

and

$$\eta_2(\tau, r^2, \lambda, \varepsilon_1, \varepsilon_2, \varepsilon_3, \alpha_4, \dots, \alpha_p) = i\tau - \beta_1\lambda - i\beta_2\lambda - \varepsilon_1 - i\varepsilon_2 + p_2.$$

We will need the following proposition, characterizing the number of solutions to (2.8) when $r = 0$ (later, we will see that these solutions correspond to points of bifurcation of 2π-periodic solutions from the origin (pure mode) or from the classical π-periodic solution (period-doubling bifurcation)):

Proposition 2.1 *For $(\varepsilon_1, \varepsilon_2, \varepsilon_3, \alpha_4, \dots, \alpha_p)$ close enough to the origin in $\mathbb{R}^p$, there are 2 and only 2 solutions in (τ, λ) (possibly a double solution) to the equation*

$$\xi(\tau, 0, \lambda, \varepsilon_1, \varepsilon_2, \varepsilon_3, \alpha_4, \dots, \alpha_p) = 0. \qquad (2.9)$$

<u>Proof</u>: Obviously, (2.9) is satisfied if and only if

$$\eta_1(\tau, 0, \lambda, \varepsilon_1, \varepsilon_2, \varepsilon_3, \alpha_4, \dots, \alpha_p) = 0, \quad \text{or}$$
$$\eta_2(\tau, 0, \lambda, \varepsilon_1, \varepsilon_2, \varepsilon_3, \alpha_4, \dots, \alpha_p) = 0.$$

It is also obvious that $\eta_1(0,\dots,0) = 0$ and $\eta_2(0,\dots,0) = 0$. It is then a simple case of applying the implicit function theorem (remembering that $\beta_1 \neq 0$) to conclude that there exist unique smooth functions

$$T_j(\varepsilon_1, \varepsilon_2, \varepsilon_3, \alpha_4, \dots, \alpha_p), \ \Lambda_j(\varepsilon_1, \varepsilon_2, \varepsilon_3, \alpha_4, \dots, \alpha_p), \ j = 1, 2$$

such that

$$\eta_1(T_1, 0, \Lambda_1, \varepsilon_1, \varepsilon_2, \varepsilon_3, \alpha_4, \dots, \alpha_p) \equiv 0, \quad \text{and}$$
$$\eta_2(T_2, 0, \Lambda_2, \varepsilon_1, \varepsilon_2, \varepsilon_3, \alpha_4, \dots, \alpha_p) \equiv 0$$

for $(\varepsilon_1, \varepsilon_2, \varepsilon_3, \alpha_4, \dots, \alpha_p)$ close enough to the origin in $\mathbb{R}^p$. $\qquad \square$

3 Singularity Theory

We will first study the case of "perfect" bifurcation; that is, we will set $\varepsilon_1 = \varepsilon_2 = \varepsilon_3 = \alpha_4 = \cdots = \alpha_p = 0$ in (2.8). Viewing the real and imaginary parts of (2.8) as the components of a real 2-dimensional mapping, we get

$$\begin{aligned}
2\tau^2 - 2\beta_2\lambda\tau - \delta_1\beta_1\lambda^2 + d_1 r^2 + q_1(\tau, r^2, \lambda) &= 0 \\
(\delta_1 + 2\beta_1)\lambda\tau - \delta_1\beta_2\lambda^2 + d_2 r^2 + q_2(\tau, r^2, \lambda) &= 0,
\end{aligned} \tag{3.1}$$

where $\delta_1 = \pm 1$, q_1 and q_2 are at least cubic in τ, λ and r, and $d_1 + i\, d_2 = b\tilde{D}(0, \ldots, 0)$.

The singularity theory we will be using is essentially the same as in [GSS 88] with a few minor modifications, and it will be assumed that the reader is familiar with the concepts of Chapters XIV and XV of that book. We also refer to [D 84]. In the case of 1:1 non-semisimple resonance, Furter ([F 92]) used a variation of the singularity theory presented here.

Let $\mathcal{E}$ denote the ring of $\mathbb{Z}_2$-invariant germs of smooth functions $g : (\mathbb{R}^{2+1}, 0) \longmapsto \mathbb{R}$ satisfying $g(\tau, -r, \lambda) = g(\tau, r, \lambda)$. We will also need $\mathcal{E}_\lambda$ which is the ring of germs of smooth functions in λ. Let $\vec{\mathcal{E}}$ denote the $\mathcal{E}$-module of maps (g_1, g_2), where g_1 and g_2 are in $\mathcal{E}$. We will denote the maximal ideal in $\mathcal{E}$ by $\mathcal{M}$.

We define a group of smooth changes of coordinates on $\vec{\mathcal{E}}$ by

$$(S, X, \Lambda)((g_1, g_2)) = S \cdot (g_1(X, \Lambda), g_2(X, \Lambda)), \tag{3.2}$$

where $S \in \overleftrightarrow{\mathcal{E}}$, the $\mathcal{E}$-module of matrices generated by

$$\begin{pmatrix} 1 & 0 \\ 0 & 0 \end{pmatrix}, \begin{pmatrix} 0 & 1 \\ 0 & 0 \end{pmatrix}, \begin{pmatrix} 0 & 0 \\ 1 & 0 \end{pmatrix}, \begin{pmatrix} 0 & 0 \\ 0 & 1 \end{pmatrix},$$

X is of the form $X(\tau, r, \lambda) = (X_1(\tau, r^2, \lambda), rX_2(\tau, r^2, \lambda))$, where $X_1(0, 0, 0) = 0$, and $\Lambda \in \mathcal{E}_\lambda$ with $\Lambda(0) = 0$, $\Lambda'(0) > 0$. These changes of coordinates will preserve the λ-direction of bifurcating branches, as well as the $\mathbb{Z}_2$-invariance of the mapping. We will not prescribe any restrictions on the signs of S and X, as we are not interested (in this paper) in the stability of the bifurcating branches. Two mappings in $\vec{\mathcal{E}}$ are said to be *strongly $\mathbb{Z}_2$-equivalent* if they are related by a change of coordinates (S, X, Λ) as in (3.2).

The mapping defined by the left-hand side of (3.1) belongs to $\vec{\mathcal{E}}$. Later, we will need the fact that the mapping (q_1, q_2) belongs to the submodule $(\mathcal{M}^3 + \mathcal{M} < u >) \cdot \vec{\mathcal{E}}$, where $u = r^2$.

Definition 3.1 *The bifurcation problem (3.1) will be called* non-degenerate *if*

a) $\beta_1 \neq 0$ *(see Hypothesis 2.2)*

b) $d_2 \neq 0$

c) $d_2\beta_1 - d_1\beta_2 \neq 0$

d) $\delta_1 + 2\beta_1 \neq 0$

e) $4\beta_1^2 a^2 + 8\delta_1\beta_1 + 8a\beta_1\beta_2 + 4\delta_1\beta_1 a^2 + 4\beta_2^2 - 4\delta_1 a\beta_2 + a^2 \neq 0$,

where $a = d_1/d_2$.

Theorem 3.1 *Suppose the bifurcation problem (3.1) is non-degenerate. Then it is strongly $\mathbb{Z}_2$-equivalent to*

$$g_{\delta_2 \delta_3} = (2\tau^2 + \theta\lambda\tau + \delta_2\lambda^2, \ \lambda\tau + \delta_3 r^2), \tag{3.3}$$

where $\delta_2 = -\delta_1 \mathrm{sgn}(\beta_1)$, $\delta_3 = \mathrm{sgn}(d_2)$ and

$$\theta = -\frac{a + 4\beta_1\beta_2 + 4\beta_1^2 a - 2\delta_1\beta_2 + 4\delta_1\beta_1 a}{\sqrt{|\beta_1| [(\delta_1 + 2\beta_1)^2 + 4\beta_2^2]}} \tag{3.4}$$

<u>Proof:</u> Consider $g + tq \in \vec{\mathcal{E}}$, where

$$g = (g_1, g_2) = (2\tau^2 - 2\beta_2\lambda\tau - \delta_1\beta_1\lambda^2 + d_1 r^2, \ (\delta_1 + 2\beta_1)\lambda\tau - \delta_1\beta_2\lambda^2 + d_2 r^2),$$

$$q = (q_1, q_2) \in (\mathcal{M}^3 + \mathcal{M} < u >) \cdot \vec{\mathcal{E}}$$

and $t \in [0, 1]$. For any $h = (h_1, h_2) \in \vec{\mathcal{E}}$, we define the *restricted tangent space* of h, $\vec{RT}(h)$, to be the submodule of $\vec{\mathcal{E}}$ generated by

$$(h_1, 0), \ (0, h_1), \ (h_2, 0), \ (0, h_2), \ r(h_{1,r}, h_{2,r}), \ \tau(h_{1,\tau}, h_{2,\tau}), \ \lambda(h_{1,\tau}, h_{2,\tau}), \ u(h_{1,\tau}, h_{2,\tau}).$$

We will first show that $\vec{RT}(g + tq) = \vec{RT}(g)$ for all $t \in [0, 1]$. This implies (slight modification of [GSS 88], Theorem 1.3, p.168) that $g + tq$ is strongly $\mathbb{Z}_2$-equivalent to g for all $t \in [0, 1]$, in particular, $g + q$ is strongly $\mathbb{Z}_2$-equivalent to g. We will need the following

Lemma 3.1 $\mathcal{M} \cdot \vec{RT}(g + tq) = (\mathcal{M}^3 + \mathcal{M} < u >) \cdot \vec{\mathcal{E}}, \ \forall t \in [0, 1]$.

<u>Proof of Lemma</u>: Since $\vec{RT}\,(g+tq) \subset (\mathcal{M}^2 + <u>)\cdot \vec{\mathcal{E}}$, it follows that

$$\mathcal{M}\cdot \vec{RT}\,(g+tq) \subset (\mathcal{M}^3 + \mathcal{M} <u>)\cdot \vec{\mathcal{E}}\,.$$

To show that $(\mathcal{M}^3 + \mathcal{M} <u>)\cdot \vec{\mathcal{E}} \subset \mathcal{M}\cdot \vec{RT}\,(g+tq)$, we will instead show that

$$(\mathcal{M}^3 + \mathcal{M} <u>)\cdot \vec{\mathcal{E}} \subset \mathcal{M}\cdot \vec{RT}\,(g+tq) + (\mathcal{M}^4 + \mathcal{M}^2)\cdot \vec{\mathcal{E}}$$

and then use Nakayama's lemma. However, since $tq \in (\mathcal{M}^3 + \mathcal{M} <u>)\cdot \vec{\mathcal{E}}$, then

$$\mathcal{M}\cdot \vec{RT}\,(g+tq) + (\mathcal{M}^4 + \mathcal{M}^2)\cdot \vec{\mathcal{E}} = \mathcal{M}\cdot \vec{RT}\,(g) + (\mathcal{M}^4 + \mathcal{M}^2)\cdot \vec{\mathcal{E}}\,.$$

We then write the 24 generators of $\mathcal{M}\cdot \vec{RT}\,(g)$ in terms of the 14 generators of $(\mathcal{M}^3 + \mathcal{M} <u>)\cdot \vec{\mathcal{E}}$. A lengthy computation shows that the resulting 24 by 14 matrix has rank 14 if g is non-degenerate. This ends the proof of the Lemma.

¿From the Lemma and the fact that $(q_1, q_2) \in (\mathcal{M}^3 + \mathcal{M} <u>)\cdot \vec{\mathcal{E}}$, we conclude that

$$\vec{RT}\,(g+tq) = W + (\mathcal{M}^3 + \mathcal{M} <u>)\cdot \vec{\mathcal{E}},$$

where W is the finite-dimensional real subspace of $\vec{\mathcal{E}}$ spanned by

$$(g_1, 0),\ (0, g_1),\ (g_2, 0),\ (0, g_2),\ r(g_{1,r}, g_{2,r}),\ \tau(g_{1,\tau}, g_{2,\tau}),\ \lambda(g_{1,\tau}, g_{2,\tau}),\ u(g_{1,\tau}, g_{2,\tau}).$$

Since this is independent of tq, we conclude that

$$\vec{RT}\,(g+tq) = \vec{RT}\,(g),\ \forall\, t \in [0,1].$$

Finally, by performing the following explicit changes of coordinates on r, τ, λ and (g_1, g_2), we can transform g into $g_{\delta_2\,\delta_3}$.

$$r \to \frac{r}{\sqrt{|\,d_2\,|}},\quad (g_1, g_2) \to (g_1 - \frac{d_1}{d_2}g_2,\ g_2),\quad \tau \to \frac{\tau + \delta_1\beta_2\lambda}{\delta_1 + 2\beta_1},$$

$$(g_1, g_2) \to ((\delta_1 + 2\beta_1)^2\, g_1,\ g_2),\quad \lambda \to \frac{\lambda}{\sqrt{|\,\beta_1\,|\,[(\delta_1 + 2\beta_1)^2 + 4\beta_2^2]}},$$

$$(g_1, g_2) \to (g_1,\ \sqrt{|\,\beta_1\,|\,[(\delta_1 + 2\beta_1)^2 + 4\beta_2^2]}\, g_2),$$

$$r \to \frac{r}{\sqrt{\sqrt{|\,\beta_1\,|\,[(\delta_1 + 2\beta_1)^2 + 4\beta_2^2]}}}.$$

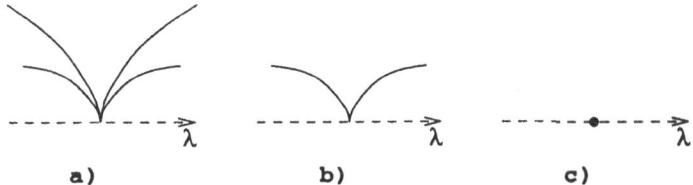

Figure 1: Perfect bifurcation diagrams

□

Notice that the non-degeneracy condition e) in Definition 3.1 is equivalent to $\theta^2 + 8 \neq 0$ if $\delta_2 = -1$ and $\theta^2 - 8 \neq 0$ if $\delta_2 = 1$. Hence, this condition specifies that the set $2\tau^2 + \theta\lambda\tau + \delta_2\lambda^2 = 0$ must be either two distinct lines through the origin in the (τ, λ)-plane, or simply the origin (i.e. hyperbolic or elliptic type). In the elliptic case, the only solution to $g_{\delta_2\,\delta_3} = 0$ is the origin. In the hyperbolic case, a solution (line) of $2\tau^2 + \theta\lambda\tau + \delta_2\lambda^2 = 0$ will yield a solution branch of $g_{\delta_2\,\delta_3} = 0$ provided that $r^2 = -\delta_3\lambda\tau$ is positive on this line (this depends on the slope of the line).

The bifurcation diagrams contained in the normal forms (3.3) are given in Figure 1. We get either 0, 1 or 2 branches of fundamental 2π-periodic solutions bifurcating from the origin on both sides of $\lambda = 0$. Specifically, Figure 1 a) corresponds to g_{++} with $\theta > \sqrt{8}$ and g_{+-} with $\theta < -\sqrt{8}$, Figure 1 b) corresponds to g_{-+} and g_{--} for any θ, and Figure 1 c) corresponds to g_{++} with $\theta < \sqrt{8}$ and g_{+-} with $\theta > -\sqrt{8}$. Note that this may appear to contrast with the result of [S 78], where 0, 2 or 4 branches were found. However, in that paper the super- and sub-critical branches are counted separately, and for a fixed value of μ (the bifurcation parameter) there are either 0, 1 or 2 solutions.

We will now study the perturbations of the bifurcation diagrams of Figure 1. This will be accomplished via unfolding theory.

Let $g \in \vec{\mathcal{E}}$. A k-parameter $\mathbb{Z}_2$-unfolding, G, of g is a germ of a smooth map $(\mathbb{R}^{2+1+k}, 0) \rightarrow \mathbb{R}^2$ such that $G(\tau, u, \lambda, 0) = g(\tau, u, \lambda)$. We will denote by $\vec{\mathcal{E}}_k$ the space of all k-parameter $\mathbb{Z}_2$-unfoldings of all elements of $\vec{\mathcal{E}}$. Let $G \in \vec{\mathcal{E}}_k$ and $\mathcal{G} \in \vec{\mathcal{E}}_\ell$ be respectively k-parameter and ℓ-parameter unfoldings of a given $g \in \vec{\mathcal{E}}$. We will say that $\mathcal{G}$ factors through G if

$$\mathcal{G}(\tau, u, \lambda, \gamma) = S(\tau, u, \lambda, \gamma) \cdot G(X(\tau, u, \lambda, \gamma), \Lambda(\lambda, \gamma), A(\gamma)),$$

where S is a 2 by 2 matrix with entries depending smoothly on τ, u, λ and γ satisfying $S(\tau, u, \lambda, 0) = Id_2$, $X(\tau, u, \lambda, \gamma) = (X_1(\tau, u, \lambda, \gamma), rX_2(\tau, u, \lambda, \gamma))$ where X_1 and X_2 are smooth functions satisfying $X_1(\tau, u, \lambda, 0) = \tau$, $X_2(\tau, u, \lambda, 0) = 1$, Λ is a smooth function satisfying $\Lambda(\lambda, 0) = \lambda$, and $A : \mathbb{R}^\ell \to \mathbb{R}^k$ is a smooth mapping satisfying $A(0) = 0$.

A $\mathbb{Z}_2$-unfolding $G \in \vec{\mathcal{E}}_k$ of $g \in \vec{\mathcal{E}}$ is said to be *versal* if every other $\mathbb{Z}_2$-unfolding of g factors through G. If G is versal and depends on the minimum number of parameters, then G is said to be *universal*. This minimum number of parameters in a universal unfolding is called the *codimension* of g in $\vec{\mathcal{E}}$.

We define the *tangent space* of $g \in \vec{\mathcal{E}}$ to be the subspace

$$T(g) = \vec{RT}(g) + \mathcal{E}_\lambda \cdot \{g_\lambda\} + \mathbb{R} \cdot \{(g_{1,\tau}, g_{2,\tau})\}.$$

Proposition 3.1 *Let* $g, \rho_1, \rho_2, \ldots, \rho_k$ *be elements of* $\vec{\mathcal{E}}$*. Then*

$$G(\tau, u, \lambda, \gamma_1, \ldots, \gamma_k) \equiv g + \sum_{i=1}^{k} \gamma_i \, \rho_i$$

is a universal unfolding of g *if and only if*

$$\vec{\mathcal{E}} = T(g) \oplus \mathbb{R} \cdot \{\rho_1, \ldots, \rho_k\}.$$

Proof: The proof which is given in [GSS 88], Chapter XV, §7 − 8 needs only a slight modification from the equivariant context there to the invariant context here.

Theorem 3.2 *Let* $g_{\delta_2 \delta_3}$ *be as in (3.3). Then*

$$G(\tau, r^2, \lambda, \gamma_1, \gamma_2, \gamma_3, \gamma_4, \tilde{\theta}) = (2\tau^2 + \tilde{\theta}\lambda\tau + \delta_2\lambda^2 + \gamma_1\lambda + \gamma_2\tau + \gamma_3, \ \lambda\tau + \delta_3 r^2 + \gamma_4)$$

is a universal unfolding of $g_{\delta_2 \delta_3}$*, where* $(\gamma_1, \gamma_2, \gamma_3, \gamma_4, \tilde{\theta})$ *is close to* $(0,0,0,0,\theta)$*.*

Proof: From Theorem 3.1, we can write

$$T(g_{\delta_2 \delta_3}) = (\mathcal{M}^3 + \mathcal{M} < u >) \cdot \vec{\mathcal{E}} + V,$$

where V is the finite-dimensional subspace of $\vec{\mathcal{E}}$ spanned by

$$(2\tau^2 + \theta\lambda\tau + \delta_2\lambda^2, 0), \ (0, 2\tau^2 + \theta\lambda\tau + \delta_2\lambda^2), \ (\lambda\tau + \delta_3 u, 0), \ (0, \lambda\tau + \delta_3 u),$$

$$(0, 2\delta_3 u), \ (4\tau^2 + \theta\lambda\tau, \lambda\tau), \ (4\tau\lambda + \theta\lambda^2, \lambda^2), \ (\theta\tau + 2\delta_2\lambda, \tau),$$

$$(\theta\lambda\tau + 2\delta_2\lambda^2, \lambda\tau), \ (4\tau + \theta\lambda, \lambda).$$

Let $\tilde{V}$ be the finite-dimensional subspace of $\vec{\mathcal{E}}$ spanned by

$$(1, 0), \ (\tau, 0), \ (\lambda, 0), \ (\lambda\tau, 0), \ (0, 1).$$

215

The conclusion of the Theorem will follow if we can show that a basis for $V + \tilde{V}$ is

$$\{(1,0), (\tau,0), (\lambda,0), (u,0), (\tau^2,0), (\lambda\tau,0), (\lambda^2,0),$$
$$(0,1), (0,\tau), (0,\lambda), (0,u), (0,\tau^2), (0,\lambda\tau), (0,\lambda^2)\}$$

This involves the verification that the 15 by 14 matrix

$$\begin{pmatrix}
0 & 0 & 0 & 0 & 2 & \theta & \delta_2 & 0 & 0 & 0 & 0 & 0 & 0 & 0 \\
0 & 0 & 0 & 0 & 0 & 0 & 0 & 0 & 0 & 0 & 0 & 2 & \theta & \delta_2 \\
0 & 0 & 0 & \delta_3 & 0 & 1 & 0 & 0 & 0 & 0 & 0 & 0 & 0 & 0 \\
0 & 0 & 0 & 0 & 0 & 0 & 0 & 0 & 0 & 0 & \delta_3 & 0 & 1 & 0 \\
0 & 0 & 0 & 0 & 0 & 0 & 0 & 0 & 0 & 0 & 2\delta_3 & 0 & 0 & 0 \\
0 & 0 & 0 & 0 & 4 & \theta & 0 & 0 & 0 & 0 & 0 & 0 & 1 & 0 \\
0 & 0 & 0 & 0 & 0 & 4 & \theta & 0 & 0 & 0 & 0 & 0 & 0 & 1 \\
0 & 0 & 0 & 0 & 0 & \theta & 2\delta_2 & 0 & 0 & 0 & 0 & 0 & 1 & 0 \\
0 & 4 & \theta & 0 & 0 & 0 & 0 & 0 & 0 & 1 & 0 & 0 & 0 & 0 \\
0 & \theta & 2\delta_2 & 0 & 0 & 0 & 0 & 0 & 1 & 0 & 0 & 0 & 0 & 0 \\
1 & 0 & 0 & 0 & 0 & 0 & 0 & 0 & 0 & 0 & 0 & 0 & 0 & 0 \\
0 & 1 & 0 & 0 & 0 & 0 & 0 & 0 & 0 & 0 & 0 & 0 & 0 & 0 \\
0 & 0 & 1 & 0 & 0 & 0 & 0 & 0 & 0 & 0 & 0 & 0 & 0 & 0 \\
0 & 0 & 0 & 0 & 0 & 1 & 0 & 0 & 0 & 0 & 0 & 0 & 0 & 0 \\
0 & 0 & 0 & 0 & 0 & 0 & 0 & 1 & 0 & 0 & 0 & 0 & 0 & 0
\end{pmatrix}$$

has full rank. The 14 by 14 matrix that results from deleting the 8th row of this matrix has determinant equal to $-16\,\delta_2\delta_3 \neq 0$. $\qquad\square$

Note that the parameter $\tilde\theta$ is a modal parameter (see [GS 85], Chapter V for a discussion).

By varying $(\gamma_1, \gamma_2, \gamma_3, \gamma_4, \tilde\theta)$ in a neighborhood of $(0,0,0,0,\theta)$, we generate all perturbations of the bifurcation diagrams in Figure 1. However, we must beware of the following technical difficulty. There are parameter values arbitrarily close to $(0,0,0,0,\theta)$ (for any θ) for which $G(\tau, 0, \lambda, \gamma_1, \gamma_2, \gamma_3, \gamma_4, \tilde\theta) = 0$ has 0, 2 or 4 distinct solutions in (τ, λ). From sections §1 and §2, we know that all possible small perturbations of a perfect 1:2 resonance bifurcation satisfying Hypotheses 2.1 and 2.2 are contained in equation (2.8). Proposition (2.1) states that there are 2 and only 2 solutions in (τ, λ) to equation (2.8) when $r = 0$. Thus, there are some perturbations contained in the universal unfolding G which do not correspond to anything in equation (2.8). This is directly related to the fact that there are some elements of $\vec{\mathcal{E}}$ which can not be expressed in the form (2.8) even after performing changes of coordinates. Therefore, when studying the bifurcation diagrams contained in G, we need only consider these which are such that there are 2 and only 2 solutions in (τ, λ) when $r = 0$:

these and these alone have relevance for the 1:2 resonance bifurcation. We can then draw the possible perturbed bifurcation diagrams by considering all cases for which the conic sections $2\tau^2 + \tilde{\theta}\lambda\tau + \delta_2\lambda^2 + \gamma_1\lambda + \gamma_2\tau + \gamma_3 = 0$ and $\lambda\tau + \gamma_4 = 0$ intersect at 2 points and only 2 points. One of these intersections corresponds to $\eta_1 = 0$ and the other to $\eta_2 = 0$ (see Proposition (2.1)). Notice that $\eta_2 = 0$ is equivalent to $w = 0$, so this intersection corresponds to a "pure" fundamental 2π-periodic solution bifurcating from the origin. On the other hand, $\eta_1 = 0$ corresponds to $P = 0$ (see equation (2.3)). Therefore this intersection corresponds to a period-doubling bifurcation from the classical π-periodic solution.

We find there are 28 persistent bifurcation diagrams $G = 0$ which satisfy the above criteria. In Figure 2, we give 14 of these, and the remaining 14 are obtained from these by transforming $\lambda \rightarrow -\lambda$. A hollow square indicates a period-doubling bifurcation from the classical solution, while a solid dot indicates the bifurcation of a pure 2π-periodic mode from the origin.

4 Conclusions

Recently in [GMSD 93], the general problem of $k : \ell$ resonance was studied for time-reversible systems. In this case, the non-degeneracy conditions a) and b) of Definition 3.1, as well as the crossing condition $\alpha_3'(0) \neq 0$ (see Hypothesis 2.2) are not satisfied. Therefore our results do not contain the results of that paper. Hamiltonian systems also violate these non-degeneracy conditions.

Our results on perfect 1:2 resonance bifurcation are in agreement with [Schmidt], with the caveat that was mentioned in section §3. Our work completes that paper in

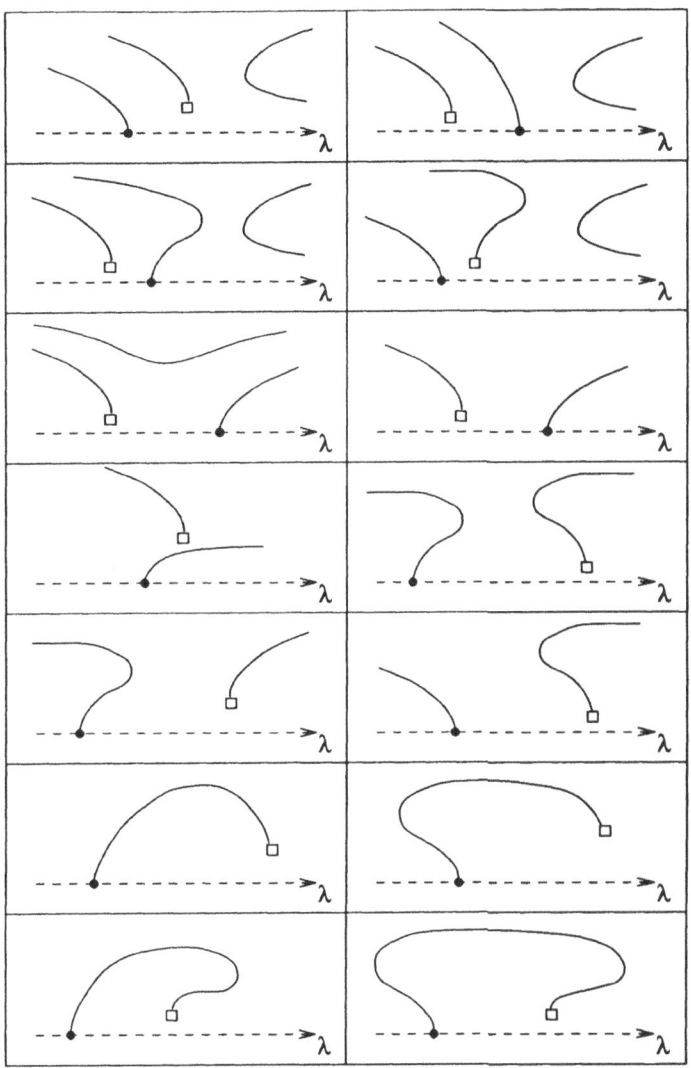

Figure 2: Persistent perturbed bifurcation diagrams

the sense that we classify the imperfect (unfolded) bifurcation diagrams.

Our approach has a few drawbacks. First, although Hypothesis 2.1 is generic, our singularity theory can not classify degeneracies where this condition is violated since the existence of the bifurcation equation (2.8) depends on this condition being satisfied. Second, it would be desirable to define a "restricted" unfolding (possibly depending on less than 5 parameters) which would describe all the perturbations contained in (2.8) and only these. It would also be advantageous to associate these unfolding parameters to the $\varepsilon_j = \alpha_j(0)$, $j = 1, 2, 3$ (see (2.7)). We are currently working on a variation of our approach which appears to be able to deal with these issues.

Acknowledgments: VGL is grateful to the Natural Sciences and Engineering Research Council of Canada (NSERC) for a 1967 Science and Engineering Scholarship, and WFL thanks NSERC for research grant support. This work was completed while both authors were visitors at the Fields Institute for Research in Mathematical Sciences, which is supported by NSERC and the Ontario Ministry of Education and Training.

References

[A 81] Arnold, V.I. 1981. On Matrices Depending on Parameters. In *Singularity Theory: Selected Papers*. London Mathematical Society Lecture Note Series, 53, pp. 46-60. Cambridge University Press.

[D 84] Damon, J. 1984. The unfolding and determinacy theorems for subgroups of $\mathcal{A}$ and $\mathcal{K}$. *Memoirs AMS*. 306. Providence.

[F 92] Furter, J.E. 1992. Hopf bifurcation at non-semisimple eigenvalues: a singularity theory approach. In *International Series of Numerical Mathematics*. Vol. 104, pp. 135-145. Birkhäuser Verlag Basel.

[GL 81] Golubitsky, M. and Langford, W.F. 1981. Classification and Unfoldings of Degenerate Hopf Bifurcations. *J. Differential Equations*. Vol. 41, No. 3, pp. 375-415.

[GMSD 93] Golubitsky, M., Marsden, J.E., Stewart, I. and Dellnitz, M. 1993. The Constrained Liapunov-Schmidt Procedure and Periodic Orbits. To appear in *Normal Forms and Homoclinic Chaos, W.F. Langford and W.K. Nagata eds.* Fields Institute Communications. AMS, Providence.

[GS 85] Golubitsky, M. and Schaeffer, D. G. 1985. *Singularities and Groups in Bifurcation Theory*. *Vol. 1* Springer-Verlag, Berlin / New York.

[GSS 88] Golubitsky, M. , Stewart , I. and Schaeffer, D. G. 1988. *Singularities and Groups in Bifurcation Theory*. *Vol. 2*. Springer-Verlag, Berlin / New York.

[S 78] Schmidt, D.S. 1978. Hopf's Bifurcation Theorem and the Center Theorem of Liapunov with Resonance Cases. *J. Math. Anal. Appl.* Vol. 63, pp. 354-370.

HAMILTONIAN STRUCTURE OF THE REVERSIBLE NONSEMISIMPLE 1:1 RESONANCE

J.C. VAN DER MEER
Faculteit Wiskunde en Informatica, Technische Universiteit Eindhoven
The Netherlands

J.A. SANDERS
Faculteit Wiskunde en Informatica, Vrije Universiteit Amsterdam
The Netherlands

A. VANDERBAUWHEDE
Vakgroep voor fundamentele en computergerichte wiskunde
Universiteit Gent
Belgium

ABSTRACT. We show that a reversible non-Hamiltonian vector field at nonsemisimple 1:1 resonance can be split into a Hamiltonian and a non-Hamiltonian part in such a way that after reduction to the orbit space for the S^1-action coming from the semisimple part of the linearized vector field the non-Hamiltonian part vanishes. As a consequence the reduced reversible vector field is Hamiltonian. We furthermore show that for vector fields in normal form on the orbit space being Hamiltonian is equivalent to being reversible.

1. Introduction

We consider a reversible - but non-Hamiltonian - vector field in the neighborhood of a 1:1 resonance. There has been a recent interest in the reversible nonsemisimple 1:1 resonance (See [5], [10], [6], [3] and other references given in this last paper) which occurs when two pairs of purely imaginary eigenvalues of the linearized system collide on the imaginary axis. From the analysis of the bifurcation in [6] it becomes clear that there is an apparent resemblance with the Hamiltonian Hopf bifurcation [12]. Part of the resemblance has been clarified in [3] where it was shown that, on the orbit space with respect to the S^1-action coming from the semisimple part of the

221

P. Chossat (ed.), Dynamics, Bifurcation and Symmetry, 221–240.
© 1994 *Kluwer Academic Publishers.*

linearized vector field, the reduced vector field was Hamiltonian with respect to a "hidden" Poissonstructure.

In [5, eqn. (46)] the equations of interest are given in complex form by

$$\frac{dA}{dt} = i(\omega_0 + \hat{P}(u, \Omega, \mu))A + B,$$

$$\frac{dB}{dt} = i(\omega_0 + \hat{P}(u, \Omega, \mu))B + \hat{Q}(u, \Omega, \mu))A, \tag{1}$$

where $(A, B) \in C^2$, $u = |A|^2$, $\Omega = \text{Im}(\bar{A}B)$. $\hat{P}$ and $\hat{Q}$ are real polynomials in u and Ω with μ-dependent coefficients, such that $\hat{P}(0, 0, 0) = \hat{Q}(0, 0, 0) = 0$.

Note that at $\mu = 0$ the linearized system has a pair of purely imaginary eigenvalues $\pm i\omega_0$ with two dimensional Jordan blocks. The reversibility is present in the symmetry $\tilde{R}$ defined by $\tilde{R}(A, B) = (\bar{A}, -\bar{B})$ which anticommutes with the vector field. The vector field is equivariant with respect to the S^1-action $\theta \cdot (A, B) = (e^{i\theta}A, e^{i\theta}B)$, which is the flow of the semisimple part of the linearized vector field. Furthermore the normalized vector field has two absolute invariants:

$$\Omega, H = -\frac{1}{2}|B|^2 + \frac{1}{2}\int_0^u \hat{Q}(s, \Omega, \mu)ds \tag{2}$$

In order to compare these equations to other results we will first change to real coordinates. Set $A = y_1 + iy_2$ and $B = x_1 + ix_2$. Then we obtain the equations

$$\frac{dx}{dt} = \omega_0 Jx + \hat{P}(u, \Omega, \mu)Jx + \hat{Q}(u, \Omega, \mu)y,$$

$$\frac{dy}{dt} = \omega_0 Jy + \hat{P}(u, \Omega, \mu)Jy + x, \tag{3}$$

where

$$J = \begin{pmatrix} 0 & -1 \\ 1 & 0 \end{pmatrix}, u = y_1^2 + y_2^2, \Omega = x_1 y_2 - x_2 y_1. \tag{4}$$

Note that the *reversor* we consider is now given by

$$R = \begin{pmatrix} -1 & 0 & 0 & 0 \\ 0 & 1 & 0 & 0 \\ 0 & 0 & 1 & 0 \\ 0 & 0 & 0 & -1 \end{pmatrix}. \tag{5}$$

In the following sections we will first show that equations (3) are actually the general normal form for reversible systems on $\mathbb{R}^4$ at nonsemisimple 1:1 resonance. After that we will show that the Poissonstructure on the orbit space is natural and that for vector fields in normal form on the orbit space being Hamiltonian is equivalent to being reversible, if the appropriate choices are made for the *reversor* and the Poisson structure. We furthermore show that the Hamiltonian structure can be identified explicitly in the original non-Hamiltonian reversible vector field. For the basics on Hamiltonian systems resp. Poisson structures we refer the reader to the textbooks [1] and [7].

2. Linear normal form

In this section we consider the linear normal form of a reversible system at non-semisimple 1:1 resonance. We show that the linear system and the *reversor* can be simultaneously normalized. A more general version of this can be found in [9].

Consider a system

$$\dot{z} = f(z, \lambda), \ z \in \mathbb{R}^4, \ \lambda \in \mathbb{R}^p, \tag{6}$$

where $f(0, \lambda) = 0$, $\forall \lambda$. Define $A(\lambda) = D_z f(0, \lambda) \in \mathcal{L}(\mathbb{R}^4)$. We call system (6) reversible if

$$f(Rz, \lambda) = -Rf(z, \lambda), \ R \in \mathcal{L}(\mathbb{R}^4), \ R^2 = I.$$

Let $A_0 = A(0)$ then obviously $A_0 R = -R A_0$.

Now suppose that $\pm i$ are non-semisimple eigenvalues of A_0, i.e. $\dim \ker(A_0^2 + I) = 2$ and $\ker(A_0^2 + I)^2 = \mathbb{R}^4$.

LEMMA 1. *Let $A_0 = A_S + A_N$ be the Jordan decomposition of A_0. Then A_S and A_N are reversible, i.e. $A_S R = -R A_S$ and $A_N R = -R A_N$.*

Proof. From $A_0 = A_S + A_N$ and $A_0 = -R A_0 R$ we find $A_0 = -R A_S R - R A_N R$. Now $-R A_S R$ is semisimple, $-R A_N R$ is nilpotent, and $-R A_S R$ and $-R A_N R$ commute. Because of the uniqueness of the Jordan decomposition it follows that $A_S = -R A_S R$ and $A_N = -R A_N R$. □

Let $U = \ker(A_0^2 + I)$, thus $\dim U = 2$. Then $K = A_0|_U \in \mathcal{L}(U)$ is semisimple, thus $K = A_S|_U$ and $A_N|_U = 0$. Furthermore $A_S^2 = -I$, that is, A_S generates an S^1-action on $\mathbb{R}^4$ given by $\{e^{A_S \phi} \mid \phi \in S^1 \cong \mathbb{R}/2\pi \mathbb{Z}\}$. This S^1-action together with

R generates an $O(2)$-action on $\mathbb{R}^4$. U is invariant under this action, actually it is irreducible under this action. Next set $R_0 = R|_U \in \mathcal{L}(U)$, then

$$R_0^2 = I_U \text{ and } KR_0 = -R_0K.$$

Now let $\mathbb{R}^4 = U \oplus V$, where V is chosen to be the $O(2)$-invariant complement to U with respect to the given $O(2)$-action. Consequently V is invariant under A_S and R. Let $\hat{K} = A_S|_V \in \mathcal{L}(V)$ and $\hat{R}_0 = R|_V \in \mathcal{L}(V)$. Then

$$\hat{K}\hat{R}_0 = -\hat{R}_0\hat{K}, \ \hat{R}_0^{\,2} = I_V, \text{ and } \hat{K}^2 = -I_V.$$

Clearly $\dim V = 2$, and V is irreducible under the S^1- and $O(2)$-actions.

LEMMA 2. $\hat{A}_N = A_N|_V \in \mathcal{L}(V,U)$ is an isomorphism.

Proof. Consider $W = A_N(V)$. W is invariant under the $O(2)$-action since $A_N A_S = A_S A_N$ and $A_N R = -RA_N$. Because A_N commutes with the S^1-action it follows from Schur's lemma that either $A_N(V) = 0$ or that $A_N|_V \in \mathcal{L}(V,W)$ is an isomorphism. If $A_N(V) = 0$ then $A_N = 0$, which is impossible since this implies $A_0 = A_S$ is semisimple. Consequently $W = \text{Im} A_N$ and $U = \ker A_N$. If $U \cap W = \{0\}$ then A_N is invertible on its image W, which is impossible since A_N is nilpotent. Thus $U \cap W$ is a nontrivial subspace of U, which is invariant under the $O(2)$-action. From this it follows that $U \cap W = U$ and consequently $U = W$. Thus we may conclude that $\hat{A}_N = A_N|_V \in \mathcal{L}(V,U)$ is an isomorphism. $\qquad\square$

From $A_S A_N = A_N A_S$ we get that $K\hat{A}_N = \hat{A}_N K$, while from $A_N R = -RA_N$ we obtain that $\hat{A}_N R_0 = -R_0 \hat{A}_N$.

LEMMA 3. There exists a linear isomorphism

$$\Phi : U \times U \to \mathbb{R}^4$$

such that

$$\Phi^{-1} A_0 \Phi(u_1, u_2) = (Ku_1 + u_2, Ku_2)$$

and

$$\Phi^{-1} R \Phi(u_1, u_2) = (R_0 u_1, -R_0 u_2).$$

Proof. Define $\Phi \in \mathcal{L}(U \times U; I\!R^4)$ by

$$\Phi(u_1, u_2) = u_1 + \hat{A}_N^{-1} u_2,$$

obviously, by lemma 2, ϕ is an isomorphism. We have

$$
\begin{aligned}
\Phi^{-1} A_0 \Phi(u_1, u_2) &= \Phi^{-1}(A_S + A_N)(u_1 + \hat{A}_N^{-1} u_2) \\
&= \Phi^{-1}(K u_1 + \hat{K} \hat{A}_N^{-1} u_2 + u_2) \\
&= \Phi^{-1}(K u_1 + u_2 + \hat{A}_N^{-1} \hat{K} u_2) \\
&= (K u_1 + u_2, K u_2),
\end{aligned}
$$

and

$$
\begin{aligned}
\Phi^{-1} R \Phi(u_1, u_2) &= \Phi^{-1} R(u_1 + \hat{A}_N^{-1} u_2) \\
&= \Phi^{-1}(R_0 u_1 + \hat{R}_0 \hat{A}_N^{-1} u_2) \\
&= \Phi^{-1}(R_0 u_1 + \hat{A}_N^{-1} R_0 u_2) \\
&= (R_0 u_1, -R_0 u_2).
\end{aligned}
$$

$\square$

Identifying $U \times U$ with $I\!R^4$ we can assume that

$$A_0 = \begin{pmatrix} K & I_U \\ 0 & K \end{pmatrix}, \text{ and } R = \begin{pmatrix} R_0 & 0 \\ 0 & -R_0 \end{pmatrix}. \tag{7}$$

LEMMA 4. *U has a basis $\{e_1, e_2\}$ such that*

$$K e_1 = e_2, \; K e_2 = -e_1, \; R_0 e_1 = e_1, \; R_0 e_2 = -e_2,$$

i.e. K and R_0 are represented by the matrices

$$K \rightarrow J = \begin{pmatrix} 0 & -1 \\ 1 & 0 \end{pmatrix} \text{ and } R_0 \rightarrow \begin{pmatrix} 1 & 0 \\ 0 & -1 \end{pmatrix}.$$

Proof. Since $R_0^2 = I_U$, R_0 has only eigenvalues $\epsilon = \pm 1$. If ϵ is an eigenvalue of R_0 with eigenvector u_0, then $R_0 u_0 = \epsilon u_0$ and $R_0(K u_0) = -K R_0 u_0 = -\epsilon K u_0$, i.e. $-\epsilon$ is an eigenvalue with eigenvector $K u_0$. Since $\dim U = 2$ we have that ± 1 are both simple eigenvalues of R_0. Let e_1 be an eigenvector of R_0 with eigenvalue $+1$, and let

$e_2 = Ke_1$. Then the lemma follows. $\qquad\qquad\qquad\qquad\qquad\qquad\qquad\qquad\square$

Consequently, on $\mathbb{R}^4$ with coordinates (x,y), we may, without loss of generality, suppose the linear system (at $\lambda = 0$) to be

$$\frac{dx}{dt} = \omega_0 Jx,$$
$$\frac{dy}{dt} = \omega_0 Jy + x, \qquad\qquad (8)$$

with R given by (5) and J given by (4).

The following theorem shows us how to obtain a normal form for the parameter dependent case.

THEOREM 1. *There exists a mapping*

$$\Psi : \mathbb{R}^p \to \mathcal{L}(U \times U; \mathbb{R}^4),$$

defined and smooth in a neighborhood of the origin, such that

(i) $\Psi(0) = \Phi$, *with Φ as in lemma 3.*

(ii) $\Psi(\lambda)^{-1} R\Psi(\lambda)(u_1, u_2) = (R_0 u_1, -R_0 u_2)$.

(iii) $\Psi(\lambda)^{-1} A(\lambda)\Psi(\lambda)(u_1, u_2) = ((1 + \beta(\lambda))Ku_1 + u_2, \alpha(\lambda)u_1 + (1 + \beta(\lambda))Ku_2)$, *for some smooth functions $\beta(\lambda)$ and $\alpha(\lambda)$, with $\alpha(0) = \beta(0) = 0$.*

That is, identifying $U \times U$ with $\mathbb{R}^4$ using $\Psi(\lambda)$ we can bring the original system in a form for which

$$A(\lambda) = \begin{pmatrix} (1 + \beta(\lambda))K & I_U \\ \alpha(\lambda)I_U & (1 + \beta(\lambda))K \end{pmatrix}, \text{ and } R = \begin{pmatrix} R_0 & 0 \\ 0 & -R_0 \end{pmatrix}. \qquad (9)$$

Proof. On U we have an $O(2)$-action generated by $\{e^{K\phi} \mid \phi \in S^1 \cong \mathbb{R}/2\pi\mathbb{Z}\}$ and R_0. Let $<,>$ denote an inner product on U for which this action is orthogonal. Then $< e^{K\phi}u_1, e^{K\phi}u_2 >=< u_1, u_2 >$, for all ϕ and thus $K^T = -K$, $R_0^T R_0 = I_U$, and $R_0^2 = I_U$. Consequently $R_0^T = R_0$. We extend this scalar product to $U \times U$ by $< (u_1, u_2), (\tilde{u}_1, \tilde{u}_2) >=< u_1, \tilde{u}_1 > + < u_2, \tilde{u}_2 >$. We may assume $A(\lambda)$ to be such that A_0 and R are given by (7). Then $R^T = R$ and

$$A_0^T = \begin{pmatrix} -K & 0 \\ I_U & -K \end{pmatrix}. \qquad (10)$$

Before continuing the proof of the theorem we will first prove a few lemma's.

LEMMA 5. *There exists a mapping*

$$\hat{\Psi} : \mathbb{R}^p \to \mathcal{L}(U \times U),$$

defined and smooth in a neighborhood of the origin, such that

(i) $\hat{\Psi}(0) = I_{U \times U}$.

(ii) $R\hat{\Psi}(\lambda) = \hat{\Psi}(\lambda)R$.

(iii) $A_0^T(\hat{\Psi}(\lambda)^{-1}A(\lambda)\hat{\Psi}(\lambda) - A_0) = (\hat{\Psi}(\lambda)^{-1}A(\lambda)\hat{\Psi}(\lambda) - A_0)A_0^T$.

Proof. (Lemma 5) Let

$$T_+ = \{\psi \in \mathcal{L}(U \times U) \mid \psi R = R\psi\},$$

$$T_- = \{\psi \in \mathcal{L}(U \times U) \mid \psi R = -R\psi\}$$

(thus $A(\lambda) \in T_-$),

$$\tilde{T}_+ = \{\psi \in T_+ \mid \psi \text{ is invertible }\}.$$

Define $F : \tilde{T}_+ \times \mathbb{R}^p \to T_-$ by $F(\psi, \lambda) = \psi^{-1}A(\lambda)\psi$. Then $F(I, 0) = A_0$ and

$$D_\psi F(I, 0)\bar{\psi} = A_0\bar{\psi} - \bar{\psi}A_0 = (\text{Ad}A_0)\bar{\psi}, \text{ for all } \bar{\psi} \in T_+,$$

with $\text{Ad}A_0 \in \mathcal{L}(T_+, T_-)$.

On $\mathcal{L}(U \times U)$ we introduce the scalar product by $< B, C > = \text{trace}(B^T C)$, where adjoint and trace are defined with respect to the inner product $<,>$ on $U \times U$. Then

$$
\begin{aligned}
< (\text{Ad}A_0)B, C > &= < A_0 B - BA_0, C > \\
&= \text{trace}((A_0 B - BA_0)^T C) \\
&= \text{trace}((B^T A_0^T - A_0^T B^T)C) \\
&= \text{trace}(B^T(A_0^T C - CA_0^T)) \\
&= < B, (\text{Ad}A_0^T)C > .
\end{aligned}
\tag{11}
$$

Moreover, from $A_0 R = -RA_0$ we get $A_0^T R = -RA_0^T$, i.e. $A_0^T \in T_-$, and $\text{Ad}A_0^T \in \mathcal{L}(T_-, T_+)$. From (11) we see that $(\text{Ad}A_0)^T = \text{Ad}A_0^T$, hence

$$T_- = \text{Im}(\text{Ad}A_0) \oplus \ker(\text{Ad}A_0^T).$$

Let $Q \in \mathcal{L}(T_-)$ be the projection in T_- such that $\operatorname{Im}Q = \operatorname{Im}(\operatorname{Ad}A_0)$, and $\ker Q = \ker(\operatorname{Ad}A_0^T)$. Define $\tilde{F} : \tilde{T}_+ \times \mathbb{R}^p \to \operatorname{Im}Q$ by $\tilde{F}(\psi, \lambda) = Q(F(\psi, \lambda) - A_0)$. Then $\tilde{F}(I, 0) = 0$ and

$$D_\psi \tilde{F}(I, 0) = Q(\operatorname{Ad}A_0) = \operatorname{Ad}A_0$$

is surjective on $\operatorname{Im}Q$. By the implicit function theorem there exists a smooth function $\hat{\psi} : \mathbb{R}^p \to \tilde{T}_+$ with $\hat{\psi}(0) = I$, such that $\tilde{F}(\hat{\psi}(\lambda), \lambda) = 0$, for all λ near zero. Thus $F(\hat{\psi}(\lambda), \lambda) - A_0 \in \ker(\operatorname{Ad}A_0^T)$. $\qquad\square$

LEMMA 6.

$$\{B \in \mathcal{L}(U) \mid BK = KB \text{ and } BR_0 = R_0B\} = \{\alpha I_U \mid \alpha \in \mathbb{R}\}$$

and

$$\{C \in \mathcal{L}(U) \mid CK = KC \text{ and } CR_0 = -R_0C\} = \{\beta K \mid \beta \in \mathbb{R}\}$$

Proof. (Lemma 6) We first observe that U is irreducible for the action generated by K. Let $B \in \mathcal{L}(U)$ be such that $BK = KB$ and let $\alpha + i\beta$ be an eigenvalue of A.

If $\beta = 0$, let $\tilde{U} = \ker(A - \alpha I_U)$. Then $\tilde{U}$ is K-invariant and non-trivial. Consequently, by irreducibility, $\tilde{U} = U$ and $B = \alpha I_U$.

If $\beta \neq 0$, let $\tilde{U} = \ker((A - \alpha I_U)^2 + \beta^2 I_U)$. By the same argument we have $\tilde{U} = U$.

Let $\tilde{K} = -\beta^{-1}(A - \alpha I_U) \in \mathcal{L}(U)$, then $\tilde{K}^2 = -I_U$ and $(\tilde{K} - K)(\tilde{K} + K) = \tilde{K}^2 - K^2 = 0$. Thus not both $\tilde{K} - K$ and $\tilde{K} + K$ are isomorphisms, but both commute with K, consequently, by Schur's lemma, either $\tilde{K} = K$ or $\tilde{K} = -K$. Therefore $B = \alpha I_U \pm \beta K$. So

$$\{B \in \mathcal{L}(U) \mid BK = KB\} = \{\alpha I_U + \beta K \mid \alpha, \beta \in \mathbb{R}\}$$

The result of the lemma now follows using $R_0K = -KR_0$. $\qquad\square$

We will now continue the proof of theorem 1. Let

$$\Psi(\lambda) = \Phi\hat{\Psi}(\lambda).$$

Set $D = \hat{\Psi}(\lambda)^{-1}A(\lambda)\hat{\Psi}(\lambda) - A_0$. Recall that we assume $A(\lambda)$ to be such that A_0 and R are given by (7). Write

$$D = \begin{pmatrix} D_1 & D_2 \\ D_3 & D_4 \end{pmatrix}, \quad D_i \in \mathcal{L}(U), \ 1 \leq i \leq 4.$$

Then by lemma 5 D has to fulfill the equations $A_0^T D = D A_0^T$ and $RD = -DR$ with R given by (7) and A_0^T given by (10). From these equations and lemma 6 we get

$$D_2 = 0, \; D_1 = D_4 = \beta K, \; D_3 = \alpha I_U,$$

which proves the theorem. □

A further transformation

$$(u_1(t), u_2(t)) = (\tilde{u}_1(1 + \beta(\lambda))t, (1 + \beta(\lambda))\tilde{u}_2((1 + \beta(\lambda))t))$$

and a choice of basis as in lemma 4 give

$$A(\lambda) = \begin{pmatrix} J & I \\ \tilde{\alpha}I & J \end{pmatrix},$$

with $\tilde{\alpha} = \alpha(\lambda)/(1 + \beta(\lambda))^2$, and J as in (4). When $\lambda \in \mathbb{R}$ and $\tilde{\alpha}'(0) \neq 0$ then we can redefine the parameter such that $\tilde{\alpha}(\lambda) = \mu$.

Remark 1. Note that with respect to the standard symplectic form ω

$$\omega = \sum_{i=1}^{2} dx_i \wedge dy_i, \tag{12}$$

(8) is Hamiltonian and that $A(\mu)$ is in normal form as a infinitesimal symplectic matrix. Furthermore R is anti-symplectic.

3. Nonlinear normal form

In [4, eqn. (8.2)] the normal form for a vector field on $\mathbb{R}^4$ in nonsemisimple 1:1 resonance, i.e. the linearized system has a pair of eigenvalues $\pm i$ and two dimensional Jordan blocks, is given as

$$\frac{dx}{dt} = (1 + F_1(u, \Omega, \mu))Jx + F_2(u, \Omega, \mu)y + F_3(u, \Omega, \mu)x + F_4(u, \Omega, \mu)Jy,$$

$$\frac{dy}{dt} = (1 + F_1(u, \Omega, \mu))Jy + x + F_3(u, \Omega, \mu)y, \tag{13}$$

where J, u, and Ω are given by (4).

This normal form is obtained by the common Lie algebraic methods. For completeness we will formulate the reversible normal form theorem which shows that formally a reversible system can be put in normal form in such a way that the *reversor* remains the same.

We will first prove the following lemma. Let X_{A_0} denote the reversible linear vector field corresponding to A_0, and let $[\,,\,]$ be the Lie bracket of vector fields.

LEMMA 7. $R[X_{A_0}, V](z) = \epsilon[X_{A_0}, V](Rz)$, where $\epsilon = +1$ if V is reversible, and $\epsilon = -1$ if V is R-equivariant, i.e. $RV(z) = V(Rz)$.

Proof.

$$
\begin{aligned}
R[X_{A_0}, V](z) &= R[X_{A_0}(z), V(z)] \\
&= RD_z V(z) \cdot X_{A_0}(z) - RD_z X_{A_0}(z) \cdot V(z) \\
&= RD_z V(z)R \cdot R X_{A_0}(z) - RD_z X_{A_0}(z)R \cdot RV(z) \\
&= \epsilon RD_z V(Rz) \cdot X_{A_0}(Rz) - \epsilon RD_z X_{A_0}(Rz) \cdot V(Rz) \\
&= \epsilon(D_u V(u) \cdot X_{A_0}(u) - D_u X_{A_0}(u) \cdot V(u)) \text{ with } u = Rz \\
&= \epsilon[X_{A_0}(u), V(u)] \\
&= \epsilon[X_{A_0}, V](Rz),
\end{aligned}
$$

where $\epsilon = +1$ if V is reversible, and $\epsilon = -1$ if V is R-equivariant. $\qquad\square$

We now come to the normal form theorem where as usual we show that given a vector field in normal form up to order $k - 1$ one can find a transformation normalizing the vector field up to order k. By $\text{ad}(X_{A_0})$ we denote the mapping defined by $\text{ad}(X_{A_0})(V) = [V, X_{A_0}]$, for V some arbitrary vector field.

THEOREM 2 (Reversible normal form theorem). *Let*

$$
f(z, \mu) = f_1(z, \mu) + f_2(z, \mu) + \dots + f_k(z, \mu) + h.o.t., \quad z \in \mathbb{R}^4, \ \mu \in \mathbb{R}^p,
$$

with $f_k(z, \mu)$ homogeneous of order k in z, be a formal power series vector field, reversible with respect to R, which is in normal form up to order $k - 1$ with respect to $f_1(z, 0) = A_0 z$. Then there exists a transformation $\exp(\text{ad}(P))$, with $P(z, \mu)$ a homogeneous vector field of order k in z and R-equivariant, such that $\exp(\text{ad}(P))f(z, \mu)$ is in normal form up to order k and reversible with respect to R.

Proof. We will start with considering the parameter independent case. Therefore let $\mu = 0$. In order to define the normal form we need to embed A_N in a subalgebra of the Lie algebra of vector fields isomorphic to $sl(2, \mathbb{R})$. We denote the generators by A_N, A_M, and A_T, A_M being the nilpotent element dual to A_N (cf. [4]). We then have a splitting of the space of vector fields into

$$[\ker(\mathrm{ad}(X_{A_S})) \cap \ker(\mathrm{ad}(X_{A_M}))] \oplus \mathrm{im}(\mathrm{ad}(X_{A_0})). \tag{14}$$

The k-th order term in $\exp(\mathrm{ad}(P)) f(z)$, $P(z)$ homogeneous of order k, is given by

$$f_k(z) + [X_{A_0}, P](z).$$

Now the reversible vector fields form a subspace, thus we may split $f_k(z)$ according to (14) as

$$f_k(z) = \bar{f}_k + \hat{f}_k, \ \bar{f}_k \in \ker(\mathrm{ad}(X_{A_S})) \cap \ker(\mathrm{ad}(X_{A_M})), \ \hat{f}_k \in \mathrm{im}(\mathrm{ad}(X_{A_0})),$$

and choose P such that

$$[X_{A_0}, P](z) = -\hat{f}_k, P \in im(ad(X_{A_S} + X_{A_M})).$$

This way, P is determined uniquely by f_k. It follows that the part of f_k in $\mathrm{im}(\mathrm{ad}(X_{A_0}))$ vanishes after applying the transformation $\exp(\mathrm{ad}(P))$. According to lemma 7 P has to be chosen R-equivariant. Consequently the *reversor* is not changed, and the remaining part $\bar{f}_k$ of f_k is in normal form and reversible.

The transformation $\exp(\mathrm{ad}(P))$ determined so far only normalizes the part of $f_k(z, \mu)$ which does not depend on μ. That is,

$$(\exp(\mathrm{ad}(P)) f_k(z, \mu))_k = \bar{f}_k(z) + \tilde{f}_k(z, \mu),$$

with $\tilde{f}_k(z, 0) = 0$. Let $\mathcal{P}_k$ denote the homogeneous polynomials of order k. We will now show that we can find an additional transformation which normalizes the μ depending part $\tilde{f}_k$. We will use an implicit function theorem argument as in lemma 5. I.e. we will show that there exists a smooth mapping

$$\hat{P} : \mathbb{R}^p \to \mathcal{P}_k,$$

with $\hat{P}(0) = 0$, $\hat{P}(\mu)(Rz) = R\hat{P}(\mu)(z)$, and such that

$$\exp(\mathrm{ad}(\hat{P})) f_k(z, \mu) \in \ker(\mathrm{ad}(X_{A_S})) \cap \ker(\mathrm{ad}(X_{A_M})).$$

Let

$$\mathcal{P}_{k,+} = \{P \in \mathcal{P}_k \mid P(Rz) = RP(z)\},$$

$$\mathcal{P}_{k,-} = \{P \in \mathcal{P}_k \mid P(Rz) = -RP(z)\}.$$

Define $F : \mathcal{P}_{k,+} \times \mathbb{R}^p \to \mathcal{P}_{k,-}$ by $F(P,\mu) = f_k(z,\mu) + [X_{A_0}, P](z,\mu)$. Then $F(0,0) = \bar{f}_k$ and $D_P F(0,0) = \mathrm{ad}X_{A_0}$. Let Q be the projection in $\mathcal{P}_{k,-}$ such that $\mathrm{Im}Q = \mathrm{im}(\mathrm{ad}(X_{A_0}))$ and $\ker Q = \ker(\mathrm{ad}(X_{A_S})) \cap \ker(\mathrm{ad}(X_{A_M}))$. Define $\tilde{F} : \mathcal{P}_{k,+} \times \mathbb{R}^p \to \mathrm{Im}Q$ by $\tilde{F}(P,\mu) = Q(F(P,\mu) - \bar{f}_k)$. Then $D_P\tilde{F}(0,0) = \mathrm{ad}X_{A_0}$ is surjective on $\mathrm{Im}Q$. By the implicit function theorem there exists a smooth function $\hat{P} : \mathbb{R}^p \to \mathcal{P}_{k,+}$ with $\hat{P}(0,0) = 0$, such that $\tilde{F}(\hat{P}(\mu),\mu) = 0$, for all μ near zero. Thus $(\exp(\mathrm{ad}(\hat{P}))f_k(z,\mu))_k = \bar{f}_k + \tilde{f}_k(z,\mu) + [X_{A_0}, P](z,\mu) \in \ker(\mathrm{ad}(X_{A_S})) \cap \ker(\mathrm{ad}(X_{A_M})).$ □

Remains to give a general expression for the normal form. Such a general expression can be obtained from (13) and is given in lemma 8.

LEMMA 8. *The equations (13) are reversible if and only if $F_3 = F_4 = 0$.*

Proof. The condition for reversibility is $f(Rz) = -Rf(z)$. This gives for (13) the following equations. The F_i are R-invariant because u and Ω are R-invariant. We will write the F_i without there arguments.

$$
\begin{aligned}
-(1+F_1)x_2 + F_2 y_1 - F_3 x_1 + F_4 y_2 &= -(1+F_1)x_2 + F_2 y_1 + F_3 x_1 - F_4 y_2, \\
-(1+F_1)x_1 - F_2 y_2 + F_3 x_2 + F_4 y_1 &= -(1+F_1)x_1 - F_2 y_2 - F_3 x_2 - F_4 y_1, \\
(1+F_1)y_2 - x_1 + F_3 y_1 &= (1+F_1)y_2 - x_1 - F_3 y_1, \\
(1+F_1)y_1 + x_2 - F_3 y_2 &= (1+F_1)y_1 + x_2 + F_3 y_2
\end{aligned}
\tag{15}
$$

Which gives

$$
\begin{aligned}
-F_3 x_1 + F_4 y_2 &= F_3 x_1 - F_4 y_2, \\
F_3 x_2 + F_4 y_1 &= -F_3 x_2 - F_4 y_1, \\
F_3 y_1 &= -F_3 y_1, \\
-F_3 y_2 &= F_3 y_2
\end{aligned}
\tag{16}
$$

The conclusion of the lemma is now straightforward. □

Thus by putting $F_3 = F_4 = 0$ in (13) we obtain a general normal form for reversible vector fields at nonsemisimple 1:1 resonance as follows.

$$
\begin{aligned}
\frac{dx}{dt} &= (1 + F_1(u,\Omega,\mu))Jx + F_2(u,\Omega,\mu)y, \\
\frac{dy}{dt} &= (1 + F_1(u,\Omega,\mu))Jy + x.
\end{aligned}
\tag{17}
$$

This is precisely the system (3) if we put $\omega_0 = 1$ (this can be done without loss of generality), $\hat{P} = F_1$, and $\hat{Q} = F_2$.

In the Hamiltonian context the normalization proceeds along the same lines. In this case we have the standard symplectic form on $I\!\!R^4$ and the normalizing transformations need to be symplectic, i.e. the generating vector field P must be a Hamiltonian vector field. In the Hamiltonian case the normalization procedure is performed with the Hamiltonian functions, which, with the Poisson bracket induced by the symplectic form, also form a Lie algebra, rather then with vector fields (see [12]). The normal form for the Hamiltonian function at nonsemisimple 1:1 resonance is given by

$$H(x, y) = \Omega + n + F(u, \Omega, \mu), \tag{18}$$

with $F(\Omega, 0) = F(0, 0) = 0$, and $n = \frac{1}{2}(x_1^2 + x_2^2)$. We obtain the vector field

$$\frac{dx}{dt} = (1 + \frac{\partial F}{\partial \Omega}((u, \Omega, \mu))Jx + 2\frac{\partial F}{\partial u}((u, \Omega, \mu)y,$$

$$\frac{dx}{dt} = (1 + \frac{\partial F}{\partial \Omega}((u, \Omega, \mu))Jy + x. \tag{19}$$

Thus we obtain

LEMMA 9. *The equations (13) are Hamiltonian with a Hamiltonian of the form (18) if and only if $F_3 = F_4 = 0$ and $(F_1, F_2) = (\frac{\partial F}{\partial \Omega}, 2\frac{\partial F}{\partial u})$ for some $F(u, \Omega, \mu)$.*

Consequently the reversible system (17) is Hamiltonian if and only if $(F_1, F_2) = (\frac{\partial F}{\partial \Omega}, 2\frac{\partial F}{\partial u})$, a remark which is also made in [5, p 243].

4. Reduction to the orbit space

Consider a system

$$\dot{z} = f(z, \mu), \ z \in I\!\!R^n, \ \mu \in I\!\!R^p, \tag{20}$$

which is equivariant with respect to a compact symmetry group G acting linearly on $I\!\!R^n$, i.e. $gf(z, \mu) = f(gz, \mu)$ for $g \in G$. according to a theorem of Hilbert the polynomials invariant under a compact group action are generated by finitely many invariants (a Hilbert basis), which can be chosen to be homogeneous polynomials, say $\sigma_1, \sigma_2, ..., \sigma_p$. We may now define the map

$$\sigma : I\!\!R^n \rightarrow I\!\!R^p; z \mapsto (\sigma_1, \sigma_2, ..., \sigma_p).$$

This map is called the orbit map, its image can be identified with $I\!\!R^n/G$ and is called the orbit space. Each point in the orbit space corresponds to precisely one G-orbit. (see [8]). A G-equivariant system on $I\!\!R^n$ can now be lifted to the orbit space, i.e. be written as a system of equations in the invariants. It seems that the orbit space is the natural setting to study symmetric systems. However, it has the disadvantage that in general the orbit space is a semi-algebraic variety.

In the context of Hamiltonian systems the concept of reduction is well known. In this case the group action is symplectic and contains in general the flow of some integral, i.e. there exists a momentum mapping. In the case of, for instance, one integral I the reduced phase spaces are given by $\sigma(I^{-1}(c))$, c some value of the integral. (For more details see [2]).

In the case of the nonsemisimple 1:1 resonance we consider a system (13) which is in normal form, that is , the system is symmetric with respect to the S^1-action given by

$$\{e^{As\phi} \mid \phi \in S^1 \cong I\!\!R/2\pi \mathbb{Z}\},$$

which is the flow of the semisimple part of the linearized vector field. By remark 1 we may consider the linearized system to be a Hamiltonian system. Consequently the S^1-action can be seen as a symplectic action. The Hamiltonian function corresponding to this symplectic S^1-action is Ω.

LEMMA 10. Ω *is an integral for the system (13) if and only if*

$$2F_3(u, \Omega, \mu)\Omega - F_4(u, \Omega, \mu)u = 0. \tag{21}$$

Proof. A straightforward calculation shows that

$$\frac{d\Omega}{dt} = 2F_3(u, \Omega, \mu)\Omega - F_4(u, \Omega, \mu)u.$$

$\square$

From this it is clear that the reversible normal form has Ω as an absolute invariant.

A Hilbert basis generating the S^1-invariant polynomials is given by $\Omega, \frac{1}{2}u, n, T$, with $T = x_1 y_1 + x_2 y_2$. Note that these are the Hamiltonian functions corresponding to A_S, A_M, A_N, and A_T respectively, i.e. $A_S z = X_\Omega(z)$, $A_M z = X_u(z)$, $A_N z = X_n(z)$, and $A_T z = X_T(z)$, where X_f denotes the Hamiltonian vector field with Hamiltonian function f.

On the C^∞ functions the symplectic form induces a Poisson bracket $\{\,,\,\}$ by

$$\omega(X_f, X_g) = \{f, g\},$$

where, writing $z = (z_1, z_2, z_3, z_4) = (x, y) = (x_1, x_2, y_1, y_2) \in \mathbb{R}^4$,

$$\{f, g\}(z) = \sum_{i=1, j=1}^{4} W_{ij} \frac{\partial f}{\partial z_i} \frac{\partial g}{\partial z_j}$$

$$= \sum_{i=1}^{2} \left(\frac{\partial f}{\partial x_i} \frac{\partial g}{\partial y_i} - \frac{\partial f}{\partial y_i} \frac{\partial g}{\partial x_i} \right).$$

The bracket $\{\,,\,\}$ defines a Poissonstructure on the C^∞ functions. The C^∞ functions together with this bracket form a Poisson algebra. The matrix

$$W_{ij} = \begin{pmatrix} 0 & 0 & 1 & 0 \\ 0 & 0 & 0 & 1 \\ -1 & 0 & 0 & 0 \\ 0 & -1 & 0 & 0 \end{pmatrix}$$

is called the structure matrix for the Poisson structure. $u, n,$ and T generate a Lie subalgebra of $(C^\infty, \{\,,\,\})$ isomorphic to $\mathrm{sl}(2, \mathbb{R})$.

The orbit map for the S^1 action is given by

$$\rho : \mathbb{R}^4 \to \mathbb{R}^4; (x, y) \mapsto (\Omega, \frac{1}{2}u, n, T). \tag{22}$$

The image is determined by the relation

$$2un - \Omega^2 - T^2 = 0. \tag{23}$$

The standard Poisson structure on $\mathbb{R}^4$ induces a natural Poisson structure on the image of the orbit map given by the $\mathrm{sl}(2, \mathbb{R})$ bracket relations.

We may now lift the equation (13) to the orbit space. First we introduce on the orbit space new coordinates by choosing a somewhat different set of generating invariants:

$$y_1 = -2\Omega, \; y_2 = 4n, \; y_3 = 2T, \; y_4 = u. \tag{24}$$

In these new coordinates the equations on the orbit space are (see [4])

$$
\begin{pmatrix} \dot{y}_1 \\ \dot{y}_2 \\ \dot{y}_3 \\ \dot{y}_4 \end{pmatrix} = \begin{pmatrix} 0 \\ 0 \\ y_2 \\ y_3 \end{pmatrix} + 2F_3(y_1, y_4, \mu) \begin{pmatrix} y_1 \\ y_2 \\ y_3 \\ y_4 \end{pmatrix} + 2F_2(y_1, y_4, \mu) \begin{pmatrix} 0 \\ y_3 \\ y_4 \\ 0 \end{pmatrix} +
$$

$$
+ 2F_4(y_1, y_4, \mu) \begin{pmatrix} y_4 \\ y_1 \\ 0 \\ 0 \end{pmatrix}. \tag{25}
$$

We call this the reduced equations. Again these equations are in normal form. On the invariants the *reversor* (5) becomes

$$
\bar{R} = \begin{pmatrix} 1 & 0 & 0 & 0 \\ 0 & 1 & 0 & 0 \\ 0 & 0 & -1 & 0 \\ 0 & 0 & 0 & 1 \end{pmatrix}. \tag{26}
$$

Obviously the vector field (25) is reversible with respect to $\bar{R}$ if and only if $F_3 = F_4 = 0$. Thus a reduced reversible vector field in normal form has the general form

$$
\begin{pmatrix} \dot{y}_1 \\ \dot{y}_2 \\ \dot{y}_3 \\ \dot{y}_4 \end{pmatrix} = \begin{pmatrix} 0 \\ 0 \\ y_2 \\ y_3 \end{pmatrix} + 2F_2(y_1, y_4, \mu) \begin{pmatrix} 0 \\ y_3 \\ y_4 \\ 0 \end{pmatrix}. \tag{27}
$$

The bracket relations among the y_i are $\{y_1, y_2\} = \{y_1, y_3\} = \{y_1, y_4\} = 0$, $\{y_2, y_3\} = 4y_2$, $\{y_2, y_4\} = 4y_3$, and $\{y_3, y_4\} = 4y_4$. Thus the Poisson structure on the orbit space is given by the structure matrix

$$
\begin{pmatrix} 0 & 0 & 0 & 0 \\ 0 & 0 & 4y_2 & 4y_3 \\ 0 & -4y_2 & 0 & 4y_4 \\ 0 & -4y_3 & -4y_4 & 0 \end{pmatrix} \tag{28}
$$

Let [,] denote the Poisson bracket for this Poisson structure. Then the vector field (27) is Hamiltonian with respect to this Poisson structure, i.e. can be written as $\dot{y} = [y, \bar{H}]$, with

$$
\bar{H} = -\frac{1}{4}y_2 + \frac{1}{2} \int_0^{y_4} F_2(y_1, s, \mu) ds. \tag{29}
$$

We obtain $\bar{H}$ as an absolute invariant for the reversible system. So far we have shown that a reduced reversible vector field in normal form is Hamiltonian. By lemma 9 a reduced Hamiltonian vector field in normal form is reversible. Thus we have shown

THEOREM 3. *Consider a vector field on $\mathbb{R}^4$ at nonsemisimple 1:1 resonance which is in normal form. Then the corresponding reduced vector field (25) is reversible with respect to (26) if and only if it is Hamiltonian with respect to (28).*

Because Ω is an invariant, a reduced reversible vector field has a natural restriction to surfaces $\Omega = constant$. On the images of the surfaces $\Omega = constant$ the Poisson structure, in a natural way, restricts to a symplectic structure. I.e. on the orbit space the surfaces $\rho(\Omega^{-1}(c))$ are the symplectic leaves for the Poisson structure. The surfaces $\rho(\Omega^{-1}(c))$ are called the reduced phase spaces. A system which on the orbit space is Hamiltonian with respect to the Poisson structure on the reduced phase spaces restricts to a genuine Hamiltonian system which is Hamiltonian with respect to the symplectic structure.

5. Hamiltonian structure in the non-reduced reversible vector field

The reversible system (17) is Hamiltonian if and only if there exists an $F(u, \Omega, \mu)$ such that $\frac{\partial F}{\partial \Omega} = F_1$ and $2\frac{\partial F}{\partial u} = \frac{1}{2}F_2$. The Hamiltonian function with respect to the standard symplectic form is then given by

$$G = \Omega - n + F, \tag{30}$$

where $n = \frac{1}{2}(x_1^2 + x_2^2)$. Now let $\hat{F}$ be such that $\hat{F}_u = \frac{1}{2}F_2$, and write $\hat{V} = F_1 - \hat{F}_\Omega$. Then it is clear that we can split our reversible vector field in a Hamiltonian (and reversible) part corresponding to $H = -n + \hat{F}$, which is precisely one of the absolute invariants, and a reversible part $(\omega_0 - \hat{V})(Jx, Jy)$, i.e. the vector field can be written as

$$\frac{dx}{dt} = (\omega_0 - \hat{V})Jx + \frac{\partial H}{\partial y},$$
$$\frac{dy}{dt} = (\omega_0 - \hat{V})Jy - \frac{\partial H}{\partial x}. \tag{31}$$

Let X_H denote the Hamiltonian vector field corresponding to H. Then the vector field (31) can be written as $(\omega_0 - \hat{V})X_\Omega + X_H$. The "non-Hamiltonian" part of the vector field $(\omega_0 - \hat{V})X_\Omega$ is thus equivalent to the Hamiltonian vector field X_Ω in the sense that the trajectories coincide in a neighborhood of the origin. However, the flow is non-Hamiltonian, which is obvious because its period, which is determined by $\omega_0 - \hat{V}$, does not depend on Ω alone.

The vector field $(\omega_0 - \hat{V})X_\Omega$ might still contain a Hamiltonian part because each vector field of the form $f(\Omega)X_\Omega$ is obviously Hamiltonian. Let $W(\mu, \Omega)$ be such that $W_\Omega = -\hat{V}(\mu, 0, \Omega)$. Furthermore let $u\tilde{V}(\mu, u, \Omega) = -\hat{V}(\mu, u, \Omega) + \hat{V}(\mu, 0, \Omega)$, and let $\hat{H} = W + H$. Then we can write the vector field as

$$u\tilde{V}X_\Omega + X_{\hat{H}}. \tag{32}$$

The vector field for the reversible 1:1 resonance can thus be seen as a non-Hamiltonian perturbation, of at least order two, of a Hamiltonian vector field. The perturbation generates a shift along the X_Ω trajectories. Consequently after reduction to the orbit space the non-Hamiltonian effect vanishes.

6. Bifurcations of periodic solutions

The non-Hamiltonian part of the reversible vector field, which is in F_1 vanishes after reduction. This is obvious from the geometry. The trajectories of the "non-Hamiltonian" part, $(\omega_0 - \hat{V})X_\Omega$, of the reversible vector field (31) coincide with the X_Ω orbits and thus will vanish on the orbit space. Thus on the reduced phase spaces we are left with the reduced Hamiltonian systems corresponding to H, i.e. which have reduced Hamiltonian $\bar{H}$. The periodic solutions are precisely the stationary points of the reduced Hamiltonian vector fields. These points correspond to the critical values of the map

$$H \times \Omega : \mathbb{R}^4 \to \mathbb{R}^2; (x, y) \mapsto (H, \Omega), \tag{33}$$

which is obvious if we factorize this map through the orbit map. The linear stability type of the periodic solution corresponds to the stability type of the stationary point of the reduced vector field (see [3]). Consequently the geometry of the bifurcation of periodic solutions at nonsemisimple 1:1 resonance of the reversible case is exactly the same as for the Hamiltonian Hopf bifurcation [12].

Because of the presence of two absolute invariants the phase space of the system (1) is fibered into invariant surfaces given by $\Omega = constant$ and $H = constant$, i.e. the fibers of the map (33). Thus we obtain exactly the same fibration as in the Hamiltonian case (see [12]). The difference lies in the fact that on these fibers the flow of the reversible system (1) is non-Hamiltonian by the fact that a small non-Hamiltonian shift is added in the direction of the X_Ω trajectories.

Like in the Hamiltonian case one can use a reversible version of Weinstein-Moser reduction ([12], [11]) to show persistence of the periodic solutions under higher order perturbations which destroy the normal form.

Acknowledgment The authors are grateful to Tom Bridges and Richard Cushman, for helpful discussions. The research in this paper was supported by the E.E.C. Science Project on "Bifurcation Theory and its Applications".

References

[1] R. Abraham and J.E. Marsden, 1978: *Foundations of mechanics*, Benjamin/Cummings Publ. Comp. Inc., Reading, Massachussets, sec. ed.

[2] J. Arms, R.H. Cushman, and M. Gotay, 1991: *A universal reduction procedure for Hamiltonian group actions*, in: The geometry of Hamiltonian systems, ed. T. Ratiu, MSRI Workshop Proceedings, Springer Verlag, pp 33-51.

[3] T.J. Bridges, 1992: *Poisson structure of the reversible 1:1 resonance*, Preprint Math. Inst. R.U.U, to appear in the proceedings of the International Conference on Bifurcations in Differential Dynamics, Diepenbeek, 1992.

[4] R.H. Cushman, and J.A. Sanders, 1986: *Nilpotent normal forms and representation theory of sl(2,$I\!R$)*, Contemporary Mathematics vol.56, pp. 31-51.

[5] G. Iooss, A. Mielke, and Y. Demay, 1989: *Theory of steady Ginzburg-Landau equation in hydrodynamic stability problems*, Eur. J. Mech. B-Fluids **8**, pp. 229-268.

[6] G. Iooss, and M.-C. Pérouème, 1993: *Perturbed homoclinic solutions in reversible 1:1 resonance vector fields*, J. Diff. Eqns. **102**, pp. 62-88.

[7] P. Libermann and C.-M. Marle, 1987: *Symplectic geometry and analytical mechanics*, D. Reidel Publ. Comp., Dordrecht.

[8] V. Poènaru, 1976: *Singularités C^∞ en présence de symétrie*, Lect. Notes in Math. vol.510, Springer Verlag, Berlin.

[9] A. Vanderbauwhede, 1990: *Hopf bifurcation for equivariant conservative and time-reversible systems*, Proc. Roy. Soc. Edinburgh **116A**, pp. 103-128.

[10] A. Vanderbauwhede, 1991: *Branching of periodic solutions in time-reversible systems*, in: *Geometry and Analysis in Nonlinear Dynamics*, eds. H. Broer and F. Takens, Pitman Research Notes in Math. Ser. vol.222, Longman, London.

[11] A. Vanderbauwhede, and J.C. van der Meer, 1993: *A general reduction principle for periodic solutions in Hamiltonian systems*, to be published.

[12] J.C. van der Meer, 1985: *The Hamiltonian Hopf bifurcation*, Lect. Notes in Math. vol.1160, Springer Verlag, Berlin.

INSTANTANEOUS SYMMETRY AND SYMMETRY ON AVERAGE IN THE COUETTE-TAYLOR AND FARADAY EXPERIMENTS

Ian Melbourne [*][†]
Institut Non Linéaire de Nice
CNRS - Université de Nice Sophia-Antipolis
FRANCE

Abstract

We describe some recent results on symmetry of attractors for dynamical systems with symmetry and consider the implications for the Couette-Taylor experiment and the Faraday surface wave experiment. In particular, we explore the relationship between symmetry of solutions at a fixed instant in time, and symmetry in the time-averaged solution. This leads to predictions that are somewhat surprising and which we believe require careful experimental exploration.

1   Introduction

Many physical interesting situations, including Rayleigh-Bernard convection and the Couette-Taylor experiment, are modeled by PDEs that have symmetries, that is they are equivariant with respect to the action of a symmetry group Γ. Equilibrium and periodic solutions to these PDEs may be invariant as a subset of phase space under some of these symmetries. In this case, there is a well understood connection between the symmetries in phase space and symmetries in physical space [16].

Recently, there has been interest in interpreting the symmetry of a solution in physical space corresponding to a chaotic attractor in phase space, see [6, 8].

---

[*]Supported in part by NSF Grant DMS-9101836, by the Texas Advanced Research Program (003652037) and by the CNRS

[†]Permanent address: Department of Mathematics, University of Houston, Houston, Texas 77204-3476, USA

P. Chossat (ed.), Dynamics, Bifurcation and Symmetry, 241–257.

Thought of as a subset A of phase space, there are (at least) two ways in which A can be symmetric. There is a subgroup Σ_A of elements of Γ that fix A as a set, and a smaller subgroup T_A of elements of Γ that fix each point in A, see Section 2 below.

There are also two ways of interpreting the symmetry of the solution in physical space, the *instantaneous symmetry* which measures the symmetry at each moment in time, and the *symmetry on average* which is the symmetry in the time-averaged solution. Clearly, we may interpret T_A to be the instantaneous symmetry. The suggestion in Dellnitz et al [8] is that Σ_A should be identified with the symmetry on average. We shall make this identification, but note that this has been justified rigorously only under the assumption that the attractor A has an SBR measure.

The motivating example for the work of Chossat and Golubitsky [6] on symmetric attractors was provided by a chaotic state in the Couette-Taylor experiment that is known as turbulent Taylor vortices, see Brandstater and Swinney [5]. A picture of turbulent Taylor vortices at an instant in time is shown in Figure 1(a). The usual interpretation for this state is that there is no instantaneous symmetry, even though on average [20] there is the symmetry of the steady state Taylor vortices, Figure 1(b). More recently, experiments have been performed by Gollub and coworkers [15] on the Faraday surface wave experiment in square and circular geometries. Instantaneous and time-averaged pictures of states that they find are shown in Figures 2 and 3. Again there is evidence of symmetry that exists only on average.

The existence of symmetry only on average has been verified numerically [8] and is well-understood theoretically [3, 13]. One purpose of the present paper is to point out that the experimental evidence for symmetry only on average is not as conclusive as has been believed. In particular, we argue that the motivating example of turbulent Taylor vortices does not exhibit symmetry only on average, but that the symmetry in the time average is already present at each instant of time. In this paper we highlight those aspects of recent results about symmetric attractors in [13, 18] that have implications for PDEs, and then use these results to make predictions for experiments. As already indicated, these predictions are often contrary to intuition.

In Section 2 we define the instantaneous symmetry and symmetry on average for a subset A. Also, we state some elementary results including a normality condition that will prove crucial in the remainder of the paper. Then in Section 3 we describe some of the results in [13] concerning attractors for ODEs that have a finite group of symmetries Γ. To make the connection with PDEs, we introduce a class of 'high-dimensional' ODEs that includes discretizations of PDEs. As a preliminary application, we interpret these results for the Faraday experiment in a square container, taking $\Gamma = \mathbf{D}_4$ the symmetry group of the square.

In Section 4 we consider some results for continuous groups of symmetries and consider the implications for experiments. In particular, we consider the Faraday experiment in a circular container, and the Couette-Taylor experiment – the latter under the assumption of periodic boundary conditions.

Returning to the Faraday experiment in a square container, it is evident that the time-averaged state in Figure 2(b) has more structure than can be explained within the context of the symmetry group $\Gamma = \mathbf{D}_4$: there appear to be additional discrete translation symmetries parallel to the sides of the square container. Using the trick of embedding Neumann boundary conditions in periodic boundary conditions we are able to explain this structure but only if the structure is already present at any instant in time. A similar story is true for turbulent Taylor vortices if we assume Neumann boundary conditions rather than periodic boundary conditions. These issues are taken up in Section 5.

We end in Section 6 with some conclusions. In particular, we discuss at some length the likelihood that turbulent Taylor vortices have no symmetry on average except for that symmetry that is already present at any instant in time.

2 Symmetry groups of sets

In this section, we define the symmetry of a set. At this point, we work quite generally, and do not require mention of any dynamics. Suppose that X is a set, and that Γ is a group acting on X. If $A \subset X$, we define

$$T_A = \{\gamma \in \Gamma;\ \gamma x = x \text{ for all } x \in A\},$$

and

$$\Sigma_A = \{\gamma \in \Gamma;\ \gamma A = A\}.$$

For reasons explained in the introduction, we call T_A the *instantaneous symmetry* and Σ_A the *symmetry on average*. Note that A is contained in the fixed-point set of T_A: $A \subset \mathrm{Fix}(T_A)$. The following result is elementary, but crucial.

Proposition 2.1 *T_A is a normal subgroup of Σ_A.*

Proof Suppose that $t \in T_A$ and $\sigma \in \Sigma_A$. If $x \in A$, $\sigma x \in A$ and hence $t\sigma x = \sigma x$. It follows that $\sigma^{-1}t\sigma x = x$ and $\sigma^{-1}t\sigma \in T_A$ as required.

If T is a subgroup of Γ, then $N(T)$ denotes the normalizer of T in Γ, namely the largest subgroup of Γ that contains T as a normal subgroup. An equivalent formulation of the normality condition in Proposition 2.1 is that

$$T_A \subset \Sigma_A \subset N(T_A).$$

For example, suppose that $\Gamma = \mathbf{O}(2)$. Up to conjugacy, the subgroups of $\mathbf{O}(2)$ are

$$\mathbf{Z}_k, \ k \geq 1, \quad \mathbf{D}_k, \ k \geq 1, \quad \mathbf{SO}(2), \quad \mathbf{O}(2).$$

(In particular, $\mathbf{Z}_1 = 1$ is the trivial group, and $\mathbf{D}_1$ is the two element group generated by any element $\kappa \in \mathbf{O}(2) - \mathbf{SO}(2)$.) The subgroups $\mathbf{Z}_k$, $\mathbf{SO}(2)$ and $\mathbf{O}(2)$ are normal in $\mathbf{O}(2)$. Hence the normality condition in Proposition 2.1 comes into effect only when $T_A = \mathbf{D}_k$ which has normalizer $\mathbf{D}_{2k}$. It then follows that either $\Sigma_A = \mathbf{D}_k$ or $\Sigma_A = \mathbf{D}_{2k}$. Proposition 2.1 can also be applied in the opposite direction. For example if $\Sigma_A = \mathbf{O}(2)$ then we deduce that $T_A \neq \mathbf{D}_k$ for any k.

3 Finite symmetry groups

In this section we recall results of Field et al [13] which classify the possible symmetry groups of attractors for flows that are equivariant with respect to the action of a finite group of symmetries Γ. The points that we emphasize here are rather different from those in [13], in particular we focus on the implications for applications modeled by PDEs. A consequence is that generally the representation-theoretic restrictions obtained in [13] (see also [19]) are not applicable.

Since the results in [13] are stated within the context of finite-dimensional flows, it is necessary to make the transition between PDEs and 'high-dimensional' ODEs. Note that if we were working with a PDE, we would expect every subgroup $T \subset \Gamma$ to be an isotropy subgroup with fixed-point subspace $\mathrm{Fix}(T)$ of infinite dimension. Moreover if I is another subgroup and $I \not\subset T$, then the intersection of $\mathrm{Fix}(I)$ with $\mathrm{Fix}(T)$ should be of infinite codimension in $\mathrm{Fix}(T)$. Passing to a discretization of such a PDE leads to a high-dimensional ODE and the dimensions and codimensions mentioned above should also be very high. In particular, we expect that the following conditions are satisfied.

(i) T is an isotropy subgroup,

(ii) $\dim(\mathrm{Fix}(T)) \geq 5$, and

(iii) $\dim(\mathrm{Fix}(T)) - \dim(\mathrm{Fix}(I) \cap \mathrm{Fix}(T)) \geq 2$ for all $I \subset \Gamma$, $I \not\subset T$.

We shall say that a representation of Γ on $\mathbf{R}^n$ is *high-dimensional* if each subgroup $T \subset \Gamma$ satisfies conditions (i)–(iii). Also a Γ-equivariant ODE or vector field is *high-dimensional* if the representation of Γ is high-dimensional. Under this assumption of high-dimensionality, the normality condition in Proposition 2.1 is necessary and sufficient.

Theorem 3.1 ([13]) *Suppose that Γ is a finite subgroup of $\mathbf{O}(n)$ and that the representation of Γ on $\mathbf{R}^n$ is high-dimensional. Let T and Σ be subgroups of Γ. Then there exists a C^∞ Γ-equivariant vector field on $\mathbf{R}^n$ possessing an Axiom A attractor A with $T_A = T$ and $\Sigma_A = \Sigma$ if and only if T is a normal subgroup of Σ.*

Remark 3.2 Axiom A attractors are structurally stable, so it follows from the theorem that whenever T is a normal subgroup of Σ, attractors with instantaneous symmetry T and symmetry on average Σ are unavoidable.

Application: the Faraday experiment in a square geometry Theorem 3.1 can be applied to the Faraday surface wave experiment performed in a square container [15], see Figure 2. For the time being we work only with the naive group of symmetries which is the symmetry group of the square $\Gamma = \mathbf{D}_4$. Later, in Section 5, we shall consider the issues that arise when trying to understand the additional structure that is evident in Figure 2(b).

It appears from the figure that we have a state with instantaneous symmetry $T = \mathbf{1}$ and symmetry on average $\Sigma = \mathbf{D}_4$. Note that this scenario is entirely consistent with Theorem 3.1 but is only one of several possibilities that is equally consistent with the theorem.

The subgroups of $\mathbf{D}_4$ are

$$\mathbf{1}, \mathbf{Z}_2, \mathbf{Z}_4, \mathbf{D}_1, \mathbf{D}_2, \mathbf{D}_4.$$

Each of these subgroups is normal in $\mathbf{D}_4$ with the exception of the subgroup $\mathbf{D}_1$ which has normalizer $\mathbf{D}_2$. Hence if $T \neq \mathbf{D}_1$, Σ can be any subgroup of $\mathbf{D}_4$ that contains T. In particular, if $T = \mathbf{1}$ then there are no restrictions on Σ. Thus we should not be surprised by symmetry on average, but neither should we expect it.

As a result of later considerations in this paper, it will become apparent that the instantaneous symmetry T is not so easy to deduce from Figure 2(a). However it is clear from Figure 2(b) that $\Sigma = \mathbf{D}_4$. Applying Theorem 3.1, we can deduce at least that $T \neq \mathbf{D}_1$.

4 Continuous groups

In this section, we consider ω-limit sets A for flows in $\mathbf{R}^n$ that are equivariant with respect to a continuous (nonfinite but compact) group of symmetries $\Gamma \subset \mathbf{O}(n)$.

Theorem 3.1 states that for high-dimensional representations of a finite group Γ, given T_A we can expect Σ_A to be any subgroup of Γ that satisfies $T_A \subset \Sigma_A \subset N(T_A)$.

Moreover, all such possibilities for Σ_A occur in a structurally stable manner (since A can be chosen to be an Axiom A attractor).

As we indicate in this section, the situation is completely different for continuous groups. Typically there are larger lower bounds for the continuous symmetries in Σ_A. These bounds depend on T_A and also in a subtle way on the dynamics in $\omega(x)$.

Recall that if G is a group, we denote by G^0 the connected component of the identity in G. Suppose that $A = \omega(x)$ is an ω-limit set for a Γ-equivariant vector field on $\mathbf{R}^n$. Then results of [2, 18] indicate that, provided the dynamics in A is 'sufficiently chaotic', typically

$$(N(T_A)/T_A)^0 \subset \Sigma_A/T_A \subset N(T_A)/T_A. \tag{4.1}$$

We do not wish to define here precisely what is meant by 'sufficiently chaotic'. It follows from [18] that at least when Γ^0 is abelian (which is the case for the applications considered in this paper) it is enough that A satisfies conditions that can be thought of as a equivariant generalization of 'Devaney's definition of chaos' [10]. However these hypotheses are unnecessarily stringent and are undergoing constant revision at present.

We note that it is certainly necessary to rule out dynamics that is too regular. For example, if A is a subset of a single group orbit (A is a relative equilibrium) then Σ_A/T_A is easily seen to be abelian, typically it is a maximal torus in $N(T_A)/T_A$, see [11, 17]. The situation is more complicated for relative periodic orbits, see [17, 12], but Σ_A/T_A remains abelian. This is not the case for general ω-limit sets. For example, the existing theory indicates that typically equation (4.1) is valid under very weak hypotheses on the irregularity of the dynamics in A. From now on we shall assume that equation (4.1) is valid at least for the states shown in Figures 1, 2 and 3.

Application: the Faraday experiment in a circular geometry We consider the Faraday surface wave experiment performed in a circular geometry [15], see Figure 3. The solution shown in the figure would appear to have no instantaneous symmetry but to have full symmetry on average. In our notation, this is $\Sigma = \Gamma = \mathbf{O}(2)$ and $T = \mathbf{1}$.

Actually, we cannot tell from the figure whether there is $\mathbf{SO}(2)$ or $\mathbf{O}(2)$ symmetry on average. This is due to the fact that all subsets $A \subset \mathbf{R}^2$ with $\mathbf{SO}(2) \subset \Sigma_A$ automatically satisfy $\Sigma_A = \mathbf{O}(2)$. In the language of [4], the observation that is being averaged is not a *detective*; it cannot distinguish between $\Sigma = \mathbf{SO}(2)$ and $\Sigma = \mathbf{O}(2)$.

It turns out that the information on symmetry on average contained in Figure 3(b) can be predicted from knowledge of the instantaneous symmetry. This follows from the entries in Table 1 where we enumerate the subgroups T of $O(2)$ and then list the possibilities for Σ that are typical according to equation (4.1). In particular, if we know that there is no instantaneous symmetry, then we can predict that there is at least $SO(2)$ symmetry on average. Hence with hindsight we can say that Figure 3(b) gives no further information on the symmetry on average. However, averaging the correct detective should determine whether the symmetry on average is $SO(2)$ or $O(2)$.

Finally, applying Table 1 in the reverse direction, we can deduce that if Σ_A contains $SO(2)$ then either there are no reflection symmetries present instantaneously or there is full symmetry instantaneously. (This is an application of the normality condition in Proposition 2.1.)

T	$N(T)$	$N(T)/T$	Σ/T	Σ
$\mathbf{Z}_k$, $k \geq 1$	$O(2)$	$O(2)$	$SO(2)$, $O(2)$	$SO(2)$, $O(2)$
$\mathbf{D}_k$, $k \geq 1$	$\mathbf{D}_{2k}$	$\mathbf{Z}_2$	1, $\mathbf{Z}_2$	$\mathbf{D}_k$, $\mathbf{D}_{2k}$
$SO(2)$	$O(2)$	$\mathbf{Z}_2$	1, $\mathbf{Z}_2$	$SO(2)$, $O(2)$
$O(2)$	$O(2)$	1	1	$O(2)$

Table 1: This table shows the interaction between instantaneous symmetry T and symmetry on average Σ for ω-limit sets in ODEs with $O(2)$ symmetry. For each subgroup $T \subset O(2)$ we list the typical possibilities for Σ as guaranteed by equation (4.1).

Application: the Couette-Taylor experiment In the Couette-Taylor experiment, the underlying symmetry group is $\Gamma = O(2) \times SO(2)$ where $SO(2)$ corresponds to the azimuthal rotations and $O(2)$ corresponds to the axial translations together with the mid-cylinder flip. Here we are assuming periodic boundary conditions at the ends of the cylinder. In Section 5 we shall consider what happens if we make the more reasonable assumption of Neumann boundary conditions at the ends of the cylinder. Our predictions are then unchanged though the analysis is completely different.

First observe that the center of Γ is $SO(2)$, and it follows that $N(T)$ contains $SO(2)$ for all subgroups T. Hence we can predict that on average all solutions will have full azimuthal symmetry, independent of the instantaneous symmetry.

Our second prediction is more interesting and concerns the solution known as turbulent Taylor vortices which is shown in Figure 1. Thanks to the previous prediction, we may as well factor out the azimuthal symmetry so that $\Gamma = O(2)$. The current trend is to believe that turbulent Taylor vortices have no instantaneous symmetry but have the symmetry of Taylor vortices on average, namely $\Sigma = \mathbf{D}_k$ (or

$D_k \times SO(2)$ if we reinstate the azimuthal symmetry). However, by Table 1, if we take $T = 1$ then we expect $\Sigma = SO(2)$ or $O(2)$. But then the averaged solution is totally homogeneous (Couette flow), which is clearly not the case. On the other hand, if we have $T_A = D_k$, then $N(T_A) = D_{2k}$ and we would expect either $\Sigma_A = D_k$ or $\Sigma_A = D_{2k}$. Based on this calculation, we propose that turbulent Taylor vortices have $T = \Sigma = D_k$.

5 Embedding Neumann boundary conditions in periodic boundary conditions

In Section 3 we considered the Faraday surface wave experiment in a square domain (Figure 2) but only taking into account the naive symmetry group $\Gamma = D_4$. On the other hand, the time-averaged solution shown in Figure 2(b) clearly has additional structure that cannot be explained in terms of elements of D_4.

Recently there has been much interest in problems satisfying certain boundary conditions, in particular Neumann boundary conditions (NBC), that admit the possibility of embedding the problem in a larger problem (on a larger domain) where the solutions satisfy periodic boundary conditions (PBC). The extra symmetry that arises in the PBC problem imposes constraints on the original NBC problem and changes the generic behavior of the associated dynamical system as well as increasing the available range of symmetries for the solutions. See Crawford et al [7] which expands upon work of Fujii, Mimura and Nishiura [14] and Armbruster and Dangelmayr [1].

In this section, we review the construction and then reconsider the Couette-Taylor and Faraday experiments in this light. Since we are interested in global dynamics, the implications for genericity do not concern us. However the additional symmetries will be of great importance.

Suppose that we have a Euclidean-equivariant PDE on the unit interval $[0, 1]$ and that we impose NBC at the ends of the interval. Then the only Euclidean transformation that preserves the domain is the flip $x \rightarrow 1 - x$ which generates the symmetry group D_1^F. By reflecting a solution across 0 we obtain a solution that satisfies the PDE on $[-1, 1]$. Moreover the solution satisfies PBC on this larger domain. (There is a technical problem concerning regularity of the solution obtained in this way but in many cases, including the ones that we shall consider, regular solutions to the NBC problem on $[0, 1]$ extend to regular solutions to the PBC problem on $[-1, 1]$.)

Conversely, solutions satisfying PBC on $[-1, 1]$ restrict to solutions satisfying NBC on $[0, 1]$ if and only if they are invariant under the reflection κ that sends x

to $-x$. Define $\mathbf{D}_1^N$ to be the group generated by κ. Now we consider the enlarged problem of PBC on $[-1,1]$ but then restrict to $\mathrm{Fix}(\mathbf{D}_1^N)$ in order to recover the NBC problem. The idea is that the PBC problem is $\mathbf{O}(2)$-equivariant and solutions in $\mathrm{Fix}(\mathbf{D}_1^N)$ may pick up symmetries in $\mathbf{O}(2)$ that do not lie in the original group $\mathbf{D}_1^F$.

Now, if we have a solution to the NBC problem, we may compute the instantaneous symmetry T and the symmetry on average Σ as subgroups of $\mathbf{O}(2)$ instead of $\mathbf{D}_1^F$. An important observation is that T must contain $\mathbf{D}_1^N$ (since solutions are assumed to satisfy Neumann boundary conditions at all times). In particular, if there is no discernible structure at an instant in time, then this should be interpreted as $T = \mathbf{D}_1^N$.

Application: the Couette-Taylor experiment revisited

In Section 4 we considered the Couette-Taylor experiment and made two predictions based upon the assumption of periodic boundary conditions. Now we show that these predictions are unchanged if we assume Neumann boundary conditions.

The prediction concerning full azimuthal symmetry on average goes through easily no matter what boundary conditions are assumed. Hence we turn to the second prediction concerning turbulent Taylor vortices. As before, we may factor out the azimuthal symmetry. Also, the radial direction plays no important role and so we may reduce to the situation of a NBC problem on the unit interval.

We now have an $\mathbf{O}(2)$-equivariant problem and we restrict attention to those solutions with instantaneous symmetry at least $\mathbf{D}_1^N$ (that is, those solutions satisfying NBC). If there is no further structure, then we have instantaneous symmetry $T = \mathbf{D}_1^N$. But then the normality condition in Proposition 2.1 guarantees that the symmetry on average Σ is isomorphic to $\mathbf{D}_1$ or $\mathbf{D}_2$, see Table 1. This is inconsistent with Figure 1(b). On the other hand if $T \cong \mathbf{D}_k$ then we have $\Sigma \cong \mathbf{D}_k$ or $\Sigma \cong \mathbf{D}_{2k}$. We are again led to our prediction that $T = \Sigma \cong \mathbf{D}_k$.

Application: the Faraday experiment revisited

Here we consider the Faraday experiment in a square container, but in such a way that we can account for additional structure in the time-averaged solution that is not accounted for by the symmetries of the square alone. The analysis here is almost identical to that for the Couette-Taylor experiment under the assumption of NBC. The corresponding questions raised by the normality condition in Proposition 2.1 were first pointed out to me by M. Golubitsky.

Solutions to the NBC problem on the square extend to solutions satisfying PBC on a square of four times the size. This problem is equivariant under the semidirect product of the naive symmetries $\mathbf{D}_4$ with translation symmetries in the directions parallel to the sides of the square. So we have $\Gamma = \mathbf{D}_4 \dotplus (\mathbf{SO}(2) \times \mathbf{SO}(2))$. Equivalently, we can write $\Gamma = \mathbf{D}_1 \dot{+} (\mathbf{O}(2) \times \mathbf{O}(2))$ which is more convenient for our purposes. Again we restrict attention to solutions that satisfy NBC on the original square: these solutions have instantaneous symmetry at least $T = \mathbf{D}_1^N \times \mathbf{D}_1^N$.

The time-averaged solution in Figure 2(b) would seem to have symmetry on average $\Sigma = \mathbf{D}_1 \dot{+} (\mathbf{D}_k \times \mathbf{D}_k)$ for a fairly large value of k. The normality condition implies that if there is no instantaneous symmetry ($T = \mathbf{D}_1^N \times \mathbf{D}_1^N$) then there is an upper bound on the symmetry on average ($\Sigma \cong \mathbf{D}_1 \dot{+} (\mathbf{D}_2 \times \mathbf{D}_2)$) which is incompatible with Figure 2(b). We propose that $T = \Sigma \cong \mathbf{D}_1 \dot{+} (\mathbf{D}_k \times \mathbf{D}_k)$.

6   Conclusions

In this paper we have made the somewhat contentious proposal that turbulent Taylor vortices have no more symmetry in the time-average than they have at any instant in time. This is surprising since much of the work on symmetry of attractors and on the existence of symmetry only in the time-average has been motivated by turbulent Taylor vortices.

The obvious objection to our proposal is that this symmetry is not very evident in Figure 1(a), and certainly not as evident as in Figure 1(b). (This is also the case for the state observed in the Faraday experiment in a square container, Figure 2.) A possible explanation is that the underlying symmetry in our model is not exact in the experiment and that the instantaneous symmetry is necessarily approximate, hence difficult to deduce from the snapshot at an instant in time. However on average the discrepancies (which are essentially random) may be expected to cancel out so that the symmetry is much clearer.

There is a related state in the Couette-Taylor experiment called turbulent wavy vortices which has a time-periodic counterpart called wavy vortices, see Figure 4. Wavy vortices have not full but discrete azimuthal symmetry (in fact the symmetry is the product of an azimuthal symmetry with the mid-cylinder flip). In addition, there is the discrete axial symmetry that is present in Taylor vortices. An intriguing question for some time has been how to distinguish on grounds of symmetry between turbulent Taylor vortices and turbulent wavy vortices.

Again, it is often claimed that turbulent wavy vortices have no instantaneous symmetry. We propose that turbulent wavy vortices have wavy vortex symmetry instantaneously. Then our prediction that there is always full azimuthal symmetry in the time average leads to the expectation that on average there is Taylor vortex symmetry. In particular, we have that symmetry on average does not distinguish between turbulent Taylor vortices and turbulent wavy vortices. However, the instantaneous symmetry does distinguish between the two states.

In principle it is not difficult to devise an experimental test of our proposal. The primary bone of contention lies in the existence or nonexistence of the mid-cylinder flip as a instantaneous symmetry. Consider the detective v consisting of the absolute value of the difference of observations taken at two reflection related points (perhaps averaged over time). For turbulent Taylor vortices, we expect the value of v to be close to zero.

The numerical difficulty of what constitutes a value close to zero can be overcome as in [4] by computing v during a transition from turbulent wavy vortices to turbulent Taylor vortices. If our predictions are correct, v should jump from a value far from zero to a value close to zero. At the same time, if we average the difference in the observations without taking the absolute value, we should obtain a value w near zero throughout the transition, indicating the presence of the mid-cylinder flip in the time-average. Note that because the symmetry on average should be much cleaner than the instantaneous symmetry, we expect w to be much closer to zero than v, even for turbulent Taylor vortices.

A second and more difficult objection to our proposal has been raised by M. Field. In experiments, there is typically a natural direction in the variation of parameters. In the case of the Couette-Taylor experiment, it is usual to consider the transitions as the Reynolds number is increased. Then one prechaotic scenario is that solutions lie in a low-dimensional fixed-point subspace (Couette flow) from which there are bifurcations to solutions lying in higher-dimensional fixed-point subspaces (such as Taylor vortices and wavy vortices) and even to solutions with no symmetry. It is rather difficult to imagine why there should then be a transition back to a (now chaotic) solution in a lower-dimensional fixed-point subspace. However, this is what we are arguing to be the case with turbulent Taylor vortices. At present, we have no answer to this line of argument except to say that this is all the more reason to obtain a better understanding of the states that occur and only then to consider the transitions as parameters are varied.

We end by giving an argument in support of our proposal that we feel is particularly compelling. In phase space, the instantaneous symmetry corresponds to the symmetry of a single point in an attractor, whereas the symmetry on average corresponds to the symmetry of the whole attractor. It is clear in this setting that the symmetry of one point yields little or no information about additional symmetry that fixes the attractor but not the point itself. Hence in physical space, the symmetries (or structure) visible in the picture of a state at an instant in time should not give any clue to additional symmetry (structure) that may be present in the time average.

Evidence for this argument is provided by Figure 3(a) where there is no sign of the symmetry that appears on average 3(b) (but note that equation (4.1) predicts

252

the time average symmetry in Figure 3(b) as a consequence of the lack of structure in Figure 3(a)). Contrast this with the picture of turbulent Taylor vortices in Figure 1(a) where it is already possible to 'see' the putative symmetry on average.

In conclusion, we suggest that whenever it is possible to guess the existence of symmetry on average by looking at the instantaneous picture, then that symmetry is probably there even instantaneously. To compute additional symmetry on average, it is necessary to appeal to theory (such as equation (4.1) and/or to use detectives [4, 9]. By the same token, even the instantaneous symmetry may not be transparent by simply observing the solution. This suggests that it is necessary in general to use detectives to compute the instantaneous symmetry as well as the symmetry on average.

Acknowledgment I am grateful to Michael Dellnitz, Mike Field and Marty Golubitsky for helpful discussions.

References

[1] D. Armbruster and G. Dangelmayr. Coupled stationary bifurcations in nonflux boundary value problems, *Math. Proc. Camb. Phil. Soc.* **101** (1987) 167-192.

[2] P. Ashwin, P. Chossat and I. Stewart. Transitivity of orbits of maps symmetric under compact Lie groups, in preparation.

[3] P. Ashwin and I. Melbourne. Symmetry groups of attractors. *Arch. Rat. Mech. Anal.*, to appear.

[4] E. Barany, M. Dellnitz and M. Golubitsky. Detecting the symmetry of attractors. *Physica D* **67** (1993) 66-87.

[5] A. Brandstater and H.L. Swinney. Strange attractors in weakly turbulent Couette-Taylor flow, *Phys. Rev. A* **35** (1987) 2207-2220.

[6] P. Chossat and M. Golubitsky. Symmetry-increasing bifurcation of chaotic attractors, *Physica D32* (1988) 423-436.

[7] J.D. Crawford, M. Golubitsky, M.G.M. Gomes, E. Knobloch and I.N. Stewart. Boundary conditions as symmetry constraints, in *Singularity Theory and its Applications* Part II, (M. Roberts, I. Stewart, eds.), Lecture Notes in Math. **1463**, Springer, Berlin, 1991.

[8] M. Dellnitz, M. Golubitsky and I. Melbourne. Mechanisms of symmetry creation. In *Bifurcation and Symmetry* (E. Allgower et al, eds.) ISNM 104, 99-109, Birkhäuser (1992).

[9] M. Dellnitz, M. Golubitsky and M. Nicol. Symmetry of attractors and the Karhunen-Loève decomposition. *Appl. Math. Sci. Ser.* **100**, Springer, to appear.

[10] R.L. Devaney. *An introduction to chaotic dynamical systems.* Benjamin/Cummings: Menlo Park, CA (1985).

[11] M. Field. Equivariant dynamical systems. *Trans. Amer. Math. Soc.* **259** (1980) 185-205.

[12] M. Field. Local structure for equivariant dynamics, in *Singularity Theory and its Applications* Part II, (M. Roberts, I. Stewart, eds.), Lecture Notes in Math. **1463**, Springer, Berlin, 1991.

[13] M. Field, I. Melbourne and M. Nicol. Symmetric attractors for diffeomorphisms and flows. In preparation.

[14] H. Fujii, M. Mimura and Y. Nishiura. A picture of the global bifurcation diagram in ecological interacting and diffusing systems. *Physica D* **5** (1982) 1-42.

[15] B.J. Gluckman, P. Marcq, J. Bridger and J.P. Gollub. Time-averaging of chaotic spatialtemporal wave patterns. *Phys. Rev. Lett.* **71** 2034-2039.

[16] M. Golubitsky, I.N. Stewart and D.G. Schaeffer. *Singularities and Groups in Bifurcation Theory*, Vol 2. Appl. Math. Sci. **69** Springer, New York, 1988.

[17] M. Krupa. Bifurcations of relative equilibria. *SIAM J. Appl. Math.* **21** (1990) 1453-1486.

[18] I. Melbourne. Generalizations of a result on symmetry groups of attractors. *Pattern Formation: Symmetry Methods and Applications* (J. Chadam, W. Langford, eds.) Fields Institute Communications, AMS, 1994, to appear.

[19] I. Melbourne, M. Dellnitz and M. Golubitsky. The structure of symmetric attractors. *Arch. Rat. Mech. Anal.* **123** (1993) 75-98.

[20] P. Umbanhowar and H.L. Swinney. Private communication.

254

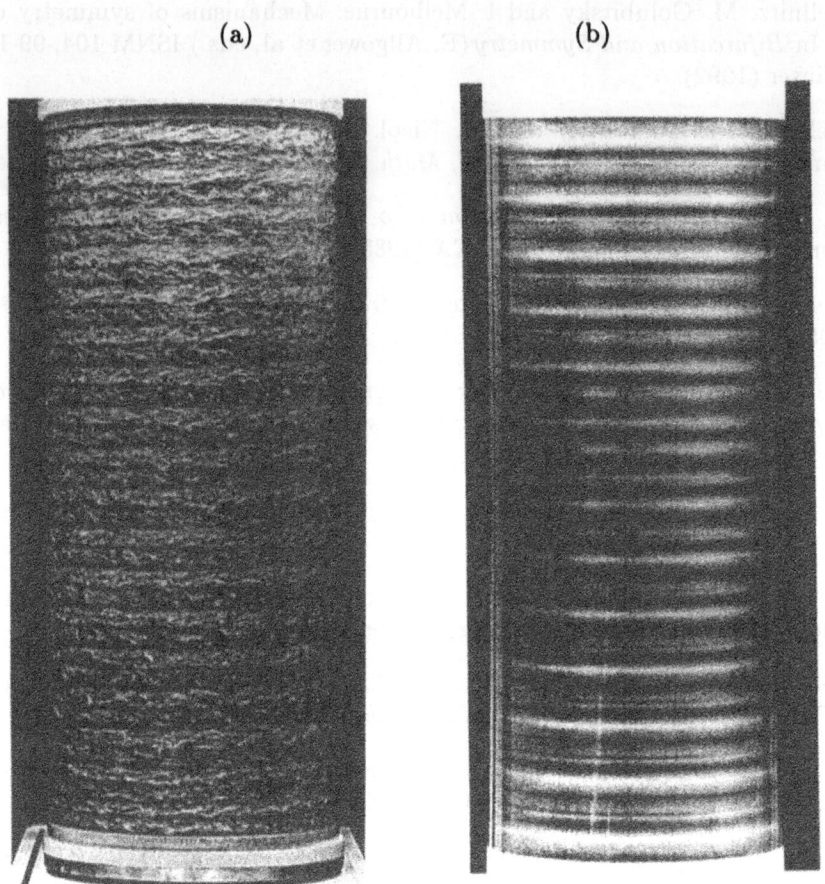

Figure 1: Turbulent Taylor vortices (a) and Taylor vortices (b) in the Couette-Taylor experiment. Pictures supplied by H.L. Swinney and R. Tagg. The state in (a), when time-averaged, looks like the state in (b) [20].

Figure 2: Instantaneous symmetry (a) and symmetry on average (b) of a state in the Faraday experiment in a square geometry. Pictures supplied by J.P. Gollub.

(a)

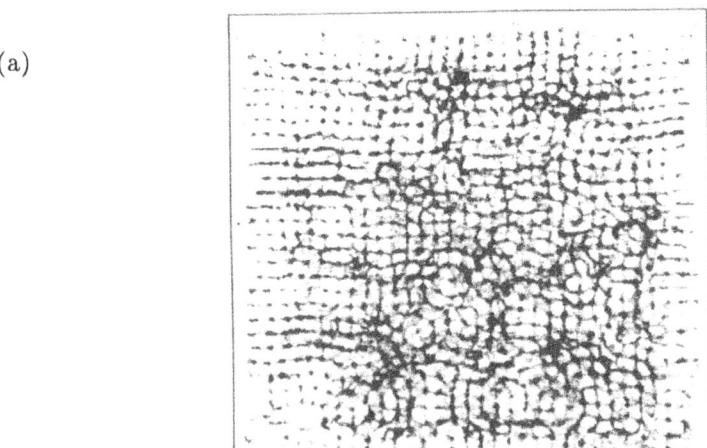

(b)

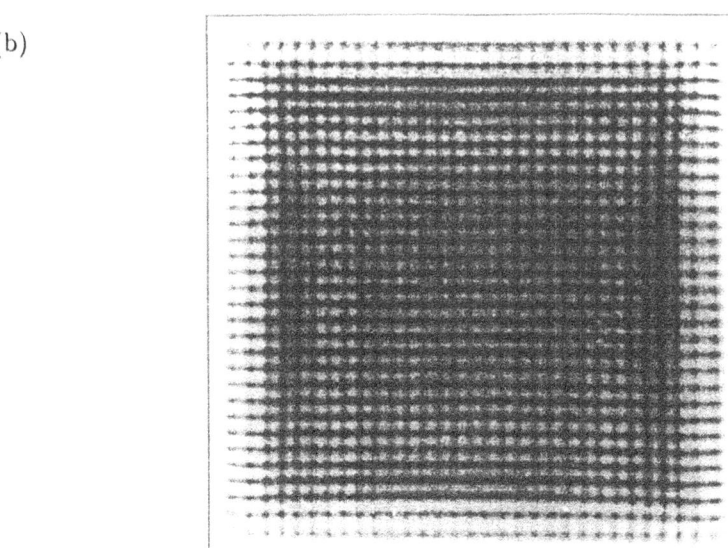

Figure 3: Instantaneous symmetry (a) and symmetry on average (b) of a state in the Faraday experiment in a circular geometry. Pictures supplied by J.P. Gollub.

(a)

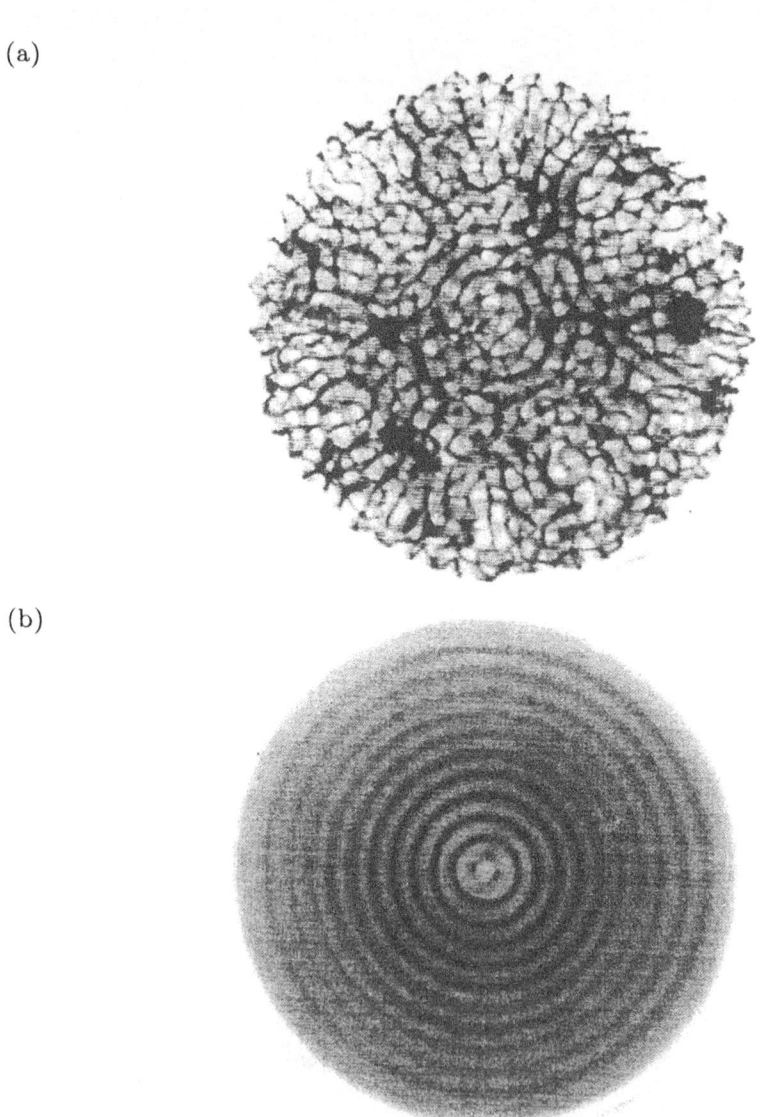

(b)

Figure 4: Wavy vortices in the Couette-Taylor experiment. Picture supplied by H.L Swinney.

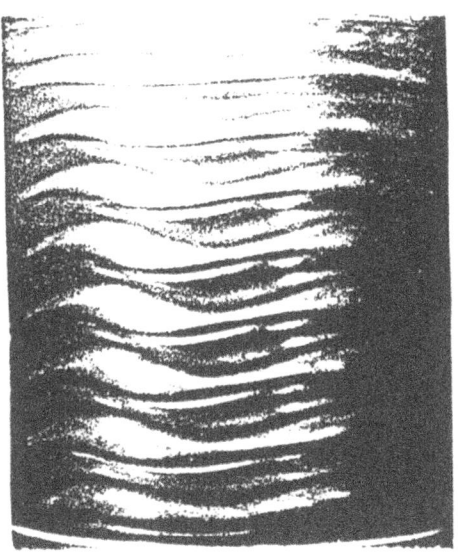

THE PATH FORMULATION OF BIFURCATION THEORY

JAMES MONTALDI
Institute Non-Linéaire de Nice
CNRS Unviversité de Nice
Sophia-Antipolis, FRANCE

ABSTRACT. We show how the path formulation of bifurcation theory can be made to work, and that it is (essentially) equivalent to the usual parametrized contact equivalence of Golubitsky and Schaeffer.

Introduction

In their original paper on imperfect bifurcation theory [GS79], Golubitsky and Schaeffer consider the so-called path formulation of bifurcation theory. However they had to abandon this approach as the calculations were mostly intractable, and they replaced it by their now standard distinguished parameter formulation. In this paper I describe how the path approach can be made to work thanks to recent advances in Singularity Theory, and I will show that it is (almost) equivalent to the distinguished parameter formulation. The new technology available, allowing the path formulation to work is two-fold: firstly, computations are facilitated by the development of computer algebra packages, and secondly the path-formulation itself is clarified by the introduction by J. Damon of $\mathcal{K}_V$-equivalence [D87].

As far as the computations go, Golubitsky and Schaeffer found the distinguished parameter approach more tractable thanks to a lemma ensuring that a bifurcation problem is finitely determined with respect to distinguished parameter contact equivalence if and only if it is finitely determined with respect to a restricted form of equivalence which is easier to compute. However, this lemma fails to hold as soon as there is more than 1 parameter, and in that case the computations of the full distinguished parameter equivalence are considerably harder than those of the path formulation. See for example [P].

There is one drawback at present to a coherent path formulation, and that is the distinction between the smooth (C^∞) and analytic theories. The problem arises as the modules of smooth vector fields tangent to certain varieties (discriminants) are not necessarily finitely generated. However, in the analytic category, all such modules are finitely generated. One can argue that this is not a problem, since one is dealing with finitely determined bifurcation problems, so that after a change of coordinates, they are analytic, and even polynomial. It seems that this shortcoming may be able to be overcome, but the details are still to be worked out.

Part of the object of this paper is to give a general description of Singularity Theory

P. Chossat (ed.), Dynamics, Bifurcation and Symmetry, 259–278.
© 1994 *Kluwer Academic Publishers.*

for the non-specialist; this is done in Section 1. This point of view which groups all the different equivalence relations together and puts "bifurcation equivalence" in a wider perspective, is not evident in Golubitsky and Schaeffer [GS85]. The remainder of the paper is organized as follows. Section 2 introduces Damon's notion of $\mathcal{K}_V$-equivalence. In Section 3, we give the main theorem (paragraph (3.3)), and in Section 4 we describe how one calculates generators of $\mathrm{Derlog}(\Delta)$, necessary for calculating $\mathcal{K}_\Delta$-tangent spaces. We conclude with some remarks and questions for symmetric bifurcations.

1 Singularities, Bifurcations and Paths

In this section we give a brief overview of the salient points of singularity theory necessary for understanding the results of this paper. We will be considering families of maps (or map-germs) from $\mathbf{R}^n$ to $\mathbf{R}^p$, and occasionally families of families of maps. The parameter spaces for the families will be denoted Λ, V or U according to the interpretation. Λ will be for the parameter space of a given bifurcation problem, U will always denote the parameter space (or base space) of a versal deformation, and V will be the base space for an arbitrary deformation. An excellent reference for the main results of singularity theory is C.T.C. Wall's survey paper [W], although much progress and consolidation has been made since then, and in particular Damon's introduction of Geometric Subgroups of $\mathcal{A}$ and $\mathcal{K}$ [D84], as the general class for singularity theoretic equivalences.

It should be borne in mind that we are really considering germs of maps and germs of deformations, so that all spaces such as $\mathbf{R}^n$ and $\mathbf{R}^p$ should really be considered as (small) neighbourhoods of the origin in $\mathbf{R}^n$ and $\mathbf{R}^p$ respectively. Similarly, although we say $\Lambda = \mathbf{R}^k$, we really mean that Λ is a neighbourhood of the origin in $\mathbf{R}^k$. We will not usually refer explicitly to germs, though there are occasional lapses — either through inconsistency or to remind the reader!

It should also be borne in mind that although we only make explicit reference in this section to real (C^∞) maps, one could equally well consider real analytic or complex analytic maps (or germs!). However, in Sections 2 and 3, there are certain results that only hold in the analytic categories.

(1.1) Bifurcation problems
For the purposes of this paper, a bifurcation problem is an equation of the form

$$g(x, \lambda) = 0,$$

where $g : \mathbf{R}^n \times \Lambda \to \mathbf{R}^p$ is a map-germ defined at $(0,0) \in \mathbf{R}^n \times \Lambda$, and $\Lambda = \mathbf{R}^k$. We view Λ as parameter space, and this distinction between Λ and $\mathbf{R}^n$ is reflected in the notion of equivalence used in bifurcation theory. Thus, a bifurcation problem is a system of p equations in n unknowns, with k parameters. In applications, it is common that $n = p$; however it makes no difference to the theory. We often refer to the map g as the bifurcation problem, with the equation $g = 0$ understood.

(1.2) Organizing centre
The organizing centre of (1.1) is obtained by putting $\lambda = 0$:

$$g_0(x) = g(x, 0),$$

so that $g_0 : \mathbf{R}^n \to \mathbf{R}^n$.

(1.3) Equivalence of bifurcation problems
Following Golubitsky and Schaeffer [GS85], two bifurcation problems

$$f, g : \mathbf{R}^n \times \Lambda \to \mathbf{R}^n$$

are said to be equivalent if there is a diffeomorphism $(x, \lambda) \mapsto (H(x, \lambda), h(\lambda))$ and an invertible $p \times p$ matrix $S(x, \lambda)$ depending on x and λ such that

$$f(H(x, \lambda), h(\lambda)) = S(x, \lambda)g(x, \lambda).$$

We say f and g are *bifurcation equivalent*, or $\mathcal{B}$-equivalent. In the special case that h is the identity, the equivalence is called *restricted bifurcation equivalence*.
    Putting $\lambda = 0$ we arrive at a natural equivalence of organizing centres,

$$f_0(H(x)) = S(x)g_0(x),$$

where now S is a $p \times p$ matrix depending only on x, and H is a change of coordinates on $\mathbf{R}^n$. This equivalence is called *contact equivalence*, or $\mathcal{K}$-equivalence; it was introduced into singularity theory by J. Mather in the late 1960's. (It is also sometimes known as V-equivalence [AGV].) Thus, bifurcation equivalence is a parametrized version of contact equivalence — see the next paragraph.

(1.4) Deformations and their equivalence
One of the important applications of singularity theory is to the study of how maps deform. One is able to deal in the same way with many types of equivalence (contact, bifurcation, right, left-right, equivariant, . . .).
    Let $f : X \to Y$ be a map (e.g. $X = \mathbf{R}^n$ for contact equivalence, $X = \mathbf{R}^n \times \Lambda$ for bifurcation equivalence). A deformation of f is a map

$$F : X \times U \to Y$$

satisfying $F(\cdot, 0) = f$. The deformed (or perturbed) map $x \mapsto F(x, u)$ is denoted F_u. If F is a deformation of f, then any map $\gamma : V \to U$ defines another deformation of f, denoted $\gamma^* F$, by

$$\gamma^* F(x, v) := F(x, \gamma(v)).$$

The deformation $\gamma^* F$ is said to be *induced from F by γ*. This idea is central to what follows. (Note tha since we are really talking about germs, we automatically have $\gamma(0) = 0$.)

Suppose now that $\mathcal{G}$ is one of the equivalence relations of singularity theory (contact, bifurcation, . . .). Then $\mathcal{G}$ defines an equivalence of deformations, sometimes denoted $\mathcal{G}_{\text{un}}$, as follows. Let $F_1 : X \times U_1 \to Y$ and $F_2 : X \times U_2 \to Y$ be two deformations of f. Then F_1 and F_2 are said to be $\mathcal{G}_{\text{un}}$-equivalent if there is a diffeomorphism $h : U_1 \to U_2$ such that $F_{1,u}$ is $\mathcal{G}$-equivalent to $F_{2,h(u)}$ for all $u \in U_1$. Moreover, the equivalences must depend smoothly on the parameter.

For contact equivalence ($\mathcal{G} = \mathcal{K}$), equivalence of deformations is precisely bifurcation equivalence. Thus $\mathcal{K}_{\text{un}} = \mathcal{B}$.

For bifurcation equivalence, equivalence of deformations is a little more complex. A deformation of a bifurcation problem $g(x, \lambda)$ is a map

$$\tilde{g} : \mathbf{R}^n \times \Lambda \times V \to \mathbf{R}^p$$

such that $g(x, \lambda) = \tilde{g}(x, \lambda, 0)$. Two deformations $\tilde{g}_1$ and $\tilde{g}_2$ of a bifurcation problem g are deformation bifurcation equivalent ($\mathcal{B}_{\text{un}}$-equivalent) if there are changes of coordinates $(x, \lambda, v) \mapsto (H(x, \lambda, v), h_1(\lambda, v), h_2(v))$ and a matrix $S(x, \lambda, v)$ such that

$$\tilde{g}_1(x, \lambda, v) = S(x, \lambda, v)\, \tilde{g}_2(H(x, \lambda, v), h_1(\lambda, v), h_2(v)).$$

(1.5) Versal deformations

One of the basic notions of singularity theory is that of a versal deformation; it applies to all the usual equivalences. A versal deformation is a deformation which contains (up to the equivalence in question) any deformation of the singularity. For contact equivalence, this reads as follows.

Let $g_0 : \mathbf{R}^n \to \mathbf{R}^p$ be given, and let $G : \mathbf{R}^n \times U \to \mathbf{R}^p$ be a deformation of g_0 (so that $g_0 = G(\cdot, 0)$). One says that G is a *versal deformation* of g_0 if for any deformation $g : \mathbf{R}^n \times V \to \mathbf{R}^p$ of g_0 there is a map $\gamma : V \to U$ such that $g(x, v)$ is parametrized contact equivalent to $G(x, \gamma(v))$.

A deformation

$$G : \mathbf{R}^n \times \Lambda \times U \to \mathbf{R}^p$$

of a bifurcation problem $g : \mathbf{R}^n \times \Lambda \to \mathbf{R}^p$ is said to be a *versal deformation* of g if for every deformation $\tilde{g} : \mathbf{R}^n \times \Lambda \times V \to \mathbf{R}^p$ of g there is a map $\gamma : V \to U$ such that $\tilde{g}(x, \lambda, v)$ and $G(x, \lambda, \gamma(v))$ are parametrized bifurcation equivalent.

There is a simple algebraic criterion for deciding whether a given deformation is versal, in terms of the tangent or normal spaces — see paragraph (1.9).

(1.6) Example

Consider the organizing centre $g_0(x) = x^3$ (here $n = p = 1$). There are several well-known bifurcation problems with this organizing centre. For example,

Pitchfork: $g(x, \lambda) = x^3 - \lambda x$;

Hysteresis: $g(x, \lambda) = x^3 - \lambda$.

A versal deformation of g_0 is given by

$$G(x, u_1, u_2) = x^3 + u_1 x + u_2,$$

where $U = \mathbf{R}^2$. The two bifurcation problems are induced by the maps $\gamma : \mathbf{R} \to \mathbf{R}^2$ given by,

Pitchfork: $\gamma(\lambda) = (-\lambda, 0)$;

Hysteresis: $\gamma(\lambda) = (0, -\lambda)$.

Versal deformations of the two bifurcation problems are given by

Pitchfork: $\tilde{G}(x, \lambda, u_1, u_2) = x^3 - \lambda x + u_1 + u_2 \lambda$;

Hysteresis: $\tilde{G}(x, \lambda, u) = x^3 - \lambda + ux$.

Versal deformations are often called universal unfoldings [GS85]. The word versal is used in singularity theory rather than universal, since the prefix 'uni' refers to uniqueness, and versal deformations are not unique. The difference between a deformation and an unfolding is mainly notational, and need not concern us here.

(1.7) Tangent spaces
Associated to any map (germ), and any equivalence relation in singularity theory, is the 'tangent space' of the map in question. It is essentially the tangent space to the equivalence class containing the map. To calculate it, one uses the given class of diffeomorphisms, and differentiates to obtain a tangent space. It is a subset of all infinitesimal deformations of the given map.

Notation: We denote by $\mathcal{E}_n$ the ring of C^∞ functions on $\mathbf{R}^n$, by $\mathcal{E}_\Lambda$ the functions on Λ, and $\mathcal{E}_{n,\lambda}$ consists of the functions on $\mathbf{R}^n \times \Lambda$. Similarly, Θ_n denotes the $\mathcal{E}_n$-module of vector fields on $\mathbf{R}^n$. The (maximal) ideal of functions vanishing at $0 \in \mathbf{R}^p$ is denoted m_p, and consequently $m_p \Theta_p$ is the $\mathcal{E}_p$-module of vector fields on $\mathbf{R}^p$ that vanish at the origin. Finally, we denote by $\Theta_{n,\lambda}$ the $\mathcal{E}_{n,\lambda}$-module of vector fields on $\mathbf{R}^n$ parametrized by $\lambda \in \Lambda$.

Let f be a map (organizing centre, bifurcation problem, or whatever), and $\mathcal{G}$ an equivalence relation (contact, bifurcation, or whatever). The space of infinitesimal deformations of f is denoted Θ_f consists of vector fields along f, that is, vector fields on $\mathbf{R}^a$ with values in $\mathbf{R}^b$ (more brutally, if $f \in C^\infty(\mathbf{R}^a, \mathbf{R}^b)$ then $\Theta_f = C^\infty(\mathbf{R}^a, \mathbf{R}^b)$).

The $\mathcal{G}$-tangent space of f is a subspace of Θ_f, denoted $T\mathcal{G}_e \cdot f$ (the $_e$ is for 'extended'[1]). Note that Θ_f is a module over the ring of smooth functions on $\mathbf{R}^a$.

For $g_0 : \mathbf{R}^n \to \mathbf{R}^p$, and $\mathcal{K}$-equivalence, one finds that

$$TK_e \cdot g_0 = tg_0(\Theta_n) + g_0^*(m_p\Theta_p).$$

The term $tg_0(\Theta_n)$ is the image of vector fields under the tangent mapping tg_0 of g_0; the term $g_0^*(m_p\Theta_p)$ is the $\mathcal{E}_n$-module generated by the pull-backs of vector fields on $\mathbf{R}^p$, that is by the set of vector fields of the form $v \circ g_0$, with $v \in m_p\Theta_p$. Such composites are vector fields along g_0.

For a bifurcation problem $g : \mathbf{R}^n \times \Lambda \to \mathbf{R}^p$, the tangent space for bifurcation equivalence is given by

$$TB_e \cdot g = t_1 g(\Theta_{n,\lambda}) + g^*(\Theta_{p,\lambda}) + t_2 g(\Theta_\Lambda).$$

Here $t_1 g$ and $t_2 g$ mean differentiating with respect to the first $(\mathbf{R}^n)$ and second (Λ) variables, respectively. Note that each of the first two terms is an $\mathcal{E}_{n,\lambda}$-module, while the third term is merely an $\mathcal{E}_\Lambda$-module. The whole is therefore only an $\mathcal{E}_\Lambda$-module. Golubtsky and Schaeffer [GS85] denote this tangent space by $T(g)$. Their restricted tangent space $RT(g)$ is given by the first two terms only (the third is omitted by forbidding changes in the parameter) and is therefore an $\mathcal{E}_{n,\lambda}$-module. In [MM], $RT(g)$ is denoted $TK_{\text{rel}} \cdot g$.

(1.8) Normal spaces

Given the (extended) tangent space $T\mathcal{G}_e \cdot f \subset \Theta_f$ one defines the normal space as the quotient:

$$N\mathcal{G} \cdot f = \frac{\Theta_f}{T\mathcal{G}_e \cdot f}.$$

This of course holds for $\mathcal{G} = \mathcal{K}, \mathcal{B}$ etc. The *codimension* of f with respect to $\mathcal{G}$-equivalence is defined to be,

$$\text{cod}_\mathcal{G} f = \dim_{\mathbf{R}} N\mathcal{G} \cdot f.$$

In the case that $\dim \Lambda = 1$, one has that $T(g)$ has finite codimension if and only if $RT(g)$ does [GS85, p. 127]. This fact allows Golubitsky and Schaeffer to make their theory computable: being an $\mathcal{E}_{n,\lambda}$-module makes $RT(g)$ much easier to compute than $T(g) = TB_e \cdot g$.

(1.9) Versality theorem

One of the basic theorems of singularity theory gives a simple criterion for determining whether a given deformation is versal (which works for all equivalence relations $\mathcal{G}$ such as contact, bifurcation, . . .all "geometric subgroups" of $\mathcal{A}$ and $\mathcal{K}$ [D84]).

---

[1] The 'unextended' tangent space $T\mathcal{G} \cdot f$ is defined in the same way, but using only the vector fields that vanish at 0; it is used in conditions for finite determinacy.

Let f be a map (germ) and G one of the singularity theory equivalences appropriate to f. Let $F = f + u_1\phi_1 + \cdots u_r\phi_r$ be a deformation of f, with $\phi_1, \ldots, \phi_r \in \Theta_f$. Then F is a versal deformation if and only if $\{\phi_1, \ldots, \phi_r\}$ spans $N\mathcal{G} \cdot f$ as a real vector space. In other words, F is versal if and only if

$$T\mathcal{G}_e \cdot f + \mathbf{R}\{\phi_1, \ldots, \phi_r\} = \Theta_f.$$

The codimension of a singularity is thus the number of parameters needed for a versal deformation. The space U is called the *base space* of the versal deformation.

(1.10) Discriminant

Let $G : \mathbf{R}^n \times U \to \mathbf{R}^p$ be a versal deformation of the map $g_0 : \mathbf{R}^n \to \mathbf{R}^p$ (with respect to $\mathcal{K}$-equivalence). For each $u \in U$, let $G_u : \mathbf{R}^n \to \mathbf{R}^p$ be the map given by

$$G_u(x) = G(x, u).$$

The following conditions on $u \in U$ are equivalent:

(i) *there is an $x \in \mathbf{R}^n$ such that $G(x, u) = 0$ and G_u is singular at x, and*

(ii) *u is a singular value of the projection $\pi_G : G^{-1}(0) \to U$ given by $\pi_G(x, u) = u$.*

This fact is easy to prove. The set of all such u is called the *discriminant* of the versal deformation G, denoted $\Delta = \Delta_G$. It is the basic geometric object for the remainder of this paper. It is a hypersurface in U, i.e. given by one equation $h(u) = 0$ with $h : U \to \mathbf{R}$.

For example, in the case $n = p$ (central to bifurcation theory), for each $u \in U$ the set $G_u^{-1}(0)$ is finite (otherwise g_0 would not be of finite codimension, and so would not have a versal deformation). The number of elements in $G_u^{-1}(0)$ is locally constant on an open dense set in U, whose complement is precisely the discriminant. Thus, the discriminant consists of those points u for which G_u has multiple roots. Over $\mathbf{C}$, the number of elements in $G_u^{-1}(0)$ is constant, for $u \notin \Delta$, not merely locally constant.

It is a central observation for this paper that:

Suppose $g = \gamma^* G$, then bifurcation points of g correspond under γ to points of image$(\gamma) \cap \Delta_G$.

2 $\mathcal{K}_\Delta$-equivalence

A new equivalence relation on maps was introduced a few years ago by J. Damon [D87], called $\mathcal{K}_V$-equivalence (or $\mathcal{K}_\Delta$-equivalence). It is a generalization of Mather's contact equivalence (see (1.3) above), which has been finding many applications. For an application to caustics see [M] and to bifurcations of periodic points see [BF].

Consider maps $\gamma : \Lambda \to U$ and a subset (subvariety) $\Delta \subset U$. The geometrical notion captured by $\mathcal{K}_\Delta$ equivalence is the *contact of γ with Δ*, which is clearly important since bifurcations correspond to points of intersection of γ with the discriminant Δ.

(2.1) Definition
Two maps $\gamma_1, \gamma_2 : \Lambda \to U$ are said to be $\mathcal{K}_\Delta$-equivalent if there exist diffeomorphisms h of Λ, and H of $\Lambda \times U$ satisfying

- $H(\lambda, u) = (h(\lambda), \bar{H}(\lambda, u))$, for some $\bar{H} : \Lambda \times U \to U$,

- $u \in \Delta \Rightarrow \bar{H}(\lambda, u) \in \Delta$, and

- $\gamma_1(h(\lambda)) = \bar{H}(\lambda, \gamma_2(\lambda))$, for all $\lambda \in \Lambda$.

In other words, H maps the graph of γ_2 to the graph of γ_1, whilst preserving Δ. In the case that $\Delta = \{0\}$, then $\mathcal{K}_\Delta$-equivalence reduces to $\mathcal{K}$-equivalence. It is clear that if γ_1 and γ_2 are $\mathcal{K}_\Delta$ equivalent, then $\gamma_1^{-1}(\Delta)$ and $\gamma_2^{-1}(\Delta)$ are diffeomorphic; however in general the converse is not true (it is true if Δ is smooth).

(2.2) Derlog(Δ)
For most varieties $\Delta \subset U$, it is not easy to characterize the set of diffeomorphisms preserving Δ. However, the infinitesimal version is often not so hard. For a vector field $\xi \in \Theta_U$ to integrate to a 1-parameter family of diffeomorphisms preserving Δ, it is necessary and sufficient that ξ be tangent to Δ. Note that if Δ is singular, then tangent to Δ means tangent to each stratum of some natural stratification of Δ.

The $\mathcal{E}_U$-module of vector fields tangent to Δ has the unfortunate name Derlog(Δ), for reasons that go well beyond this paper [S]．

A few words about the structure of Derlog(Δ) are in order. Firstly, Δ is a hypersurface, given by the equation $h(u) = 0$, so that

$$\mathrm{Derlog}(\Delta) = \{\theta \in \Theta_U \mid \theta(h) \in \langle h \rangle\},$$

where $\langle h \rangle$ is the ideal generated by h. This is because if θ is tangent to Δ and as h is constant on Δ, then $\theta(h) = 0$ on Δ. In other words, $\theta \in \mathrm{Derlog}(\Delta)$ if and only if there exists $f \in \mathcal{E}_U$ for which $\theta(h) = fh$.

We can define a submodule Derlog(h) $\subset$ Derlog(Δ), by

$$\mathrm{Derlog}(h) = \{\theta \in \Theta_U \mid \theta(h) = 0\}.$$

It consists of those vector fields that are tangent to all level sets of h, and not just to the zero level set Δ. Clearly, Derlog(h) depends on the choice of function used to define Δ, whereas Derlog(Δ) does not.

Suppose now that h is weighted homogeneous, so that there are integers $w_1, \ldots, w_\ell$ (where $\ell = \dim U$), such that

$$h(t^{w_1} u_1, \ldots, t^{w_\ell} u_\ell) = t^d h(u_1, \ldots, u_\ell),$$

for some d — the *degree* of h with respect to the given weights. Then Euler's formula states

$$\sum_{j=1}^{\ell} w_j u_j \frac{\partial}{\partial u_j} h = d.h.$$

The vector field

$$e = \sum_{j=1}^{\ell} w_j u_j \frac{\partial}{\partial u_j},$$

is called the Euler vector field for these weights, and we have $e(h) = d.h$.

Suppose now that $\theta \in \mathrm{Derlog}(\Delta)$, with $\theta(h) = fh$. Then the vector field $\bar{\theta} = \theta - d^{-1}fe$ satisfies $\bar{\theta} \in \mathrm{Derlog}(h)$, as is easy to see. Consequently, there is a natural projection $\mathrm{Derlog}(\Delta) \longrightarrow \mathrm{Derlog}(h)$, $\theta \to \bar{\theta}$ whose kernel is precisely $\mathcal{E}_U.e$. Thus, in the weighted homogeneous case,

$$\mathrm{Derlog}(\Delta) = \mathcal{E}_U.e \oplus \mathrm{Derlog}(h),$$

a direct sum of $\mathcal{E}_U$-modules. It is generally easier to calculate $\mathrm{Derlog}(h)$ than $\mathrm{Derlog}(\Delta)$.

(2.3) Liftable vector fields

There is an important geometric characterization of elements of $\mathrm{Derlog}(\Delta)$ when Δ is the discriminant of a map $\pi_G : G^{-1}(0) \to U$, namely they are the *liftable* vector fields. However, this only holds in the analytic categories (real and complex), and not in general for C^∞ maps and vector fields.

In general, let $f : X \to U$ be a map. A vector field $\eta \in \Theta_U$ is said to be *liftable over f* (or *via f*) if there is a vector field $\xi \in \Theta_X$ such that $df_x(\xi_x) = \eta_{f(x)}$. It is not hard to show that any liftable vector field must be tangent to the discriminant $\Delta(f)$ of f (integrating ξ and η give diffeomorphisms r of X and ℓ of U such that $f \circ r = \ell \circ f$, so that ℓ must preserve $\Delta(f)$).

For certain maps the converse is also true. In particular, Looijenga proved [L] that if $G : \mathbf{R}^n \times U \to \mathbf{R}^p$ is a versal deformation, and $\pi_G : G^{-1}(0) \to U$ the associated projection, then a vector field $\eta \in \Theta_U$ is liftable over π_G if and only if $\eta \in \mathrm{Derlog}(\Delta_G)$ (recall that Δ_G is the discriminant of π_G, see (1.10)).

More recently the general relationship between liftable vector fields and vector fields tangent to a discriminant has been clarified by Bruce, du Plessis and Wilson [BdPW].

(2.4) Example

Let $U = \mathbf{R}^2$ and Δ be defined by the equation $h(u_1, u_2) = 4u_1^3 + 27u_2^2 = 0$ (this is the equation for the discriminant of the versal deformation of $g_0(x) = x^3$ given in (1.6)). Then $\mathrm{Derlog}(\Delta)$ is generated over $\mathcal{E}_U$ by the two vector fields

$$e = \begin{pmatrix} 2u_1 \\ 3u_2 \end{pmatrix} \quad \text{and} \quad \begin{pmatrix} 9u_2 \\ -2u_1^2 \end{pmatrix}.$$

It is easy to show that any vector field annihilating h is a multiple of the second generator. Here $\begin{pmatrix} \alpha \\ \beta \end{pmatrix} = \alpha \frac{\partial}{\partial u_1} + \beta \frac{\partial}{\partial u_2}$.

Note that as discussed in the previous paragraph, these two vector fields are indeed liftable. The Euler field e lifts to $\begin{pmatrix} x \\ 2u_1 \\ 3u_2 \end{pmatrix}$, while the other generator lifts to $\begin{pmatrix} 3x^2 + 2u_1 \\ 9u_2 \\ -2u_1^2 \end{pmatrix}$. These vector fields on $\mathbf{R} \times U$ are both tangent to $G^{-1}(0)$, as is easily checked.

For futher examples, see Section 4 below.

(2.5) $\mathcal{K}_\Delta$ tangent and normal spaces

Let $\gamma : \Lambda \to U$, and $\Delta \subset U$ a subvariety. Then the extended $\mathcal{K}_\Delta$-tangent space to γ is

$$TK_{\Delta,e} \cdot \gamma = t\gamma(\Theta_\Lambda) + \gamma^* \operatorname{Derlog}(\Delta).$$

Notice how this is very similar to the ordinary $\mathcal{K}$-tangent space, except that $m_p \Theta_p$ has been replaced by $\operatorname{Derlog}(\Delta)$. This is because instead of diffeomorphisms preserving the origin in $\mathbf{R}^p$, here we are considering diffeomorphisms preserving Δ.

The $\mathcal{K}_\Delta$-normal space is of course defined by

$$NK_\Delta \cdot \gamma = \Theta_\gamma / TK_{\Delta,e} \cdot \gamma.$$

(2.6) Example

For the paths defining the pitchfork and hysteresis bifurcations (1.6), and the discriminant $\Delta \subset U$, we can compute the $\mathcal{K}_\Delta$ tangent and normal spaces. Generators of the module $\operatorname{Derlog}(\Delta)$ are given in (2.4).

For the pitchfork $\gamma(\lambda) = (-\lambda, 0)$, so that

$$t\gamma(\Theta_\Lambda) = (\mathcal{E}_\Lambda, 0),$$

while

$$\gamma^*(\operatorname{Derlog}(\Delta)) = (\lambda \mathcal{E}_\Lambda, \lambda^2 \mathcal{E}_\Lambda).$$

A similar calculation for the hysteresis bifurcation gives

$$t\gamma(\Theta_\Lambda) = (0, \mathcal{E}_\Lambda),$$

$$\gamma^*(\operatorname{Derlog}(\Delta)) = (\lambda \mathcal{E}_\Lambda, \lambda \mathcal{E}_\Lambda).$$

Pitchfork: $TK_{\Delta,e} \cdot \gamma = (\mathcal{E}_\Lambda, \lambda^2 \mathcal{E}_\Lambda)$,

Hysteresis: $TK_{\Delta,e} \cdot \gamma = (\lambda \mathcal{E}_\Lambda, \mathcal{E}_\Lambda)$.

The normal spaces are thus given by

Pitchfork: $N\mathcal{K}_\Delta \cdot \gamma \simeq \mathbf{R}\{(0,1), (0,\lambda)\}$,

Hysteresis: $N\mathcal{K}_\Delta \cdot \gamma \simeq \mathbf{R}\{(1,0)\}$.

The $\mathcal{K}_\Delta$-codimension of the first path is thus 2, while that of the second is only 1.

Generators for Derlog(Δ) for organizing centres of low codimension: see text for explanations

Type	$G(x,u)$	Generators of Derlog(Δ)
A_1	$x^2 + a$	a.
A_2	$x^3 + ax + b$	$\begin{pmatrix} 2a \\ 3b \end{pmatrix}; \begin{pmatrix} 9b \\ -2a^2 \end{pmatrix}$.
A_3	$x^4 + ax^2 + bx + c$	$\begin{pmatrix} 2a \\ 3b \\ 4c \end{pmatrix}; \begin{pmatrix} 6b \\ 8c - 2a^2 \\ -ab \end{pmatrix}, \begin{pmatrix} 48c - 4a^2 \\ 12ab \\ 16ac - 9b^2 \end{pmatrix}$.
A_4	$x^5 + ax^3 + bx^2 + cx + d$	$\begin{pmatrix} 2a \\ 3b \\ 4c \\ 5d \end{pmatrix}; \begin{pmatrix} 15b \\ 20c - 6a^2 \\ 25d - 4ab \\ -2ac \end{pmatrix}, \begin{pmatrix} 40c - 6a^2 \\ 50d - 17ab \\ 8ac - 12b^2 \\ 15ad - 6bc \end{pmatrix}, \begin{pmatrix} 50d - 2ab \\ -4ac - 3b^2 \\ 30ad - 10bc \\ 15bd - 8c^2 \end{pmatrix}$.
A_5	$x^6 + ax^4 + bx^3 + cx^2 + dx + e$	$\begin{pmatrix} 2a \\ 3b \\ 4c \\ 5d \\ 6e \end{pmatrix}; \begin{pmatrix} 9b \\ 12c - 4a^2 \\ 15d - 3ab \\ 18e - 2ac \\ -ad \end{pmatrix}, \begin{pmatrix} 40c - 8a^2 \\ 50d - 22ab \\ 60e + 4ac - 15b^2 \\ 10ad - 10bc \\ 16ae - 5bd \end{pmatrix}, \begin{pmatrix} 75d - 6ab \\ 90e - 10ac - 9b^2 \\ 45ad - 27bc \\ 15bd + 30ae - 20c^2 \\ 27be - 10cd \end{pmatrix}, \begin{pmatrix} 180e - 4ac \\ -6bc - 10ad \\ 120ae - 8c^2 - 15bd \\ 90be - 30cd \\ 48ce - 25d^2 \end{pmatrix}$.
$I_{2,2}$	$(x^2 - y^2 + az + by + c,\, zy + d)$	$\begin{pmatrix} a \\ b \\ 2c \\ 2d \end{pmatrix}; \begin{pmatrix} 3b \\ -3a \\ 4d + 2ab \\ -c \end{pmatrix}, \begin{pmatrix} 16c - 3a^2 \\ 32d + ab \\ 2ac - 24bd \\ 6ad \end{pmatrix}, \begin{pmatrix} 32d + ab \\ -16c - 3b^2 \\ 2bc + 24ad \\ 6bd \end{pmatrix}$.
$II_{2,2}$	$(x^2 + ay + c,\, y^2 + bx + d)$	$\begin{pmatrix} a \\ b \\ 2c \\ 2d \end{pmatrix}; \begin{pmatrix} 3a \\ -3b \\ 2c \\ -2d \end{pmatrix}, \begin{pmatrix} 0 \\ 8d \\ 3a^2b \\ -8bc \end{pmatrix}, \begin{pmatrix} 8c \\ 0 \\ -8ad \\ 3ab^2 \end{pmatrix}$.

(2.7) $\mathcal{K}_\Delta$-versal deformations

As with other equivalence relations, a deformation $\Gamma : \Lambda \times V \to U$ of γ is said to be $\mathcal{K}_\Delta$-versal if any deformation of γ is equivalent to one induced from Γ. Also as with other equivalence relations (of the singularity theory type) one has the following result:

The deformation Γ of γ given by

$$\Gamma(\lambda, v_1, \ldots, v_r) = \gamma(\lambda) + \sum_j v_j \phi_j(\gamma)$$

is versal if and only if the ϕ_j span $N\mathcal{K}_\Delta \cdot \gamma$.

In the example above, versal deformations of γ are given by

Pitchfork: $\Gamma(\lambda, u_1, u_2) = (-\lambda, u_1 + u_2\lambda)$,

Hysteresis: $\Gamma(\lambda, u) = (u, -\lambda)$.

These expressions should be compared to the versal deformations of the two bifurcation problems given in (1.6)

(2.8) Finite determinacy

Another property of maps considered in singularity theory is finite determinacy. For $\mathcal{K}_\Delta$-equivalence, this reads as follows.

A path $\gamma : \Lambda \to U$ is k-determined with respect to $\mathcal{K}_\Delta$-equivalence if

$$m_\Lambda^{k+1} \Theta_\gamma \subset T\mathcal{K}_\Delta \cdot \gamma,$$

where $T\mathcal{K}_\Delta \cdot \gamma \subset T\mathcal{K}_{\Delta,e} \cdot \gamma$ is the tangent space given by

$$T\mathcal{K}_\Delta \cdot \gamma = t\gamma(m_n \Theta_n) + \gamma^* \operatorname{Derlog}_0(\Delta).$$

Here $\operatorname{Derlog}_0(\Delta) = \operatorname{Derlog}(\Delta) \cap m_U \Theta_U$ consists of those vector fields tangent to Δ that vanish at 0. Note that if G is a miniversal deformation (i.e. $\dim U$ is as small as possible) then $\operatorname{Derlog}_0(\Delta) = \operatorname{Derlog}(\Delta)$, since then $\{0\}$ is a stratum of Δ. The proof of this is similar to the standard proofs of finite determinacy for $\mathcal{R}$- and $\mathcal{K}$-equivalence using the homotopy method and Nakayama's lemma, see for example [AGV].

3   Paths and bifurcation problems

To recapitulate, let $g_0 : \mathbf{R}^n \to \mathbf{R}^p$ be a $\mathcal{K}$-finite map (germ), and $G : \mathbf{R}^n \times U \to \mathbf{R}^p$ be a versal deformation of g_0. Any path (map) $\gamma : \Lambda \to U$, induces a deformation (or bifurcation problem) $\gamma^* G$ of g_0 given by

$$(\gamma^* G)(x, \lambda) = G(x, \gamma(\lambda)).$$

Moreover, the bifurcation points of $\gamma^* G$ are the points $\lambda \in \Lambda$ for which $\gamma(\lambda) \in \Delta_G$. For this section, we assume that all maps are (real or complex) analytic.

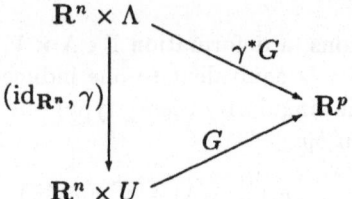

Since G is versal, any deformation $g(x, \lambda)$ of g_0 is (bifurcation) equivalent to one of the form $\gamma^* G$ for some path γ, as explained in (1.5).

Thus, for a given organizing centre g_0, we have a map

$$\text{paths } [\Lambda \to U] \quad \longrightarrow \quad \text{bifurcation problems with organizing centre } g_0.$$

The precise form of this map depends, of course, on our choice of versal deformation.

(3.1) Deformations of paths and bifurcation problems
Suppose now that we deform the path γ. Let $\Gamma : \Lambda \times V \to U$ be such a deformation (V is the parameter space), so that $\Gamma(\lambda, 0) = \gamma(\lambda)$. This then induces a deformation $\Gamma^* G$ of the bifurcation problem $\gamma^* G$ by

$$\Gamma^* G : \mathbf{R}^n \times \Lambda \times V \quad \longrightarrow \quad \mathbf{R}^p$$
$$(x, \lambda, v) \quad \mapsto \quad G(x, \Gamma(\lambda, v)).$$

(3.2) The morphism Ψ_γ
This correspondence from deformations of γ to deformations of $\gamma^* G$ can be infinitesimalized, to obtain a map associating to any infinitesimal deformation of γ an infinitesimal deformation of the bifurcation problem $g = \gamma^* G$:

$$\Psi_\gamma : \Theta_\gamma \longrightarrow \Theta_g,$$

where $g = \gamma^* G$. If $G(x, u) = g_0(x) + \sum_{j=1}^d u_j \phi_j(x)$ then it is easy to see that

$$\Psi_\gamma(\xi_1(\lambda), \ldots, \xi_d(\lambda)) = \xi_1(\lambda)\phi_1(x) + \cdots + \xi_d(\lambda)\phi_d(x).$$

The map Ψ_γ is $\mathcal{E}_\Lambda$-linear; in other words, it is a morphism of $\mathcal{E}_\Lambda$-modules. Note that *a priori* Θ_g is an $\mathcal{E}_{n,\lambda}$-module, and can therefore be considered as an $\mathcal{E}_\Lambda$-module, although as such it is not finitely generated.

As has already been pointed out, the important geometry of a perturbation of γ is how it meets the discriminant Δ: bifurcation points of $\gamma^* G$ correspond to points of $\gamma^{-1}(\Delta)$. It is thus reasonable to consider $\mathcal{K}_\Delta$-equivalence of paths γ, as an alternative to bifurcation equivalence of bifurcation problems g. The following theorem shows that given γ and $g = \gamma^* G$, the notions of codimension of the two coincide, and moreover Γ is a versal deformation of γ if and only if $\Gamma^* G$ is a versal deformation of g.

(3.3) Isomorphism Theorem

Suppose $g_0 : \mathbf{R}^n \to \mathbf{R}^p$ is an analytic $\mathcal{K}$-finite map-germ (at 0), and that $G : \mathbf{R}^n \times U \to \mathbf{R}^p$ is a versal deformation of g_0. Let $\gamma : \Lambda \to U$ be an analytic map-germ, and let $g = \gamma^* G$ be the bifurcation problem induced by γ (with organizing centre g_0). The morphism $\Psi_\gamma : \Theta_\gamma \to \Theta_g$ of $\mathcal{E}_\Lambda$-modules (defined above) induces an isomorphism

$$\psi_\gamma : N\mathcal{K}_\Delta \cdot \gamma \xrightarrow{\cong} N\mathcal{B} \cdot g.$$

(3.4) Discussion of proof

To begin with of course it is necessary to prove that the map ψ is well-defined; that is, that $\Psi_\gamma(T\mathcal{K}_{\Delta,e} \cdot \gamma) \subset T\mathcal{B}_e \cdot g$.

Firstly, it is clear from the definitions that $\Psi_\gamma(t\gamma(\Theta_\Lambda)) = t_2 g(\Theta_\Lambda)$. It therefore remains to show that $\Psi_\gamma(\gamma^* \mathrm{Derlog}(\Delta)) \subset T\mathcal{K}_{\mathrm{rel}} \cdot g$, where $T\mathcal{K}_{\mathrm{rel}} \cdot g = RT(g)$ is the sum of the first two terms in the expression for $T\mathcal{B}_e \cdot g$ given in (1.7). One can actually show more, see [MM, Lemma 3.2]:

$$\gamma^* \mathrm{Derlog}(\Delta) = \Psi_\gamma^{-1}(T\mathcal{K}_{\mathrm{rel}} \cdot g).$$

This relies heavily on the characterization of elements of $\mathrm{Derlog}(\Delta)$ as liftable vector fields over the map $\pi_G : G^{-1}(0) \to U$, as described in paragraph (2.3).

The map Ψ_γ thus descends to an injective map

$$\psi_\gamma : N\mathcal{K}_\Delta \cdot \gamma \longrightarrow N\mathcal{B} \cdot g.$$

The surjectivity of ψ follows from the preparation theorem. See [MM, Section 3].

(3.5) Example

Consider the organizing centre $g_0(x) = x^3$, its versal deformation $G(x, u_1, u_2) = x^3 + u_1 x + u_2$ and the pitchfork and hysteresis bifurcations (paragraphs (1.6), (2.6) and (2.7)).

Applying Ψ_γ to each of the $\mathcal{K}_\Delta$-versal deformations in (2.7) (i.e. substituting for Γ in G) we get:

Pitchfork: $(x, \lambda, u_1, u_2) \mapsto x^3 - \lambda x + u_1 + u_2 \lambda$,

Hysteresis: $(x, \lambda, u) \mapsto x^3 - \lambda + u$.

These agree with the versal deformations $\tilde{G}$ of the bifurcation problems given in (1.6).

(3.6) Equivalence of path and parametrized-contact formulations

We have been concentrating on the equivalence between the unfolding theories for g and for γ. However there is a more fundamental question that we have not addressed. Namely, whether $\mathcal{K}_\Delta$-equivalence of paths is *equivalent to* bifurcation equivalence of the induced bifurcation problems.

Suppose that γ_1 and γ_2 induces two bifurcation problems g_1 and g_2 from a versal deformation G, with all maps assumed to be analytic. One can show the following.

If γ_1 and γ_2 are K_Δ-equivalent, then g_1 and g_2 are bifurcation equivalent.

The proof of this fact is based on the fact that a diffeomorphism of U that preserves the discriminant Δ of π_G is liftable over π_G, which is a particular case of general results of du Plessis, Gaffney and Wilson. See for example [dPGW].

On the other hand, although it is probably true, I do not have a proof of the converse.

4   Calculations of $\mathrm{Derlog}(\Delta)$

There are some theoretical results giving more or less explicit generators for $\mathrm{Derlog}(\Delta)$, where Δ is the discriminant of a versal deformation, *without* calculating an equation for the discriminant. In the case $p = 1$, this is due to Bruce [B], and in the general case (with $n \geq p$) to Goryunov [G]. The basic result in these cases is Looijenga's theorem that $\mathrm{Derlog}(\Delta)$ is a *free $\mathcal{E}_U$-module* [L], and so has $\dim(U)$ generators. (It has at least that many, otherwise one could not obtain all vector fields away from Δ. If it had more, then there would be relations between the generators, and the module would not be free.)

In spite of the existence of theoretical results, calculations of $\mathrm{Derlog}(\Delta)$ are more easily done by brute force using computer algebra packages. The two most adapted to the sort of calculations necessary are *Macaulay* and *Singular*[2], though with some extra work it is possible to adapt other packages to do this type of computation.

The calculation proceeds as follows. First calculate the (an) equation $h(u) = 0$ for Δ. This is done by eliminating x from the equations $G(x, u) = \partial G(x, u)/\partial x_j = 0$. Using Grobner bases, this can be done very efficiently (finding the Grobner bases though can use a great deal of computer time).

To find elements of $\mathrm{Derlog}(\Delta)$, one uses the fact that a vector field

$$\theta(u) = \sum_j^d a_j(u)\frac{\partial}{\partial u_j}$$

is in $\mathrm{Derlog}(\Delta)$ if and only if

$$\sum_j^d a_j(u)\frac{\partial h}{\partial u_j} - fh = 0$$

for some function $f \in \mathcal{E}_U$. The $(d+1)$-tuple

$$(a_1, \ldots, a_d, -f) \in \mathcal{E}_U^{d+1}$$

---

[2]Both *Macaulay* and *Singular* are free and can be obtained by anonymous ftp; the first from zariski.harvard.edu, and the second from Kaiserslautern

defines a relation between the elements $\left(\frac{\partial h}{\partial u_1}, \ldots, \frac{\partial h}{\partial u_d}, h\right)$. Thus relations between the partial deriviatives of h and h itself correspond to elements of $\text{Derlog}(\Delta)$; the correspondence being given by omitting the last term (here called f). It is also possible to take advantage of the decomposition

$$\text{Derlog}(\Delta) = \mathcal{E}_U.e \oplus \text{Derlog}(h)$$

described in (2.2), by omitting the last term throughout.

Finding relations between elements of a ring (here $\mathcal{E}_U$) is also easy once Grobner bases have been calculated, and Macaulay and Singular are both purpose built for this type of task.

Some results of such calculations are listed in the table below. A column vector is identified with a vector field in alphabetical order, so that the first row is the coefficient of $\frac{\partial}{\partial a}$ the second of $\frac{\partial}{\partial b}$ and so on. The first four singularities are all corank 1, while the other two are of corank 2. Note that the first of the generators of $\text{Derlog}(\Delta)$ is the Euler field, while the others are generators of $\text{Derlog}(h)$ (for a suitable choice of h, namely for h quasihomogeneous).

5   Symmetric Bifurcations

Throughout this section we assume that H is a finite group acting linearly on $\mathbf{R}^n$ and $\mathbf{R}^p$ and that $g_0 : \mathbf{R}^n \to \mathbf{R}^p$ is H-equivariant. We also assume that g_0 is of finite $\mathcal{K}$-codimension as usual (which is the reason we assume H is finite).

One can then choose the versal deformation $G : \mathbf{R}^n \times U \to \mathbf{R}^p$ to be equivariant, with a suitable action on U. The discriminant $\Delta_G \subset U$ is then invariant under H.

If H acts on a set S, we let S^H denote $\text{Fix}(H; S)$, the subset of S consisting of all points fixed by H.

The actions of H on $\mathbf{R}^n$, $\mathbf{R}^p$ and U induce actions on each of the spaces Θ_n, Θ_p, and on Θ_γ if $\gamma : \Lambda \to U$ is equivariant, as well as on the tangent spaces $T\mathcal{K} \cdot g_0$, $T\mathcal{B}_e \cdot g$, $T\mathcal{K}_{\Delta,e} \cdot \gamma$, and consequently on the normal spaces $NB \cdot g$ and $N\mathcal{K}_\Delta \cdot \gamma$. The proof of Theorem 3.3 shows that provided γ is equivariant, then the isomorphism ψ_γ is also H-equivariant.

Consider the situation where the H-action on Λ is trivial, so that for γ to be equivariant it is necessary (and sufficient) that $\gamma(\Lambda) \subset U^H$. If one only considers perturbations of $g = \gamma^* G$ that are equivariant, then it is natural to consider the subspace

$$(T\mathcal{B}_e \cdot g)^H = T\mathcal{B}_e \cdot g \cap \Theta_g^H \subset T\mathcal{B}_e \cdot g,$$

which is isomorphic to $(T\mathcal{K}_{\Delta,e} \cdot \gamma)^H$.

However, it usually happens that equivariant organizing centres have high codimension, so that $\dim(U)$ is large, and the calculations of $\text{Derlog}(\Delta)$ become impractical. It is therefore natural to ask whether it is possible to restrict to U^H before calculating the normal spaces. This comes down to the following:

Question For γ as above, are $(T\mathcal{K}_{\Delta,e}\cdot\gamma)^H$ and $T\mathcal{K}_{\Delta^H,e}\cdot\gamma$ isomorphic?

It is known (simple linear algebra) that $(T\mathcal{K}_{\Delta,e}\cdot\gamma)^H$ is that part of $T\mathcal{K}_{\Delta,e}\cdot\gamma$ obtained by using only equivariant vector fields: $(T\mathcal{K}_{\Delta,e}\cdot\gamma)^H = t\gamma(\Theta_\Lambda^H) + \gamma^*\,\mathrm{Derlog}(\Delta)^H$. Moreover, any element of $\mathrm{Derlog}(\Delta)^H$ (i.e. any equivariant vector field tangent to Δ) restricts to a vector field on U^H tangent to $\Delta \cap U^H = \Delta^H$. Thus there is a natural map given by restriction to U^H,

$$\mathrm{Derlog}(\Delta)^H \longrightarrow \mathrm{Derlog}(\Delta^H),$$

and the question would be answered if one knew that this map was surjective.

Similar questions arise if the action of H is not trivial, corresponding to forced symmetry breaking bifurcation problems. Further problems arise if g_0 does not have a finite dimensional versal deformation, but it does have a finite dimensional equivariant deformation.

References

[AGV] V.I. Arnol'd, S. Gusein-Zade, A. Varchenko, *Singularities of Differentiable Mappings, Vol. I*, Birkhauser, Boston, 1985.

[BF] T. Bridges, J. Furter, *Singularity Theory and Equivariant Symplectic Maps*. Springer Lecture Notes in Math. **1558**, 1993.

[B] J.W. Bruce, Vector fields on discriminants and bifurcation theory. *Bull. London Math. Soc.* **17** (1985), 257–262.

[BdPW] J.W. Bruce, A.A. du Plessis, L.C. Wilson, Discriminants and liftable vector fields. *Preprint, University of Liverpool* (1992).

[D84] J. Damon, *The Unfolding and Determinacy Theorems for Subgroups of $\mathcal{A}$ and $\mathcal{K}$*. Memoirs A.M.S. **50**, no. 306 (1984).

[D87] J. Damon, Deformations of sections of singularities and Gorenstein surface singularities. *Am. J. Math.* **109** (1987), 695–722.

[D91] J. Damon, $\mathcal{A}$-equivalence and the equivalence of sections of images and disriminants. In *Singularity Theory and its Applications, Part I*, ed D. Mond, J. Montaldi, Springer Lecture Notes in Mathematics **1462** (1991), 93–121.

[DM] J. Damon, D. Mond, $\mathcal{A}$-codimension and the vanishing topology of discriminants. *Invent. math.* **106** (1991), 217–242.

[dPGW] A. du Plessis, T. Gaffney, L. Wilson, Map-germs determined by their discriminants. *Preprint, Aarhus University* (1992).

[GS79] M. Golubitsky, D. Schaeffer, A theory for imperfect bifurcation theory via singularity theory. *Comm. Pure Appl. Math.* **32** (1979), 21–98.

[GS85] M. Golubitsky, D. Schaeffer, *Singularities and Groups in Bifurcation Theory, Vol. I*. Springer-Verlag, New York etc., 1985.

[G] V.V. Goryunov, Projection and vector fields tangent to the discriminant of a complete intersection. *Func. An. Appl.* **22** (1988), 104–113.

[L] E.J.N. Looijenga, *Isolated Singular Points of Complete Intersections*. L.M.S. Lecture Notes, **17**. C.U.P., 1984.

[MM] D. Mond, J. Montaldi, Deformations of maps on complete intersections, Damon's $\mathcal{K}_V$-equivalence and bifurcations. To appear in *Proceedings of the Congrès Singularités de Lille* ed. J.P. Brasselet. L.M.S. Lecture Notes, 1994.

[M] J. Montaldi, Caustics in time reversible Hamiltonian systems. In *Singularity Theory and its Applications, Part II*, ed M. Roberts, I. Stewart, Springer Lecture Notes in Mathematics **1463** (1991), 266–277.

[P] M. Peters, *Classification of two-parameter bifurcation problems*. Thesis, University of Warwick, 1991. [See also in, *Singularity Theory and its Applications, Part II*, ed M. Roberts, I. Stewart, Springer Lecture Notes in Mathematics **1463** (1991), 294–300.]

[S] K. Saito, Theory of logarithmic differential forms and logarithmic vector fields. *J. Fac. Sci., Univ. Tokyo* **27** (1980), 265–291.

[W] C.T.C. Wall, Finite determinacy of smooth map-germs. *Bull. L.M.S.* **13** (1981), 481–539.

e-mail: james@doublon.unice.fr

CODIMENSION TWO LOCAL ANALYSIS OF SPHERICAL BÉNARD CONVECTION

J.D. RODRIGUEZ, C. GEIGER AND G. DANGELMAYR
Institut für Informationsverarbeitung
Universität Tübingen
Germany

ABSTRACT. Bifurcation of planforms in spherical Rayleigh-Bénard convection is studied by direct center manifold reduction of the Boussinesq equations. Codimension two local analysis of the reduced bifurcation equations is then used to classify the symmetry and stability of steady-state patterns near the onset of instability.

1. Introduction

The study of convection in spherical shells has attracted considerable research interest over a number of years as a simplified model for convective phenomena in geophysics and astrophysics. Applications of particular importance include the convective flow within the earth's mantle [2], stellar convection zones, planetary atmospheres, and the convective portion of the geophysical magnetic dynamo [8].

In a number of important applications, *e.g.* mantle convection, the rotation of the shell may be neglected. We thus assume our model to be non-rotating. In the absence of rotation the governing Boussinesq equations are invariant under the O_3 symmetry group and the linear problem is degenerate. For a given irreducible representation l of O_3 the linear problem has $2l + 1$ unstable modes. The critical value of l is determined by the aspect ratio, η, of the inner and outer radii. We shall focus on $l = 4$ which is relevant for convection in the lower earth mantle.

Near the onset of instability, the local behavior is governed by bifurcation equations which we calculate by center manifold reduction of the Boussinesq equations. The bifurcation equations are determined up to constants only by the particular representation of the O_3 group. When $l = 4$ the codimension one problem yields only unstable solutions [7], [3]. The codimension two case has only recently been completely analyzed [5], [6] and will permit us to classify stable patterns near onset.

The goal of this paper is to perform a full classification of steady state patterns for the

279

P. Chossat (ed.), Dynamics, Bifurcation and Symmetry, 279–289.
© 1994 *Kluwer Academic Publishers.*

range of the aspect ratio in which the $l = 4$ mode is critical and where the codimension two analysis is valid. One such codimension two point occurs when the Boussinesq equations are self adjoint [9]. In addition to the self adjoint case, additional codimension two points may also occur as functions of the Prandtl number [4]. We find that a unique codimension two point occurs as a function of Prandtl number for all physically relevant parameter values of the gravity and heat source terms. Finally, we investigate further possible codimension two points by fixing the Prandtl number and varying the gravity and heat source terms. The results of [5], [6] show that steady state patterns are completely classified by a normal form which includes terms of order three. Using center manifold reduction we calculate the local bifurcation equations to third order and apply the results of the local analysis [6] to determine the stability and symmetry of patterns as a function of the gravity and heat source terms as well as the Prandtl number. The remainder of this paper is organized as follows, in § 2 we introduce the governing equations, in § 3 we reduce the partial differential equations to a finite dimensional dynamical system on the nine dimensional center manifold, in § 4 we present the results of the classification scheme.

2. Governing equations

We consider a spherical shell of fluid with inner and outer radii r_1 and r_2 and boundary temperatures T_1 and T_2. The fluid is internally heated by a radially symmetric distribution of heat sources $\epsilon(r)$ and a radially symmetric gravity field $\gamma(r)$, is assumed to act upon the fluid. We assume that the conditions of the Boussinesq approximation are satisfied, *i.e.* the material properties are assumed constant except for density variations in the buoyancy term.

The temperature field is decomposed as a sum of the temperature contribution for the radially symmetric conduction state $T_0(r)$, and a perturbation to allow for convective effects. The conduction state satisfies the dimensionless diffusion equation,

$$\nabla^2 T_0(r) = -\frac{h^2 \epsilon(r)}{c_p \kappa \Delta T},$$

where c_p is the constant pressure specific heat. The thermal diffusivity is denoted by κ, h is the shell thickness and $\Delta T = T_1 - T_2$.

In nondimensional form the equations governing perturbations of velocity $\mathbf{v}$, pressure p and temperature Θ from the conduction state then become [1],

$$\frac{1}{P}\frac{\partial \mathbf{v}}{\partial t} = -\nabla\Pi + R\gamma(r)\mathbf{r}\Theta + \nabla^2\mathbf{v} + \frac{1}{P}\mathbf{v} \times (\nabla \times \mathbf{v}) \tag{2.1}$$

$$\nabla \cdot \mathbf{v} = 0 \tag{2.2}$$

$$\frac{\partial}{\partial t}\Theta = \tau(r)\mathbf{v} \cdot \mathbf{r} + \nabla^2\Theta - \mathbf{v} \cdot \nabla\Theta. \tag{2.3}$$

where $\nabla\Pi$ includes all terms which can be written as gradients.

The heat source term, $\tau(r)$ is defined as,

$$\tau(r) = -\frac{2}{r}\frac{d}{dr}T_0(r),$$

with $T_0(r)$ the steady state temperature distribution. The dimensionless parameters which appear in (2.1)-(2.3) are the Rayleigh number,

$$R = \frac{\alpha h^4 \Delta T \mathcal{G} \rho}{\kappa \nu}$$

and the Prandtl number,

$$P = \frac{\nu}{\kappa}.$$

Here α, ρ and ν are, respectively, the coefficient of thermal expansion, density and kinematic viscosity of the fluid and $\mathcal{G}$ is the gravitational constant. The Rayleigh number represents buoyancy forcing of the system and will serve as a bifurcation parameter. The Prandtl number measures the relative importance of thermal and kinematic diffusive effects.

The solonoidal property of the velocity field permits the decomposition of the velocity field into solonoidal Φ and toroidal Ψ scalar velocity fields related by,

$$\mathbf{v}(\mathbf{r}) = \nabla \times (\nabla \times \mathbf{r}\Phi(\mathbf{r})) + \nabla \times \mathbf{r}\Psi(\mathbf{r}).$$

The governing equations for the poloidal and toroidal velocity fields and the temperature perturbation Θ may then be derived from (2.1)-(2.3) by operating on the momentum equation with, in turn, $\mathbf{r} \cdot \nabla \times (\nabla \times)$ and $\mathbf{r} \cdot \nabla \times$ to give,

$$\frac{1}{P}\mathcal{L}^2 \nabla^2 \frac{\partial}{\partial t}\Phi = \mathcal{L}^2 \nabla^4 \Phi - R\gamma(r)\mathcal{L}^2\Theta - \frac{1}{P}\mathbf{r} \cdot \nabla \times \left(\nabla \times (\mathbf{v} \times (\nabla \times \mathbf{v}))\right) \quad (2.4)$$

$$\frac{1}{P}\mathcal{L}^2 \frac{\partial}{\partial t}\Psi = \nabla^2 \mathcal{L}^2 \Psi + \frac{1}{P}\mathbf{r} \cdot \nabla \times (\mathbf{v} \times (\nabla \times \mathbf{v})) \quad (2.5)$$

$$\frac{\partial}{\partial t}\Theta = \nabla^2 \Theta + \tau(r)\mathcal{L}^2\Phi - \mathbf{v} \cdot \nabla\Theta, \quad (2.6)$$

where $\mathcal{L}^2$ is the Laplacian in the angular variables,

$$\mathcal{L}^2 = -\frac{1}{\sin\vartheta}\frac{\partial}{\partial\vartheta}\sin\vartheta\frac{\partial}{\partial\vartheta} - \frac{1}{\sin^2\vartheta}\frac{\partial^2}{\partial\varphi^2}.$$

Together with the Boussinesq equations in potential form (2.4)-(2.6) we have the boundary conditions: $\Phi = 0$ and $\Theta = 0$ at $r = r_1, r_2$. We assume stress-free boundaries which yields the remaining boundary conditions as $\Phi_{rr} = 0$ and $\Psi_r - \frac{1}{r}\Psi = 0$ at $r = r_1, r_2$.

For constant internal heating and for a shell of constant density, the heat source and gravity terms may be written as [1],

$$\tau(r) = b_2 + b_1\frac{r_1^3}{r^3} \quad \text{and} \quad \gamma(r) = g_2 + g_1\frac{r_1^3}{r^3},$$

respectively. By a rescaling, b_2 and g_2 may be set to unity without loss of generality. In the following sections the constants b_1 and g_1 are considered as free parameters and are allowed to vary over $0 \leq g_1 \leq \infty$ and $0 \leq b_1 \leq \infty$. We recall that the boundary value problem (2.1)-(2.4) is self adjoint only for $g_1 = b_1$. We regard the Prandtl number as a third free parameter which may vary as $0 \leq P \leq \infty$.

Center manifold reduction

We derive bifurcation equations near the critical Rayleigh number, R_c, by center manifold reduction. The approach is similar to that used by Friedrich and Haken [4] to study one-two mode interactions. The center manifold technique allows the á priori infinite dimensional phase space of the Boussinesq equations to be reduced to a finite dimensional subspace of dimension equal to the number of critical modes. The center manifold may then be computed approximately as a Taylor series about the critical point. In order to classify time independent patterns we must calculate the coefficients of the Taylor series expansion to third order [6].

The critical mode, l_0, and critical Rayleigh number, R_c, are determined by linear stability analysis [1] as a function of the aspect ratio, η. The $l = 4$ mode is critical for $0.556 \leq \eta \leq 0.630$. This value of η is particularly important for convection in the lower earth mantle where $\eta = 0.61$. For the critical mode corresponding to $l_0 = 4$ the center manifold is of dimension $2l_0 + 1 = 9$. We note however, that the general reduction technique described below is not specific to $l_0 = 4$.

Setting $\lambda = R - R_c$, we write the Boussinesq equations in the form,

$$S\frac{\partial}{\partial t}\xi = (L + \lambda\hat{L})\xi + N[\xi,\xi], \tag{3.1}$$

where $\xi = (\Phi_l, \Psi_l, \Theta_l)^T$ and the operators S and L are defined as,

$$L = \begin{pmatrix} \nabla^4 \mathcal{L}^2 & 0 & -R_c\gamma(r)\mathcal{L}^2 \\ 0 & \nabla^2\mathcal{L}^2 & 0 \\ \tau(r)\mathcal{L}^2 & 0 & \nabla^2 \end{pmatrix} \quad \text{and} \quad S = \begin{pmatrix} \frac{1}{P}\nabla^2\mathcal{L}^2 & 0 & 0 \\ 0 & \frac{1}{P}\mathcal{L}^2 & 0 \\ 0 & 0 & 1 \end{pmatrix}.$$

The adjoint operator, $L^\dagger$, and the linear operator $\hat{L}$ are respectively,

$$L = \begin{pmatrix} \nabla^4\mathcal{L}^2 & 0 & \tau(r)\mathcal{L}^2 \\ 0 & \nabla^2\mathcal{L}^2 & 0 \\ -R_c\gamma(r)\mathcal{L}^2 & 0 & \nabla^2 \end{pmatrix} \quad \text{and} \quad \hat{L} = \begin{pmatrix} 0 & 0 & -\gamma(r)\mathcal{L}^2 \\ 0 & 0 & 0 \\ 0 & 0 & 0 \end{pmatrix}.$$

The transition to instability of the conduction state, $\xi = 0$, occurs when λ passes through zero from below, thus λ is considered as the bifurcation parameter which we assume to be small. The nonlinear operator $N[\xi,\xi]$, defined as,

$$N[\xi,\xi] = (N_1[\xi,\xi], N_2[\xi,\xi], N_3[\xi,\xi])^T$$

is a bilinear operator with components

$$
\begin{aligned}
N_1[\xi,\xi] &= -\frac{1}{P}\mathbf{r}\cdot\nabla\times\left(\nabla\times(\nabla\times(\nabla\times\Phi)+\nabla\times\Psi)\times(\nabla\times(\nabla\times(\nabla\times\Phi)\right.\\
&\left.+\nabla\times\Psi))\right)\\
N_2[\xi,\xi] &= \frac{1}{P}\mathbf{r}\cdot\nabla\times(\nabla\times(\nabla\times\Phi)+\nabla\times\Psi)\times(\nabla\times(\nabla\times(\nabla\times\Phi)\\
&+\nabla\times\Psi))\\
N_3[\xi,\xi] &= -(\nabla\times(\nabla\times\Phi)+\nabla\times\Psi)\cdot\nabla\Theta.
\end{aligned}
$$

Let H be the Hilbert space of vector fields $\xi=(\Phi,\Psi,\Theta)^T$ square-integrable over the domain $D=\{r_1\leq r\leq r_2, 0\leq\varphi\leq 2\pi, -\pi/2\leq\vartheta\leq\pi/2\}$ such that $\xi=0$ for $r=r_1$ and $r=r_2$ with the inner product defined by,

$$
<\xi_1,\xi_2>=\int_D\xi_1\cdot\bar{\xi}_2d\mathbf{r}.
$$

We shall also need the radial and angular scalar products defined by,

$$
<\xi_1,\xi_2>_r=\int_{r_1}^{r_2}\xi_1\cdot\bar{\xi}_2r^2dr,\quad\text{and}\quad<\xi_1,\xi_2>_a=\int_0^{2\pi}d\varphi\int_0^\pi\xi_1\cdot\bar{\xi}_2\sin\vartheta d\vartheta,
$$

repectively.

Let $H_0\subset H$ represent the null space of the linear operator L. A complete basis for H_0 is provided by the eigenfunctions $\{\varphi_m|-l_0\leq m\leq l_0\}$ with $\varphi_m\equiv f_{l_0}^0(r)Y_{l_0m}$, where $f_{l_0}^0$ is the radial null eigenvector of L, with $L_{l_0}f_{l_0}^0=0$ and $Y_{l_0m}(\vartheta,\varphi)$ are the spherical harmonics of order l_0. Here L_{l_0} is the radial part of L for $l=l_0$.

Then a solution $\xi\in H$ of (3.1) may be written as

$$
\xi=v+w,
$$

with $v\in H_0$ and $w\in H'$ where H' is the range of L.

We define the projection operator $P\colon H\to H_0$ for $\xi\in H$ as

$$
P=\sum_{m=-l}^{l}<\xi,\varphi_m^*>\varphi_m,
$$

where $\varphi_m^*=f_{l_0}^{0\dagger}\bar{Y}_{l_0m}$ is the null eigenvector of the adjoint linear operator $L_{l_0}^\dagger$. We denote by $Q\colon H\to H'$ the projection on the complementary space H', i.e., $Q=I-P$. Then $v=P\xi\in H_0$ may be expanded in spherical harmonics as

$$
P\xi=\sum_{m=-l_0}^{l_0}z_m(t)f_{l_0}^0(r)Y_{l_0m}(\vartheta,\varphi).
$$

The $z_m \in \mathbf{C}$ are complex coordinates for h_0 and satisfy the reality condition $z_{-m} = (-1)^m \bar{z}_m$. An appropriate ansatz for $w \in H'$ is given by

$$w(z, \lambda) = \sum_{l=1}^{\infty} \sum_{m=-l}^{l} \sum_{i,j,k} \lambda^k Z_{lm}^{ij}(z) f_l^{ijk}(r) Y_{lm}(\vartheta, \varphi).$$

For each $i = 1, 2, \ldots$ the $Z_{lm}^{ij}(z)$ consist of a finite set of homogeneous polynomials of degree i of the coordinates $z = \{z_{-m}, \ldots, z_m\}$ with which the center manifold is parameterized. The functions

$$f_l^{ijk}(r) = (f_{l,1}^{ijk}(r), f_{l,2}^{ijk}(r), f_{l,3}^{ijk}(r))^T$$

must be determined numerically and are required to be orthogonal to $f_{l_0}^{0\dagger}$ for $l = l_0$.

The vector field $g_m(z, \lambda)$ on the center manifold is obtained by projecting (3.1) onto H_0,

$$\dot{z}_m = g_m(z, \lambda) = \lambda z_m < \hat{L}_{l_0} f_{l_0}^0, f_{l_0}^{0\dagger} >_r + < N[v + w(z, \lambda), v + w(z, \lambda)], f_{l_0}^{0\dagger} \bar{Y}_{l_0 m} > \quad (3.2)$$

where we have normalized the eigenvectors such that $< S_{l_0} f_{l_0}^0, f_{l_0}^{0\dagger} >_r = 1$ and where $\hat{L}_l$ is the radial part of $\hat{L}$.

The O_3 symmetry of the Boussinesq equations has the consequence that the $l_0(l_0 + 1)$ dimensional vector field on the center manifold,

$$g(z, \lambda) = (g_{-l_0}(z, \lambda) \ldots g_{l_0}(z, \lambda))$$

is *equivariant* under the irreducible representation, T_{l_0}, of O_3, *i.e.*, the vector field commutes with the group action,

$$T_{l_0}(A)g(z, \lambda) = g(T_{l_0}(A)z, \lambda),$$

for all $A \in O_3$. The successive approximations to the center manifold are calculated by expanding the vector field in a Taylor series in z. The equivariance property permits the series expansion to be decomposed at each order i in z into a restricted set of t_i equivariant homogeneous polynomials $G_{is}(z)$, independent over $\mathbf{R}$ where $1 \leq s \leq t_i$. Since t_i is small for lower orders i, this results in a dramatic simplification of the Taylor series. The explicit form of the dominant $G_{is}(z)$ for the $l_0 = 4$ representation of O_3 may be found in [5], [6]. When $i \geq 2$ we are only interested in the λ independent terms. The vector field $g(z, \lambda)$ is therefore approximated as

$$g(z, \lambda) = a_0 \lambda z + \sum_{i \geq 2} \sum_{s=1}^{t_i} a_{is} G_{is}(z). \quad (3.3)$$

and the sum over the nonlinear terms is truncated at an appropriate order which is dependent on the degeneracy of the problem. The main task is to calculate the (real) coefficients in the expansion (3.3).

The lowest order coefficients depend solely on the critical radial basis functions whose components we shall write as $f_{l_0}^0 = (f_1, f_2, f_3)$ and $f_{l_0}^{0\dagger} = (f_1^\dagger, f_2^\dagger, f_3^\dagger)$. The coefficient a_0 is determined by

$$a_0 = <\hat{L}_{l_0} f_{l_0}^0, f_{l_0}^{0\dagger}>_r = -\int_{r_1}^{r_2} \gamma(r) f_{l_0,3} f_{l_0,1}^\dagger r^2 dr. \tag{3.4}$$

The sign of (3.4) is important in the bifurcation classification and is positive for the cases we consider.

For even values of l_0 there is a unique quadratic equivariant G_{21}. We wish to consider a codimension two vector field and hence we require that a_{21} vanishes. To lowest order $w = 0$ and a_{21} can be explicitly calculated as an integral involving the critical eigenvector,

$$a_{21} = \int_{r_1}^{r_2} \left[f_{l_0,3}^\dagger (2f_{l_0,1} f_{l_0,3}' r + f_{l_0,3}(f_{l_0,1} + r f_{l_0,1}')) \right.$$

$$+ \frac{l_0(l_0+1)}{Pr^2} f_{l_0,1}^\dagger (l_0(l_0+1)(f_{l_0,1}^2 - 3f_{l_0,1} f_{l_0,1}' r) + (4f_{l_0,1}'^2 + 3f_{l_0,1} f_{l_0,1}'')r^2$$

$$\left. + 2(f_{l_0,1}' f_{l_0,1}'' + f_{l_0,1} f_{l_0,1}''')r^3) \right] dr \tag{3.5}$$

In the self-adjoint case this may be integrated to give,

$$r f_{l_0,3}^2 f_{l_0,1} + \frac{l_0(l_0+1)}{P} f_{l_0,1}^2 (f_{l_0,1}'' r - l_0(l_0+1)\frac{f_{l_0,1}}{r} + 2f_{l_0,1}').$$

With the boundary conditions that $f_{l_0,1}(r_1) = f_{l_0,1}(r_2) = 0$ (3.5) vanishes identically and thus the self-adjoint case is an exact codimension two point.

For $l_0 = 4$ there exist two cubic equivariant functions G_{31} and G_{32} [5], [6]. The calculation of G_{31} and G_{32} requires the computation of the second approximation to the component of the vector field $w \in H'$ which may be calculated by applying the projection operator Q to (3.1) ,

$$QS\dot{u} = Q(S\dot{v} + S\dot{w}) = Q(L + \hat{L})w + QN \tag{3.6}$$

The explicit time dependence of the component $v \in H_0$ is then eliminated using (3.3) to give a series of linear, inhomogeneous partial differential equations which yield successive approximations to w. The O_3 symmetry of the Boussinesq system allows the angular part of the calculations to be carried out analytically using MAPLE. The remaining set of coupled ordinary differential equations for the radial component f_l^{210} of w is,

$$L_l f_l^{210} = a_{21} \delta_{ll_0} S_{l_0} f_{l_0}^0 - N_{l_0 l_0}^l [f_{l_0}^0, f_{l_0}^0], \tag{3.7}$$

with $< f_l^{210}, f_{l_0}^{0\dagger} >= 0$ if $l = l_0$ and where $N_{l_0 l_0}^l [f_{l_0}^0, f_{l_0}^0]$ is the bilinear, radial operator defined by

$$N_{l_0 l_0}^l [f_{l_0}^0, f_{l_0}^0] = < N[f_{l_0}^0 Y_{l_0 m_1}, f_{l_0}^0 Y_{l_0 m_2}], \bar{Y}_{l_0 m} >_\alpha < Y_{l_0 m_1} Y_{l_0 m_2}, \bar{Y}_{l_0 m} >_\alpha^{-1} \tag{3.8}$$

for all $m = -l_0 \ldots l_0$.

When $l_0 = 4$ the equations (3.7) constitute a set of four boundary value problems on the interval $r_1 \leq r \leq r_2$ for $l = 2, 4, 6$ and 8. The equations for all remaining values of l vanish after the integration over the angular variables in (3.8) is carried out. In the second approximation the toroidal velocity field component f_l^{210} is again zero so the equations (3.7) consist of only two ordinary differential equations each. The integration of (3.7) must be carried out numerically and a multiple shooting algorithm was used for this purpose. The boundary value problem for $l = l_0 = 4$ is singular and numerical solution will not generally yield a unique solution. Uniqueness is enforced by requiring that the orthogonality condition $< f_4, f_4^{0\dagger} >_r = 0$ be satisfied which ensures that $w \in H'$. The second approximation to w was then inserted into (3.2) and a numerical integration over the radial variable then gave the required third order coefficients.

4. Results and discussion

The codimension two singularity is determined by the condition that the coefficient a_{21} of the second order equivariant polynomial G_{21} vanishes. We have shown explicitly that a_{21} vanishes identically when the Boussinesq equations are self-adjoint. The classification of patterns near the onset of instability for the self-adjoint case has been analyzed in [9]. Inspection of (3.5) indicates that a_{21} may also vanish in the general non-self-adjoint case for certain values of the Prandtl number. Here we investigate numerically the existence of codimension two points in the general non-self-adjoint case in the range $10^{-2} \leq g_1, b_1 \leq 10^{2.5}$ and Prandtl number $10^{-2} \leq P \leq 10^2$. The above ranges were found to be sufficient to determine the behavior over the complete ranges of $0 \leq g_1, b_1, P \leq \infty$ as the asymptotic limits are approached quite rapidly.

Since we are interested in stable patterns we restrict the range of η to those values for which $l = 4$ is critical. We find that the critical range of η is only weakly dependent on the parameters g_1 and b_1. The critical range is also, of course, completely independent of the Prandtl number since P does not appear in the linear problem. We thus consider the range $0.556 \leq \eta \leq 0.631$.

The numerical investigation proceeds by calculating the coefficient a_{21} for fixed values of the parameters g_1, b_1 and η. The zeros of a_{21} are then investigated as a function of the Prandtl number for each triple (b_1, g_1, η) with $0 \leq P \leq \infty$. We find that for all parameter values a_{21} has a single zero at $P = P_c(g_1, b_1, \eta)$. The numerical value of P_c is relatively insensitive to the parameter values with $P_c \approx 0.24$ and varies only approximately 10% for all g_1, b_1 and η. By fixing the Prandtl number we may also investigate further codimension two points as functions of g_1 and b_1. Here, however we find no further codimension two points in the specified range of η, b_1, g_1 and P.

The coefficient a_{21} may be shown [9] to be of the form $a_{21} = c_1 + c_2/P$ where c_1 and c_2 are constants resulting from projecting the nonlinear terms of (3.1) onto the critical adjoint eigenvector. The calculation of the linear eigenvalue problem shows that the critical eigenvector is not strongly dependent on the parameters g_1 and b_1 which accounts for the

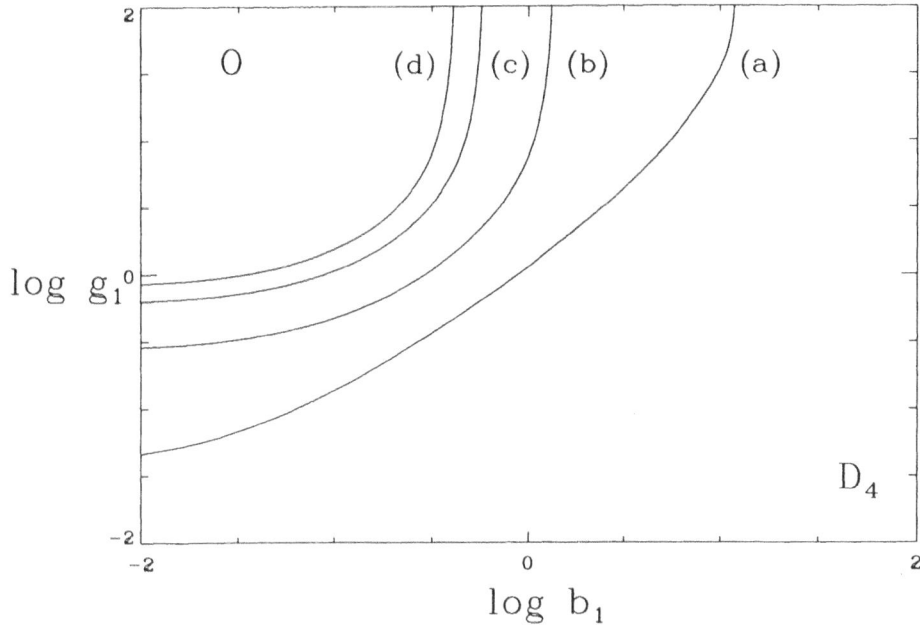

Figure 1: Regions of stable patterns with octahedral O and D_4 symmetry at $P = P_c$ vs. η. These regions are separated by codimension three singular lines shown for (a) $\eta = 0.560$, (b) $\eta = 0.575$, (c) $\eta = 0.594$, and (d) $\eta = 0.627$.

relatively simple functional dependence of a_{21}.

The second step of the analysis consists of the application of the results of the local analysis in [6], [5] in order to determine the stability and symmetry of the patterns at $P = P_c$. The analysis is codimension two and the two third order equivariants completely determine the bifurcation behavior. To apply the classification scheme we calculated a_{31} and a_{32} for five values of η over the $l = 4$ critical range and for ten values each of g_1 and b_1 distributed logarithmically over the interval $10^{-2} \leq g_1, b_1 \leq 10^{2.5}$. The classification scheme [5] predicts stable persistent patterns with D_4 symmetry for $0 < a_{31}/|a_{32}| < 35/16$ when $a_{32} < 0$. Stable and persistent patterns with octahedral (O) symmetry occur for $a_{31}/|a_{32}| < -21/4$ if $a_{32} > 0$ and for $a_{31} < 0$, if $a_{32} < 0$.

The results of the classification scheme indicate that stable patterns are always present for $P = P_c$. The calculations indicate that both stable and persistent patterns with D_4 symmetry as well as stable and persistent patterns with octahedral symmetry appear. These two cases occur in two regions of the (g_1, b_1, η) parameters space separated by a

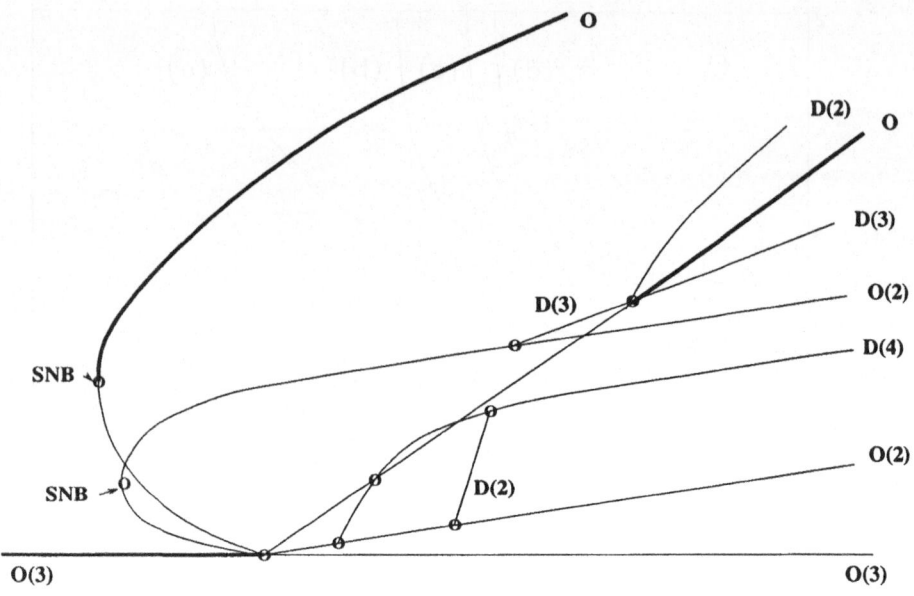

Figure 2: Bifurcation diagram for the unfolding of the codimension two singularity with stable and persistent O symmetric patterns. Bold lines indicate stable branches and the symmetry of each pattern is indicated.

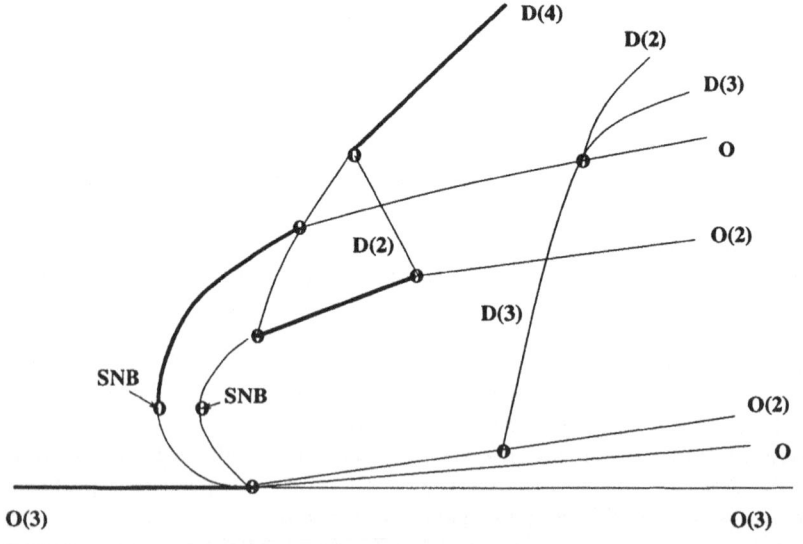

Figure 3: As above with stable and persistent D_4 symmetric patterns.

codimension three singular surface. The regions of stable patterns as determined by g_1 and b_1 are shown in Fig. 1 for four values of η. The analogous results for the self-adjoint case [9] indicate that stable and persistent patterns occur only for small Prandtl numbers with stable D_4 patterns appearing in the limit of vanishing Prandtl number.

The local analysis remains valid in a sufficiently small neighborhood of the codimension two point. Nevertheless, the weak dependence on g_1 and b_1 suggest that the general non-self-adjoint case with arbitrary Prandtl number may be qualitatively understood in terms of the codimension two calculation. For weak perturbations the bifurcation structure for the general case can then be determined by unfolding the codimension two singularity. The unfolding results in the appearance of additional non-persistent solution branches in the neighborhood of the instability. Bifurcation diagrams for the O and D_4 symmetric cases are shown in Figure 2 and Figure 3 respectively. A complete discussion of the properties of these solution branches may be found in [5], [6].

Acknowledgements:

Support for JDR was provided by the Alexander von Humboldt Stiftung and for CG by the Deutsche Forschungsgemeinschaft.

References

[1] S. Chandrasekhar, *Hydrodynamic and Hydromagnetic Stability*, Clarendon, Oxford, 1961.

[2] C.G. Chase, *Nature* **282** 464-8.

[3] P. Chossat, R. Lauterbach and I. Melbourne, Arch. Rat. Mech. Anal. **113**, 313 1990.

[4] R. Friedrich and H. Haken, Phys. Rev. A, **34**, No. 3 1986.

[5] C. Geiger, *Strukturbildung in nichtlinearen dissipativen Systemen mit sphärischer Symmetrie*, dissertation 1993.

[6] C. Geiger, G. Dangelmayr, J.D. Rodriguez and W. Güttinger, *On a degenerate bifurcation with O_3 symmetry*, Proc. of the AMS (to appear).

[7] M. Golubitsky, I. Stewart and D.G. Schaeffer, *Singularities and Groups in Bifurcation Theory*, Vol.II, Springer Verlag, New York, 1988.

[8] E.N. Parker, *Cosmical Magnetic Fields*, Clarendon Press, Oxford, 1979.

[9] J.D. Rodriguez, C. Geiger, G. Dangelmayr, and W. Güttinger, *Symmetry Breaking Bifurcations in Spherical Bénard Convection*, Proc. of the AMS (to appear).

A Geometric Hamiltonian Approach to the Affine Rigid Body

E. SOUSA DIAS
Mathematics Institute
University of Warwick

England

ABSTRACT. A short account of the modern treatment of symmetric Hamiltonian systems by methods of Poisson and symplectic geometry for the affine rigid body is presented, with emphasis on the presence of symmetry.

The formulation in this set up of the so-called Dedekind theorem on ellipsoidal figures of equilibrium for the Dirichlet problem is shown to be a straightforward consequence of $\mathbf{Z}_2$ symmetry and of the relative equilibrium theorem.

Introduction

An affine rigid body can be defined as a system of points which is allowed to have two kinds of motion, rigid rotations and homogeneous deformations, during which all affine relationships are preserved.

Affine rigid body problems can be found in several areas of Physics and Mechanics. It is worth mentioning, for example, the so-called Dirichlet problem of a self gravitating homogeneous fluid mass whose internal motions are linear functions (in some inertial frame) of the position, and which maintains at all times an ellipsoidal form, which can vary.

The Classical treatment of Dirichlet's problem by an analytical study of the differential equations governing it can be traced back to Newton with the study of the earth's motion. Over a period of centuries the problem has interested many authors such as Jacobi, Maclaurin, Dirichlet, Dedekind, Riemann and others. One of the most important contributions is due to Riemann, with his work on the classification of the possible ellipsoidal figures of equilibrium and their stability. A historical background on Dirichlet's problem and on the studies made by several authors on ellipsoidal figures of equilibrium can be found in Chandrasekhar [2] and references therein.

Although these kinds of problem have a Hamiltonian structure, their study is far from being completely explored within the modern treatment of Hamiltonian systems,

P. Chossat (ed.), Dynamics, Bifurcation and Symmetry, 291–299.

which has as its main characteristic the exploitation of geometric properties: not only the configuration and phase space structures but in particular the level sets of the momentum map, by using Poisson and symplectic geometry. The symmetries often play an important role in the analysis of a particular system, either as a symmetry of the system or of a particular equilibrium, or as the generator of the trajectory under consideration.

The study of an affine rigid body in this context is far from being completly exhausted, although some works can be found which explore particular cases or apply results from the geometric approach just mentioned (see D. Lewis *et al.* [5, 4] and Slawianowski [7]).

Affine rigid body

All these kinds of problem, as the formulation suggests, deal with two kinds of reference frame: one fixed in the space (spatial or inertial frame) and the other fixed in the body and moving with it (body frame). We can formulate this by considering two smooth Riemannian manifolds $(\mathcal{B}, v)$ and $(\mathcal{S}, V)$, where v, V are metrics and $\mathcal{B}, \mathcal{S}$ are respectively the reference configuration and the ambient space in which the body $\mathcal{B}$ moves.

The configuration space $\mathcal{C}$ is the set of (orientation preserving) embeddings $\phi :$ $\mathcal{B} \to \mathcal{S}$, that is $\mathcal{C} = \text{Emb}(\mathcal{B}, \mathcal{S})$, and the set $\phi(\mathcal{B})$ is called the current configuration. A motion is a curve of configurations $\phi_t(X) = \phi(t, X)$ which denotes a configuration at time t. The deformation gradient P_t is the tangent map $P_t = T\phi_t$.

In the affine rigid body the reference manifold $\mathcal{B}$ is a subset of $\mathcal{S} = \mathbf{R}^3$ and the embedding is of the form

$$\phi_t(X) = Q(t)X, \quad Q(t) \in \mathcal{C} , \ X \in \mathcal{B}.$$

That is, $Q(t)$ is the deformation gradient and it acts by matrix multiplication.

The configuration space $\mathcal{C}$ in the affine rigid body case is the linear group $GL^+(3)$, the set of 3×3 invertible matrices with positive determinant. This corresponds in the Dirichlet problem to the condition that internal motions be a linear function of the position (in some inertial frame).

The phase space is taken to be the cotangent bundle of the configuration space, $T^*\mathcal{C} = \mathcal{P}$, which is a symplectic manifold with the canonical symplectic form.

Comparing with the rigid body case, where the configuration space is the special orthogonal group $SO(3)$ and the phase space its cotangent bundle, we see that a rigid body problem is a particular case of an affine rigid body problem.

Any matrix $Q(t) \in GL^+(3)$ can be decomposed into its bipolar decomposition,

$$Q(t) = R(t)A(t)S(t)^T \tag{1}$$

where $A(t)$ is a time dependent diagonal matrix and R, S^T are elements of $SO(3)$. The physical interpretation of the matrices R, S (in Dirichlet's problem) is linked, respectively, to the rate of rotation of the body frame with respect to the spatial frame, and to the vorticial motion of the fluid. That is, respectively, with the vector identification of the skew-symmetric matrices $\Omega = \dot{R}R^T$ and $\Lambda = \dot{S}S^T$ (which belong to the Lie algebra $so(3)$ of $SO(3)$) given by the Lie algebras isomorphism

$$l : (\mathbf{R}^3, \times) \to (so(3), [\,,\,])$$

$$\mathbf{R}^3 \ni (x_1, x_2, x_3) \mapsto \begin{bmatrix} 0 & -x_3 & x_2 \\ x_3 & 0 & -x_1 \\ -x_2 & x_1 & 0 \end{bmatrix} \in so(3).$$

Here $\times$ and $[\,,\,]$ denote respectively the cross product in $\mathbf{R}^3$ and the Lie algebra bracket.

¿From (1) it is clear that there is an $SO(3) \times SO(3)$-action, say Φ, on the configuration space $GL^+(3)$, given by

$$\Phi : (SO(3) \times SO(3)) \times GL^+(3) \to GL^+(3)$$

$$\Phi\left((\lambda, \rho), Q\right) = \lambda Q \rho^T = \Phi_{(\lambda, \rho)}(Q).$$

This action induces an action on the phase space $T^*GL^+(3)$, called the lifted action.

For symmetric Hamiltonian systems, i.e for systems where the Hamiltonian H is invariant under a group action, the symmetries give rise to conserved quantities. These quantities are expressed via a map, the momentum map, which is a first integral for the symmetric Hamiltonian system (see Abraham & Marsden [1] p.277).

The momentum map J, a function from the phase space $\mathcal{P}$ into the dual of the Lie algebra γ of the Lie group Γ acting on $\mathcal{P}$, is an equivariant momentum map when $\mathcal{P}$ is a cotangent bundle.

For the action induced by Φ on $T^*GL^+(3)$, the momentum map $J : T^*GL^+(3) \to so(3)^* \times so(3)^*$ is given by

$$\langle J(\alpha_Q), (\xi, \eta) \rangle = \langle \alpha, (\xi, \eta)_{GL^+(3)}(Q) \rangle. \tag{2}$$

Here $(\xi, \eta)_{GL^+(3)}(Q)$ denotes the infinitesimal generator associated to $(\xi, \eta) \in so(3) \times so(3)$ for the $SO(3) \times SO(3)$-action on $GL^+(3)$, $\alpha_Q \in T^*_Q GL^+(3)$ and $\langle,\rangle$ on the left hand side denotes the pairing between $so(3)^* \times so(3)^*$ and $so(3) \times so(3)$, while the one on the right hand side denotes the pairing between $TGL^+(3)$ and $T^*GL^+(3)$.

By definition, the infinitesimal generator $(\xi, \eta)_{GL^+(3)}(Q)$ is just

$$(\xi, \eta)_{GL^+(3)}(Q) = \frac{d}{dt}\Phi_{(\exp t\xi, \exp t\eta)}(Q)\Big|_{t=0} = \xi Q + Q\eta^T = \xi Q - Q\eta$$

(note that $\xi = -\xi^T$ and $\eta = -\eta^T$).

Defining the pairings in (2) by $\langle u, v \rangle = tr\,(u^T v)$ we have for (2) the form

$$\langle J(\alpha), (\xi, \eta) \rangle = tr(\alpha^T \xi Q + \alpha^T Q\eta^T) = tr(Q\alpha^T \xi - \alpha^T Q\eta). \tag{3}$$

In this expression the elements $Q\alpha^T$ and $\alpha^T Q$ are not skew-symmetric matrices; but replacing each of them in (3) by its symmetric and skew-symmetric parts we have

$$J(\alpha) = \frac{1}{2}(Q\alpha^T - \alpha Q^T, Q^T\alpha - \alpha^T Q) = (J_1, J_2). \tag{4}$$

The introduction of the so-called body and space coordinates in the generalized rigid body, i.e coordinates of $\Gamma \times \gamma^*$ which correspond to the left or right identification of $T^*\Gamma$ with $\Gamma \times \gamma^*$ (where Γ is a Lie group) have been proved of special importance in the treatment of these kind of problems. They play a key role in the normal form of the momentum map and very often the momentum map takes a particularly simple form in these coordinates.

Let $i_L, i_R : T^*_Q GL^+(3) \to GL^+(3) \times gl(3)^*$ denote the left and right identifications, where

$$i_L(\alpha) = (Q, \langle \alpha, T_e L_Q(\xi) \rangle)$$

and

$$i_R(\alpha) = (Q, \langle \alpha, T_e R_Q(\xi) \rangle)$$

for $\alpha \in T^*_Q GL^+(3)$, that is $i_L(\alpha) = (Q, Q^T\alpha)$ and $i_R(\alpha) = (Q, \alpha Q^T)$. Then, in coordinates $(Q, \beta) \in GL^+(3) \times gl(3)^*$, the momentum map becomes

$$J_L(Q, \beta) = J \circ i_L^{-1}(Q, \beta) = J(Q, Q^{-T}\beta) = \frac{1}{2}(Q\beta^T Q^{-1} - Q^{-T}\beta Q^T, \beta - \beta^T)$$

$$J_R(Q, \beta) = J \circ i_R^{-1}(Q, \beta) = J(Q, \beta Q^{-T}) = \frac{1}{2}(\beta^T - \beta, Q^T\beta Q^{-T} - Q^{-1}\beta^T Q)$$

where J_R, J_L denote the momentum map when the phase space identification with $GL^+(3) \times gl(3)^*$ is performed using the isomorphisms i_R, i_L respectively.

Consider the following $\mathbf{Z}_2$-actions :

1. $\mathbf{Z}_2 = \{1, \tau\}$ acts on $GL^+(3)$ by

$$\Theta_\sigma(Q) = \begin{cases} Q & \text{for} \quad \sigma = 1 \\ Q^T & \text{for} \quad \sigma = \tau \end{cases}.$$

2. The $\mathbf{Z}_2$-action on $(x, y) \in SO(3) \times SO(3)$

$$\tau \cdot (x, y) = (y, x)$$

for the nontrivial element τ of $\mathbf{Z}_2$.

The $\mathbf{Z}_2$-action on $SO(3) \times SO(3)$ induces an action on $so(3)^* \times so(3)^*$, say $\tau_{so(3)^* \times so(3)^*}$ given by

$$\tau_{so(3)^* \times so(3)^*}(\Omega, \Lambda) = \frac{d}{dt}\Big|_{t=0} \tau \cdot (\exp t\Omega, \exp t\Lambda) = (\Lambda, \Omega) = \tau \cdot (\Omega, \Lambda) \tag{5}$$

The next proposition will show that the momentum map for the $SO(3) \times SO(3)$-action on $\mathcal{P}$ is $\mathbf{Z}_2$ equivariant. Denote by Θ_σ^* the lift of the Θ_σ-action to $T^*GL^+(3)$ identified with $GL^+(3) \times gl(3)^*$ via i_L.

Proposition 1 *The lifted action of $\mathbf{Z}_2$ to $\mathcal{P}$ identified with $GL^+(3) \times gl(3)^*$ is*

$$\Theta_\sigma^*(Q,\beta) = \begin{cases} (Q,\beta) & \text{for} \quad \sigma = 1 \\ (Q^T, Q\beta^T Q^{-1}) = (Q^T, Ad_{Q^T}^*\beta^T) & \text{for} \quad \sigma = \tau \end{cases}.$$

The momentum map J for the $(SO(3) \times SO(3))$-action is $\mathbf{Z}_2$-equivariant in the following sense:

$$J(\Theta_\sigma^*(Q,\beta)) = \sigma \cdot J(Q,\beta) \quad \text{for} \quad \sigma = 1, \tau. \tag{6}$$

Proof: By the definition of the lifted action (see [1] p. 283) we have

$$\text{for} \quad v \in T_{\Theta_\sigma(Q)}GL^+(3) = \begin{cases} T_Q GL^+(3) & \sigma = 1 \\ T_{Q^T} GL^+(3) & \sigma = \tau \end{cases}$$

$$\langle \Theta_\sigma^*(\alpha), v \rangle = \langle T^*\Theta_{\sigma^{-1}}(\alpha), v \rangle = \langle \alpha, T\Theta_{\sigma^{-1}}(v) \rangle = \langle (T\Theta_{\sigma^{-1}})^*(\alpha), v \rangle.$$

That is, $\Theta_\sigma^*(\alpha) = (T\Theta_{\sigma^{-1}})^*(\alpha)$. As we are dealing with the left identification of the phase space $\Theta_\sigma^*(Q,\beta) = (i_L \circ \Theta_\sigma^* \circ i_L^{-1})(Q,\beta)$ then:

$$\Theta_\sigma^*(Q,\beta) = \begin{cases} i_L\left((T\Theta_{\sigma^{-1}})^*(Q^{-T}\beta)\right) = (Q, \langle (T\Theta_{\sigma^{-1}})^*(Q^{-T}\beta), T_e L_Q \rangle) & \sigma = 1 \\ i_L\left((T\Theta_{\sigma^{-1}})^*(Q^{-T}\beta)\right) = (Q^T, \langle (T\Theta_{\sigma^{-1}})^*(Q^{-T}\beta), T_e L_{Q^T} \rangle) & \sigma = \tau, \end{cases} \tag{7}$$

where the last equalities on the right hand side follow from the definition of i_L and from the fact that the cotangent projection of $(T\Theta_{\sigma^{-1}})^*(Q^{-T}\beta)$ is Q^T and Q, respectively for $\sigma = \tau$ and $\sigma = 1$. For $\sigma = \tau$, we have

$$\langle (T_{Q^T}\Theta_{\tau^{-1}})^*(Q^{-T}\beta), T_e L_{Q^T}(\xi) \rangle = \langle Q^{-T}\beta, \left(T_{Q^T}\Theta_{\tau^{-1}}^* \circ T_e L_{Q^T}\right)(\xi) \rangle$$

$$= \langle Q^{-T}\beta, T_{Q^T}\Theta_{\tau^{-1}}(Q^T\xi) \rangle$$

$$= \langle Q^{-T}\beta, \xi^T Q \rangle = tr(\beta^T Q^{-1}\xi^T Q)$$

$$= tr(\beta Ad_{Q^T}\xi) = \langle Ad_{Q^T}^*\beta^T, \xi \rangle.$$

Therefore $\Theta_\tau^*(Q,\beta) = (Q^T, Q\beta^T Q^{-1}) = (Q^T, Ad_{Q^T}^*\beta^T)$ for $\sigma = \tau$. For $\sigma = 1 = id$ the same kind of calculations follow, since

$$\langle (T_Q\Theta_{id^{-1}})^*(Q^{-T}\beta), T_e L_Q(\xi) \rangle = \langle Q^{-T}\beta, (T_Q\Theta_{id^{-1}}^* \circ T_e L_Q)(\xi) \rangle$$

$$= tr(\beta^T Q^{-1}Q\xi) = \langle \beta, \xi \rangle.$$

Then $\Theta_1^*(Q,\beta) = (Q^T, Q\beta^T Q^{-1}) = (Q,\beta)$ for $\sigma = 1$.

For the equivariance of J, as in (6), it is sufficient to prove equivariance for $\sigma = \tau$, since for $\sigma = 1$ it is a straighforward consequence of the fact that Θ_1^* is the identity map. Now

$$
\begin{aligned}
J(\Theta_\tau^*(Q,\beta)) &= J(Q^T, Q\beta^T Q^{-1}) \\
&= \tfrac{1}{2}\left(Q^T(Q\beta^T Q^{-1})^T - (Q^T)^{-T}(Q\beta^T Q^{-1})Q, Q\beta^T Q^{-1} - (Q\beta^T Q^{-1})^T\right) \\
&= \tfrac{1}{2}\,\tau\cdot\left(\beta - \beta^T, Q\beta^T Q^{-1} - Q^{-T}\beta Q^T\right) = \tau\cdot J(Q,\beta).
\end{aligned}
$$

$\square$

Note that it is possible to prove that J is also the momentum map for the semidirect product $\mathbf{Z}_2 \times_s (SO(3) \times SO(3))$ with the following multiplication rule:

$$
(\sigma_1,(\lambda_1,\rho_1))\,(\sigma_2,(\lambda_2,\rho_2)) = (\sigma_1\sigma_2,(\lambda_1,\rho_1)(\sigma_1\cdot(\lambda_2,\rho_2))) \tag{8}
$$

for $(\sigma_i,(\lambda_i,\rho_i)) \in \mathbf{Z}_2 \times_s (SO(3) \times SO(3))$ for $i,j = 1,2$ (see Sousa Dias [3]).

Dedekind's Theorem

Dedekind's theorem is concerned with the existence of a congruent (or adjoint) ellipsoid to a given ellipsoid of equilibrium for the Dirichlet problem. We will show that these ellipsoids of equilibrium are consequences of the $\mathbf{Z}_2$ symmetry and of the relative equilibrium theorem.

Let us formulate Dedekind's theorem as given in Chandrasekhar [2], §28, chapter 4. It states that if a motion determined by $X(t) = Q(t)x_0$, with $x_0 = I$ and $Q(t)$ an ellipsoid of equilibrium, is admissible under Dirichlet's conditions, then the motion determined by Q^T is also admissible, and the configurations determined by Q and Q^T are called adjoint configurations. Due to the bipolar decomposition of Q, see (1), this can also be formulated as follows: if there is a solution for some $(A; \Omega, \Lambda)$ then there is another solution for the same A but with the roles of Ω and Λ interchanged, i.e $(A; \Lambda, \Omega)$.

Under a Hamiltonian formulation of Dirichlet's problem, Dedekind's theorem will be equivalent to saying that if $X(t) = Q(t)x_0$ is a dynamic orbit of the Hamiltonian vector field X_H, generated by $H : \mathcal{P} \to \mathbf{R}$, then $X(t) = Q(t)^T x_0$ is also a dynamic orbit.

The relative equilibrium theorem provides us with a tool for finding dynamic orbits of a given Hamiltonian vector field with prescribed momentum values (see Marsden [6] p. 77 for the relative equilibrium theorem and Lewis & Simo [5] for its application to dynamical orbits). Briefly, it states that z_0 being a relative equilibrium is equivalent to the existence of a Lie algebra element ξ for which z_0 is a critical point of the augmented Hamiltonian

$$H_\xi(z) = H(z) - \langle J(z) - \mu, \xi \rangle,$$

where $\mu = J(z_0)$. The solution through z_0 is then given by $z(t) = \exp(t\xi)z_0$.

With this in mind, represent Q by

$$Q(t) = (\exp t\Omega)^T A(\exp t\Lambda).$$

(Since the matrices R, S of the bipolar decomposition are considered independent of time, that is they can be represented respectively by $R = (\exp t\Omega)$ and $S = (\exp t\Lambda)$).

Let $\tau_{\tilde{C}}^*$ be the cotangent bundle projection, $\tau_{\tilde{C}}^* : GL^+(3) \times gl(3)^* \to GL^+(3)$, $\tau_{\tilde{C}}^*(Q, \beta) = Q$, and $\tau_{\tilde{C}}^*(Q_0, \beta_0) = A$ (where (Q_0, β_0) is the value of (Q, β) at $t = 0$), $(J_1, J_2) = J(Q_0, \beta_0)$. Under these notations let us define an admissible configuration for the Dirichlet problem:

Definition 1 *An admissible configuration for the Dirichlet problem with angular velocity Ω and vorticity Λ is a triplet*

$$(A; \Omega, \Lambda) = (\tau_{\tilde{C}}^*(Q_0, \beta_0); \Omega, \Lambda) \in GL^+(3) \times so(3) \times so(3),$$

298

or (equivalently) a triplet

$$(A; J_1, J_2) = (\tau_{\mathbb{C}}^*(Q_0, \beta_0); J_1, J_2) \in GL^+(3) \times (so(3)^* \times so(3)^*),$$

provided that $(A, \beta_0) = (Q_0, \beta_0) \in GL^+(3) \times gl(3)^*$ *is a critical point of the augmented Hamiltonian*

$$H_{(\Omega, \Lambda)}(Q, \beta) = H(Q, \beta) - \ll J(Q, \beta) - J(Q_0, \beta_0), (\Omega, \Lambda) \gg \tag{9}$$

where J is the momentum map for the $SO(3) \times SO(3)$ action on $\mathcal{P}$.

Dedekind's theorem now follows:

Theorem 1 (Dedekind) *If $(A; J_1, J_2) \in GL^+(3) \times (so(3)^* \times so(3)^*)$ is an admissible configuration for the Hamiltonian system generated by the $\mathbb{Z}_2$- invariant Hamiltonian function $H : GL^+(3) \times gl(3)^* \to \mathbb{R}$, then $(A^T; J_2, J_1)$ is also an admissible configuration where $(J_1, J_2) = J(Q_0, \beta_0)$ is the momentum map value at time zero.*

Before proving this theorem note that the $\mathbb{Z}_2$-action on $so(3)^* \times so(3)^*$, say $\tau_{so(3)^* \times so(3)^*}$ can be easily deduced from (5) as the adjoint action of $\tau_{so(3) \times so(3)}$, i.e

$$\langle \tau_{so(3)^* \times so(3)^*} \cdot (\alpha, \beta), (\Omega, \Lambda) \rangle = \langle (\alpha, \beta), \tau_{so(3) \times so(3)}(\Omega, \Lambda) \rangle = \langle (\alpha, \beta), (\Lambda, \Omega) \rangle$$

$$= \langle \tau \cdot (\alpha, \beta), (\Omega, \Lambda) \rangle.$$

That is, $\tau_{so(3)^* \times so(3)^*} \cdot (\alpha, \beta) = \tau \cdot (\alpha, \beta)$.

Proof :

By the invariance of H, equation (6), and proposition 1 we have

$$
\begin{aligned}
H_{\tau_{so(3) \times so(3)}(\Omega, \Lambda)} \left(\Theta_\tau^*(Q, \beta) \right) &= H \left(\Theta_\tau^*(Q, \beta) \right) - \langle J(\Theta_\tau^*(Q, \beta)), \tau_{so(3) \times so(3)}(\Omega, \Lambda) \rangle \\
&\quad + \langle (J_2, J_1), \tau_{so(3) \times so(3)}(\Omega, \Lambda) \rangle \\
&= H(Q, \beta) - \langle \tau \cdot J(Q, \beta) - (J_2, J_1), \tau \cdot (\Omega, \Lambda) \rangle \\
&= H(Q, \beta) - \langle J(Q, \beta), (\Omega, \Lambda) \rangle + \langle (J_1, J_2), (\Omega, \Lambda) \rangle \\
&= H(Q, \beta) - \langle J(Q, \beta) - (J_1, J_2), (\Omega, \Lambda) \rangle = H_{(\Omega, \Lambda)}(Q, \beta).
\end{aligned}
$$

Since by hypothesis $(A; \Omega, \Lambda)$ is a critical point of (9), it follows that $(\tau_{\mathbb{C}}^*(\Theta_\tau^*(Q, \beta); (J_2, J_1)$ $(Q^T; J_2, J_1)$ is an admissible configuration . □

ACKNOWLEDGMENTS: This work has been carried out as part of the author's Ph.D research. I am grateful to my Ph.D supervisor Dr. Mark Roberts, not only for directing me towards this subject but for many helpful conversations. Special thanks to Prof. Ian Stweart for his helpful reading of the draft of this work.

References

[1] R. Abraham and J. E. Marsden. *Foundations of Mechanics*. Addison-Wesley, New York, 2nd edition, 1978.

[2] S. Chandrasekhar. *Ellipsoidal Figures of Equilibrium*. Dover, New York, 1987.

[3] E. Sousa Dias. *Local Dynamics of Symmetric Hamiltonian Systems and the Affine Rigid Body*. PhD thesis, University of Warwick, Mathematics Institute, University of Warwick, 1993. To be submitted.

[4] D. Lewis. Bifurcation of liquid drops. Sudam Report 91-7, Department of Mechanics, Stanford University, California, December 1991.

[5] D. Lewis and J.C. Simo. Nonlinear stability of rotating pseudo-rigid bodies. *Proc. R. Soc. London*, A(427):281–319, 1990.

[6] J. E. Marsden. *Lectures on Mechanics*. London Mathematical Society. Cambridge University Press, Cambridge, 1992.

[7] J. Slawianowski. Affinely rigid body and hamiltonian systems on $GL(n, \mathbf{R})$. *Reports on Mathematical Physics*, 26(1):73–119, 1988.

A NOTE ON DISCONTINUOUS VECTOR FIELDS AND REVERSIBLE MAPPINGS

MARCO A. TEIXEIRA
IMECC - UNICAMP, BRAZIL

ABSTRACT. The purpose of this note is to present an elementary discussion of the dynamic behavior of both reversible mappings and ordinary differential equations with discontinuous right-hand sides. We explore the main ideas and results needed for the qualitative study of such ODE through reversible mappings. The key of the problem is that, under certain conditions, reversible mappings work as first return mappings associated to the equations.

0. Introduction

The purpose of this note is to present an elementary discussion of the dynamic behavior of both reversible mappings and ordinary differential equations with discontinuous right-hand sides. We explore the main ideas and results needed for the qualitative study of such ODE through reversible mappings contained in [T2], [T3], [T4] and [T5]. The key of the problem is that, under certain conditions, reversible mappings work as first return mappings associated to the equations.

We consider $Z : \mathbb{R}^3 \to \mathbb{R}^3$ as being a discontinuous vector field on $\mathbb{R}^3$, in such a way that its discontinuities are concentrated on a 2-dimensional submanifold S. Systems of such kind arise in various applications (see for instance [AVK], [B] and [F]).

Our formulation of the problem needs some concepts involving the Filippov's rule for defining the solution orbits of the discontinuous ODE, the contact between a smooth vector field on $\mathbb{R}^3$ and S, and the reversible mappings associated to discontinuous systems. This is done in Section 1.

In Section 2, we analyse those discontinuous systems Z having the following property: "the orbit of Z refracts when it meets the manifold S".Those points where Z is

301

P. Chossat (ed.), Dynamics, Bifurcation and Symmetry, 301–310.

tangent to S are singularities of the vector field. Each one of these singularities can be expressed as a fixed point of a reversible mapping. We derive the generic normal forms of such reversible systems and characterise the local structrutal stability of Z. Moreover, we exhibit a topological invariant (in the global sense) for this class of discontinuous systems as well as for the associated reversible mappings. We observe that the class of reversible mappings treated in this section is in a certain way very degenerate in the world of all reversible mappings.

In Section 3, the local behavior of generic discontinuous systems is analysed. The basic ingredient for such study is the dynamics of generic reversible mappings. We restrict our discussion to the asymptotic stability of the system.

Section 4 is devoted to present some final comments.

The references are given in Section 5.

1. Preliminaries

For simplicity we assume that S is the canonical sphere in $\mathbb{R}^3$. Let N be a small neighborhood of S in $\mathbb{R}^3$, N^- (resp N^+) be the part of N contained in the interior (resp. exterior) of S.
Denote by V the space of C^∞ vector fields on N.

1.1. FILIPPOV'S RULE

Let W be the set of vector fields Z defined on N by

$$Z(q) = \begin{cases} X(q) & \text{if } q \in N^+ \\ Y(q) & \text{if } q \in N^- \end{cases} \qquad \text{where } X, Y \in V,$$

and the solutions curves of Z at points of S are defined by the Filippov's rule which we briefly recall in what follows.

Denote $Z = (X, Y)$. We have to distinguish the following regions in S:

a) Sewing Region (SR): in this region, the vector field X (resp. Y) is directed away from S (resp. toward) S. If a point of the phase space is moving in an orbit of Z

falls onto S then it crosses S over the another part of space.

b) Escaping Region (ER): the vector fields X and Y point toward N^+ and N^- respectively. The solution through a point p e S follows the orbit of one of the vector fields X or Y, according to which has the largest normal component with respect to M.

c) Sliding Region (SLR): both vector fields X and Y point toward S. In this case the solution of Z through points of S follows the orbit of the vector field $F = F(X,Y)$ (called the SL-vector field associated to $Z = (X,Y)$). F is tangent to S and is defined at p e S by the vector $F(p) = m - p$, where m is the point where the segment joining $p + X(p)$ and $p + Y(p)$ is tangent to S.

All "curves" in S separating the above named regions are constituted by points where X or Y are tangent to M. Call by T_X (resp. T_Y) the set of tangency points between X (resp. Y) and S.

1.2. INVOLUTION ASSOCIATED TO A C^∞ VECTOR FIELD

Let $p \in S$ and $f : \mathbb{R}^3, S \to \mathbb{R}, 0$ be a C^∞ local representation of S at p (i.e. $f^{-1}(0) = S$) with $df(p) \neq 0$, such that $N^- = \{f < 0\}$ and $N^+ = \{f > 0\}$.

We say that $p \in S$ is a fold point of $X \in V$ if $Xf(p) = 0$ and $XXf(p) \neq 0$. This means that the contact between X and S at p is quadratic.

We list some facts:

i) If p is a fold point of X then the connected component of T_X containing p is a smooth curve in S.
ii) The orbit $\gamma_X(q)$ of X through a point $q \in S$ close to p, intercepts S either : (1) at two distinct points q and $\tilde{q}$ (provided that $q \notin T_X$) or (2) just at q (provided that $q \in T_X$).

We define the C^∞ (local) diffeomorphism $\phi_X : S, p \Rightarrow S, p$ by:

$$f(q) = \begin{cases} \tilde{q} & \text{if } q \notin T_X \\ q & \text{if } q \in T_X \end{cases}$$

It is easy to recognise that ϕ_X is an involution. One knows that it is C^∞ equivalent to the canonical involution $\phi_0(x, y) = (x, -y)$.

1.3. FIRST RETURN MAPPING

As our treatment in this section is local, S and p are identified to $\mathbb{R}^2$ and 0 respectively.

Let $Z = (X, Y)$ in W and $p \in S$ be such that $Xf(p) = Yf(p) = 0, XXf(p) < 0$ and $YYf(p) > 0$ with f as above; in this case we say that p is a focus of Z.

In this case the reversible mapping $\phi = \phi_X \circ \phi_Y$ works as a 1th return mapping associated to Z and p where ϕ_X and ϕ_Y are the involutions associated to X and Y respectively. In $[T_5]$ this construction is made in details and it is illustrated in Figure 1.

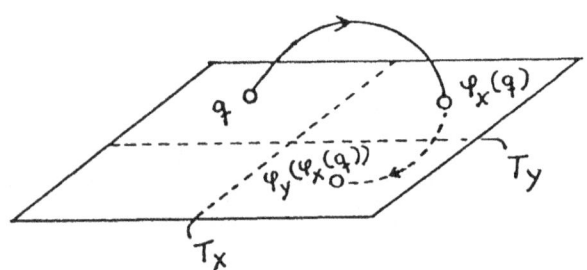

Figure 1- The first return mapping

2. Topological invariance

2.1. LOCAL THEORY

In this section, we deal with those $Z = (X, Y)$ satisfying $X(q).\vec{n} = Y(q).\vec{n}$ where $q \in S$ and $\vec{n} = \vec{0q}$.

Call by ξ the subset of W constituted by such Z.

We observe that if $Z = (X, Y)$ is in ξ then $T_X = T_Y$.

The following result is proved in $[T_2]$:

Theorem 1 - Assume that p is a focus of $Z = (X, Y) \in \xi$. Then Z is, at p, locally struturally stable if and only if $\phi = \phi_X \circ \phi_Y$ is C^0 conjugated to one of the following generic normal forms:
I- regular case: $\phi_0(x, y) = (x + y, y)$
II- singular case: $\phi_1(x, y) = (x + xy, y + \varepsilon y^2)$ with $\varepsilon = \pm 1$.

This result can be transfered to the following form:

Theorem 1' - Assume that $\phi : \mathbb{R}^2, 0 \Rightarrow \mathbb{R}^2, 0$, is a local reversible mapping satisfying: $\phi = L \circ K$, $L \circ L = Id$, $K \circ K = Id$, $Det L'(0) = Det K'(0) = -1$ and $Fix(L) = Fix(K)$. Then the generic C^0 normal forms of ϕ are as above.

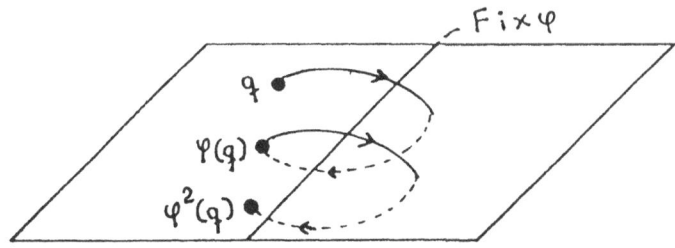

Figure 2. The field Z: regular case

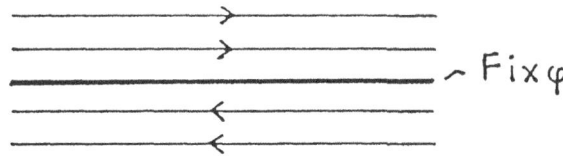

Figure 3. Dynamics of the reversible mapping: regular case

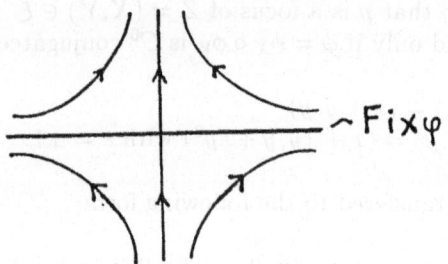

Figure 4. Dynamics of the reversible mappings: singular case ($\varepsilon = -1$).

We remark that due to the presence of symmetry to relate the dynamics of the vector field and the associated mapping it is enough to concentrate the attention to the up side of the line of tangency. Of course, the tangency set of Z coincide with the fixed point set of the associated mapping.

The singular cases say that Z has an invariant 2-dimensional manifold transverse to the tangency set of Z.

2.2. EXAMPLE

In $\mathbb{R}^3$ consider :
i) the vector fields defined by:
$X(x, y, z) = (1, -1, \cos\, y + \sin\, y - 1)$,
$Y(x, y, z) = (0, 1, \cos\, y + \sin\, y - 1)$.

ii) the plane $S = \{z = 0\}$ and the line $r = \{(x, 0, 0)\}$.

We define the vector field $Z = (X, Y)$, descontinuous on S.
Observe that r belongs to the tangency set between Z and S. Moreover, any point p in r is a focus of Z. We can deduce (see T_4) that the involutions associated to X and Y at $(x, y, 0) \in S$ have the forms:

$$\phi_X(x, y) = (x + 2y - (2y^2/3), -y - (2y^2/3)) + \text{h.o.t.}$$
$$\phi_Y(x, y) = (x, -y - (y^2/3)) + \text{h.o.t.}$$

We proceed with some calculations, which will be very usefull in the sequel (see for instance Section 2.3))   The reversible mapping $\phi = \phi_Y \circ \phi_X$ imbedds formally in a flow H having the form:

$$H(x,y) = (2y - (2y^2/3), y^2/3) + \text{h.o.t.}$$

Take now the sections $\Sigma_0 = \{(0,y)\}$ and $\Sigma_1 = \{(2\pi, y)\}$ and the Poincare transformation

$$\rho : \Sigma_0 \Rightarrow \Sigma_1$$

associated to (H/y). It is given by:

$$\rho(y) = \text{ by } + \text{h.o.t}$$

where $b = (3 + \pi)/3$.

We recall that this number b plays an important role in the global structural stability of $Z = (X, Y)$ in the following sense.

If $\tilde{Z}$ is a small perturbation of Z in ξ then we can get similar constructions and objects as above. In particular we find a similar number $\tilde{b}$. We can deduce that if Z is C^0 equivalent to Z then $b = \tilde{b}$. In another words the number b is a topological invariant for the structural stability in ξ.

2.3. TOPOLOGICAL INVARIANT

Let ϕ be a reversible mapping on $\mathbb{R}^2$ such that: Fix $\phi = S^1$ and at any point $p \in S$, ϕ is locally equivalent to $\phi_0(x,y) = (x + y, y)$.

We do the following construction:

(i) ϕ is , locally around any $p \in S$, formally imbedded in a flow;

(ii) we consider ϕ restricted to the strip $B = \{(x,y) : 0 \leq x \leq 1, \text{ and } -\delta < y < \delta\}$ in such a way that $r = \{(x,0)\}$ represents the fixed set of the diffeomorphism;

(iii) we may consider finite retangles Q_i, $i = 1, 2, \ldots, k$ satisfying $r \subset Q_1 \cup Q_2 \cup \ldots \cup Q_k$ with $Q_i \cap Q_j (i \neq j)$ being the boundary of the retangles and such that ϕ restricted to Q_i is formally imbedded in a flow H_i, $i = 1, 2 \ldots, n$.

(iv) Let H be a vector field on B satisfying $H_{|Q_i} = H_i$. Observe that H can be non smooth;

(v) Define the vector field $B = H/y$ and sections transverse to B given by $S = \{0\} \times \delta$ and $S = \{1\} \times \delta$;

(vi) The first return mapping associated to B, Σ_0 and S has the form

$$\rho(y) = by + \text{h.o.t.}$$

The following result is proved in [T4]:

Theorem 2: Let $Z = (X, Y)$ be in ξ and ϕ be the reversible mapping associated to it. Then the number b is a topological invariant for the structural stability in ξ.

3. Generic reversible mappings

Our treatement in this section will be local and the proofs are contained in [T_5].
Let $Z = (X, Y) \in W, p \in S$ be such that:
(i) p is a focus of Z;
(ii) The manifolds T_X and T_Y are in general position in S.

We list some facts and concepts:
a) T_X and T_Y determine four quadrants (see Figure 5) in S in such a way one always has: a escaping region (Q_1), sewing regions (Q_2 and Q_3) and a sliding region (Q_4);
b) the sliding vector field $F(X, Y)$ can be smoothly extended to a full neighborhood of p; moreover, this point is a critical point of F;
c) we say that p is an stable singularity of $F = F(X, Y)$ if the eigenvalues associated to $dF(p)$ are real, negative, distinct and the strong stable manifold W^{uu} does not meet SLR whereas the weak stable manifold W^u meets SLR; moreover W^u e W^{uu} are transverse to both T_X and T_Y (see Figure 6).

d) let $\phi = \phi_X \circ \phi_Y$ be the reversible mapping associated to $Z = (X, Y)$. Then generically p is either a saddle fixed point or a elliptic fixed point of ϕ. We say that p is an s-singularity of ϕ if one of the followings conditions is satisfied: d1) p is a saddle fixed point, the eigenvalues of $d\phi(p)$ are negative and the respective eigenspaces meet SLR and they are transverse to T_X and T_Y; d2) p is a elliptic fixed point and the rotation angle associated to the eigenvalues is irrational. We have the following result:

Theorem 3. Let $Z = (X, Y) \in W$ and $p \in S$ be a focus point of Z. Then Z is asymptotically stable at p provided that p is an stable singularity of the associated SVF and an s-singularity of the associated reversible mapping.

In $[T_5]$ there are others results concerning structutral stability in W.

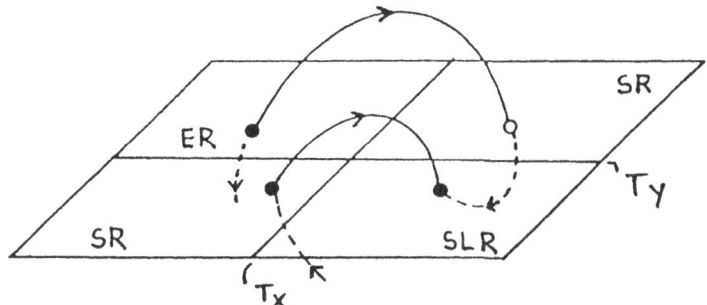

Figure 5. Generic singularity of $Z = (X, Y)$

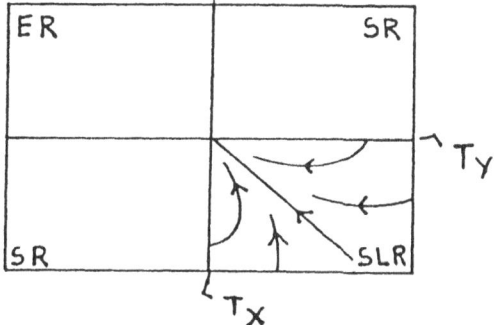

Figure 6. Stable sliding vector field

4. Final Remarks

There are some problems involving reversible mappings and classification of divergent diagrams of mappings. See for instance $[T_1], [T_2], [T_3]$ and mainly the list of problems contained in [V].

Concerning the refracted discontinuous vector fields discussed in Section 2, there are some open problems involving cubic contact between the vector fields and the surface of discontinuity, and reversible mappings (see $[T_4]$).

In Section 3, a connection between generic reversible mappings and stability of vector fields has been studied. Similar study can be done (in our point of view) for the case when the eigenvalues associated to the mappings are ±1.

5. References

[A,K,V] - A.A. Andronov et al, Theory of oscillators, Dover, NY, 1987.

[B] - E. Barbashin, Introduction to the theory of stability, Noordhoff, Groningen, 1970.

[F] - A.F. Filippov, Differencial equations with discontinuous righ-hand sides, Kluwer, 1988.

[T1] - M.A. Teixeira, Generic bifurcation of discontinuous vector fields, An. Acad. Bras. Cienc., V. 53 (2) 1981.

[T2] -..............., Local and simultaneous structural stability of certain diffeomorphisms, Springer-Verlag Lecture Notes , V. 898, 1982.

[T3] -..............., On topological stability of divergent diagrams of folds, Math. Z., V. 180, 1982.

[T4] - ...., A topological invariant for discontinuous vector fields, Nonlinear Anal.: TMA, V. 9, 1985.

[T5] - ...., Stability conditions for discontinuous vector fields, J. Diff. Eq., V. 88 (1), 1990.

[V] - S. M. Voronin , Analytic classification of pairs of involutions and its applications, Funktional'nyi Anaiz Prilozheniya, V. 16, N. 2, 1982.

BIFURCATION OF SINGULARITIES NEAR REVERSIBLE SYSTEMS

MARCO A. TEIXEIRA   A. JACQUEMARD
IMECC - UNICAMP   Laboratoire de Topologie
BRAZIL   Université de Bourgogne
                                    FRANCE

ABSTRACT. In this paper we study generic unfoldings of certain singularities in the class of all C^∞ reversible systems on R^2.

1. Introduction

This paper studies generic bifurcations of certain singularities of C^∞ reversible mappings on the plane. There is a rapidly growing literature about such systems and they have been studied from different points of view; it should be mention that this problem has been treated recently by Seivryuk [S] and by some physicists (see for example [P,C,Q,W]). Our analysis follows the main ideas of these works but with different formulation of the problem. The development of this work has been influenced by the chapter 6 of [A2] and our results, in certain sense, continue those ones contained in [P,C,Q,W].   A mapping $f : I\!R^2, 0 \to I\!R^2, 0$ is reversible if it can be written as the composition of two involutions $f = \phi_0 \circ \phi_1$. We shall just deal with those involutions ϕ which are germs of C^∞ diffeomorphisms (at 0) satisfying $\phi^2 =$ Id and $\text{Det}(\phi'(0)) = -1$.

Denote by G the space of germs of involutions at 0. The space of all reversible mappings is identified, in a natural way, with $W = G \times G$. To simplify the notation, we will make no distinction between a germ of an involution (at 0) and any one of its representatives.

Given (ϕ_0, ϕ_1) in W, we may choose coordinates (see [T1]) (x, y) around 0 in $I\!R^2$ such that $\phi_0(x, y) = (-x, y)$ and for some a, b and c in $I\!R$ we have

$$d\phi_1(0) = \begin{bmatrix} a & b \\ c & -a \end{bmatrix}$$

with $a^2 + cb = 1$.

We fix, troughout the paper, the above coordinates.

The eigenvalues of $df(0)$ are $\lambda = a \pm (a^2 - 1)^{1/2}$. Moreover, 0 is a hyperbolic fixed point (saddle point) of f if and only if $a^2 > 1$. If $a^2 < 1$, then 0 is an ellyptic fixed point

P. Chossat (ed.), Dynamics, Bifurcation and Symmetry, 311–325.
© 1994 *Kluwer Academic Publishers*.

of f (the linearization of the mapping at 0 is a conventional rotation). For $a^2 = 1$ both eigenvalues coincide with either 1 or -1 (double eigenvalue).

It may be clear that, taking "a" as a parameter, the orbit structure of f can change drastically when the real number $|a|$ cross generically the value $|a_0| = 1$. To study such changes, we deal with 1$-$ and 2-parameter families of diffeomorphisms f_α such that: i) f_0 is a reversible mapping;

ii) the 1$-$jet of f_α at 0 (the linear part) is a reversible mapping;

iii) either $df_0(0) = \begin{bmatrix} 1 & 0 \\ c & 1 \end{bmatrix}$ or $df_0(0) = \begin{bmatrix} -1 & 0 \\ c & -1 \end{bmatrix}$.

Our results will concern the qualitative behavior of generic parameter families of such diffeomorphisms. An outstanding question in this approach is the case when $c = 0$.

We observe that a natural question to be exploited in this context is the analysis of symetric homoclinic and the heteroclinic loop bifurcations (see [AS]) as well as the emergence or disappearance of limit cycles.

We recall that, given (ϕ_0, ϕ_1) in W, we know, from [T], that it is simultaneously C^0 conjugate to (L_0, L_1) provided that 0 is a hyperbolic fixed point of $f = \phi_0 \circ \phi_1(L_1$ is the linear part of $\phi_1, i = 0, 1)$.

We will use the following definitions and notations:

$I\!M(2)$: the space of real matrices 2×2;

Diff(2) : the space of germs of C^∞ diffeomorphisms in 0 e $I\!R^2$;

$V(2)$: the space of germs of C^∞ vector fields (in 0), vanishing in 0;

Ω: the space of reversible mappings in Diff(2);

Σ : the space of f e Diffeo(2) such that $j^1 f(0)$ is a reversible mapping;

Σ_0: the space of f e Σ such that the eigenvalues of $df(0)$ are different from ± 1;

$\Omega_0 = \Omega \cap \Sigma_0$;

$W_1 = \Sigma / \Sigma_0$;

Σ_1 : the space of $f \in W_1$ such that $df(0)$ is not diagonalizable (non degenerate case);

$\Omega_1 = \Omega \cap \Sigma_1$;

Σ_2 : the space of f e W such that df(0) is diagonalizable (degenerate case);

$\Omega_2 = \Omega \cap \Sigma_2$.

The main aim of this paper is to study generic "arcs" in Σ passing through a generic element f_0 of Ω. Our strategy is as follows:

(i) find conditions on the coefficients of $j_k f_0(0)$ for the reversibility of f_0 (for some k); this is made via the Mathematica Program in a PC-386 and the Axiom Program in a Risc workstation (see bellow).

(ii) determination and study of the singular points of a perturbation f of f_0 in Σ; we do this by taking suitable generic parameter families in Σ, which are in fact, universal unfoldings of f_0.

(iii) analysis of the bifurcation set (of fixed points) of each family.

In this paper we consider just the cases where the eigenvalues of df(0) are positive.

Before developing the results, we turn to discuss how the algebraic and analytic computings were carried out using the axiom software. We give here a sketch of the programs written in AXIOM. The key parts involving formal computing are

1) finding reversibility conditions of order 2.

Writing a general Taylor expansion of F, we compute the Taylor expansion of the square of $s \circ F$, where s is the involution $(x, y) \to (x, -y)$ and identify the expression with the identity.

We get then for each coefficient of degree $[i, j]$ in $[x, y]$ a condition between the coefficients of the Taylor expansion of F, denoted by V.

This set E of equations generates an ideal, and applying the Groebner basis algorithm with respect to the set of variables V (nothing more than the "solve" function of AXIOM), we obtain that exactly the half of these equations define that ideal. (see Appendix 1 for the AXIOM code)

2) finding fixed points

The fixed points of a given perturbation are solution of a system of two algebraic equations. We apply on these two equations again the Groebner basis algorithm with respect to variables $[y, x]$ and we get there a splitting in:

– the solution $[x, y] = [0, 0]$

– one equation in y of degree 3 with coefficients in the parameters named $sy = \{Py = 0\}$ and one equation sx there x is given as an explicit function in y.

3) studying the number and the type of fixed points

First of all the number of fixed points is directly given via the study of the sign of the discriminant of Py for small values of the parameters.

The location of the fixed points is possible with exact formulas involving radicals but their expression is too complicated to be handled with benefit. So we deliberately chose to solve $Py = 0$ in y with the Puiseux method. Since we only need here the first terms of the Puiseux expansion of the solution in order to decide the type of the fixed points, we write a simple program for this particular purpose.

The first step consists in finding the Puiseux exponents, and the second step solves the equations avoiding as long as possible radicals (characterization of coeficients by anulating polynomials).

Note that a slight modification of this program is able to give at once the number, the localization, and the type of the fixed points for given fixed coefficients.

2. Linear Mappings

In this section, we discuss the behavior of linear reversible systems and prepare the way for the proof of the main results.

We fix some notations:

S : the space of $A \in I\!M(2)$ such that $A = A_1 A_2$ with A_1, A_2 in $I\!M(2)$, $A_1^2 = A_2^2 =$ Id and $\mathrm{Det} A_1 = \mathrm{Det} A_2 = -1$;

$S_0(s)$: the space of $A \in S$ such that its eigenvalues are real and distinct;

$S_0(e)$: the space of $A \in S$ such its eigenvalues have non zero imaginary parts;

$S_0 = S_0(s) \cup S_0(e)$;

S_1 : the space of $A \in S/S_0$ such that A is not diagonalizable;

S_2 : the space of $A \in S/S_0$ such that A is diagonalizable.

We identify $I\!M(2)$ with $I\!R^4$; so in our coordinates we have:

$S = \{(x, y, z, w) : x - w = 0 \text{ and } xw - yz = 1\}$, $S_0 = \{(x, y, z, w) \in S : x^2 \neq 1\}$,

$S_1 = \{(x, y, z, w) \in S : x^2 = 1, y = 0 \text{ and } z \neq 0\}$ and

$S_2 = \{\mathrm{Id}\} \cup \{-\mathrm{Id}\}$. Observe that $S = S_0 \cup S_1 \cup S_2$.

For a general $A \in S$, denote $\sigma(A)$ and $\mathrm{Det} A$ the trace and the determinant of A, respectively.

2.1. Lemma. The sequence $S \supset S_1 \cup S_2 \supset S_2$ defines a Whitney stratification on S, where $\dim S_0 = 2$, $\dim S_1 = 1$ and $\dim S_2 = 0$.

Proof:

First of all, consider the mapping h : $I\!M(2) \rightarrow I\!R^2$ given by $h(x, y, z, w) = (x - w, xw - yz - 1)$. We have that this mapping is regular at a point $A = (a, -b, c, a)$, provided that $c^2 + b^2 \neq 0$.

Let now $A_0 = \mathrm{Id}$.

Define the mapping $g : S \rightarrow I\!R$ by

$g(A) = \sigma^2(A) - 4\,\mathrm{Det} A$.

In coordinates, we have that $A = (x, y, z)$, $A_0 = (1, 0, 0)$ $g(A) = 4x^2 - 4$ and $g(A_0) = 0$. A straighforward computation shows that A_0 is a non degenerate critical point of $g_{|S}$ (in this case a saddle point). Similar proceeding can be done around $B_0 = -\mathrm{Id}$. Now the conclusion of the lemma is direct.

The proof of next lemmas are straigthforward.

2.2. Lemma: Let B_α be a 1-parameter family of matrices in S transverse to S_1 at $A_0 = (1, 0, 1)$. Then B_α is linearly equivalent to A_α where

$$A_\alpha = \begin{bmatrix} 1+\alpha & (\alpha^2 + 2\alpha) \\ 1 & 1+\alpha \end{bmatrix}$$

Call $H_\alpha = A_\alpha - \mathrm{Id}$.

2.3. **Lemma:** Let $B_{\alpha\beta}$ be a 2-parameter family of matrices in S "transverse to S_2" at $A_0 = (1,0,0)$ (this means that the mapping : $(\alpha,\beta) \Rightarrow B_{\alpha\beta} \in S$ is surjective). Then $B_{\alpha\beta}$ is equivalent to $A_{\alpha\beta}$ where

$$A_{\alpha\beta} = \begin{bmatrix} 1+\alpha\beta & 2\alpha + 2\beta\alpha^2 \\ \beta & 1+\alpha\beta \end{bmatrix}$$

Call $H_{\alpha\beta} = A_{\alpha\beta}$ - Id.

2.4. **Remark.** The two above families, characterise the generic transition "$S_0(s)$ (saddle point) $\Longleftrightarrow S_0(e)$ (elliptical point)" in S.

2.5. **Remark.** Let us consider the projection $\pi : J_\infty \Rightarrow J_1$ (J_1 being the space of the i-th jet of mappings in 0) given by
$\pi(j^\infty f)(0) = j^1 f(0)$. It is evident that $\sum = \pi^{-1}(S), \sum_i = \pi^{-1}(S_i)$, $(i = 0,1,2)$ and it induces, in a natural way, a stratification in $\sum$ given by $\sum \supset \sum_1 \cup \sum_2 \supset \sum_2$. Moreover, $\sum_0 \cup \sum_1$ is a codimension 2 submanifold of Diff(2), $\sum_0$ is open in $\sum, \sum_1$ is a codimension 1 submanifold of $\sum$ and $\sum_2 \subset Cl(\sum_0 \cup \sum_1)$.

3. Generic Singularities

3.1. Regular Case : $f \in \Omega_1$

In our coordinates we have that
$f(x,y) = L(x,y) + Q(x,y) + C(x,y) + (o|x,y|^2)$ where
$L(x,y) = (x, x+y)$,
$Q(x,y) = (a_{20}x^2 + a_{11}xy + a_{02}y^2, b_{20}x^2 + b_{11}xy + b_{02}y^2)$ and $C(x,y) = (a_{30}x^3 + a_{21}x^2y + a_{12}xy^2 + a_{03}y^3, b_{30}x^3 + b_{21}x^2y + b_{12}xy^2 + b_{03}y^3)$.
The reversibility conditions on f implies that $a_{02} = 0, a_{11} = 2a_{20}$ and $a_{20} - b_{11} + b_{02} = 0$.
Define the following subsets of Ω_1:
i) G_{11} is the set of all f in Ω_1 such that $a_{11} \neq 0$ and $b_{02} \neq 0$.
ii) G_{12} is the set of all f such that $a_{11} = 0$, $b_{02} \neq 0$ and $a_{03} \neq 0$ (in this case the reversibility conditions imply that $a_{12} = 2a_{21}$ and $a_{20} = a_{02} = 0$).

Denote $G_1 = G_{11} \cup G_{12}$.

We have then :
I1) if $f \in G_{11}$ then it is imbedded in the flow map $X = (a_{20}x^2 + a_{11}x\,y,\ x + B_{20}x^2 + B_{11}x\,y + b_{02}\,y^2) +$ higher order terms, where both B_{20}, B_{11} depend linearly on B_{20}, B_{11}, b_{02} and a_{20}.

By changing coordinates $u = x + B_{20}x^2 + B_{11}xy + b_{02} y^2 +$ h.o.t and $v = y$ and doing a suitable rescaling we arrive to the following general form of the vector field:

$$X(u,v) = ((a_{11} + 2b_{02})uv - (a_{03} - a_{11}b_{02})v^3 - a_{11}(b_{02} + b_{11})u\, v^2, u) + \text{h.o.t}$$

We know, from [DSRZ, p.2], that the topological type of the singularity falls into one of the following categories:

i) The saddle case: $\varepsilon = a_{03} - a_{11}b_{02} > 0$;

ii) The focus case: $\varepsilon < 0$ and $\zeta^2 - 8\varepsilon < 0$, where $\zeta = (a_{11} - 2b_{02})$. We are assuming that $\zeta \neq 0$.

iii) The elliptic case: $\varepsilon < 0$ and $(\zeta - 8\varepsilon) > 0$ (see Figure 1). We are assuming generically that $(\zeta^2 - 8\varepsilon) \neq 0$.

I2) if $f \in G_{12}$ we arrived, as above, to the following general form of X:

$$X(u,v) = (buv + \varepsilon v^3 + \zeta uv^2, u) + \text{h.o.t.} \quad \text{where } b = 2b_{02}, \varepsilon = a_{03}.$$ This case is very similar to the first one and we get the same categories as above.

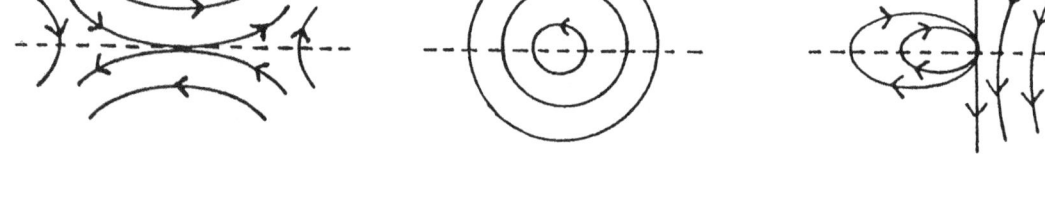

| a) Saddle Case | b) Focus Case | c) Elliptic Case |

Figure 1. The G_{11}−singularity;the picture is qualitative

The cases I1 and I2 are singularities of codimension 3 in the class of all germs of singularities.

3.2. Singular Case: $f \in \Omega_0$

In this case the linear part L of f is the identity and the reversibility conditions on f imply that $a_{20} = a_{02} = b_{11} = 0$. Define G_2 as the subset of elements f of Ω_2 such that $a_{11}^2 + b_{02}^2 + b_{02}^2 \neq 0$, $a_{03} \neq 0$ and $(a_{11} + 2b_{02}) \neq 0$ (for technical reasons we are assuming that $(3a_{03} + b_{21}) \neq 0$.

We know by [Ta] and [DRR] that if $f \in G_2$ then it is formally imbedded in a flow defined by a vector field X. Moreover the first non vanishing jet of X at 0 is $\tilde{X} =$

$(a_{11}xy, b_{20}x^2 + b_{02}y^2)$.

By means of a blowing up in polar coordinates we get that the phase portraits of a generic X as illustrated in Figure 3. The results obtained in [K] are useful in this section.

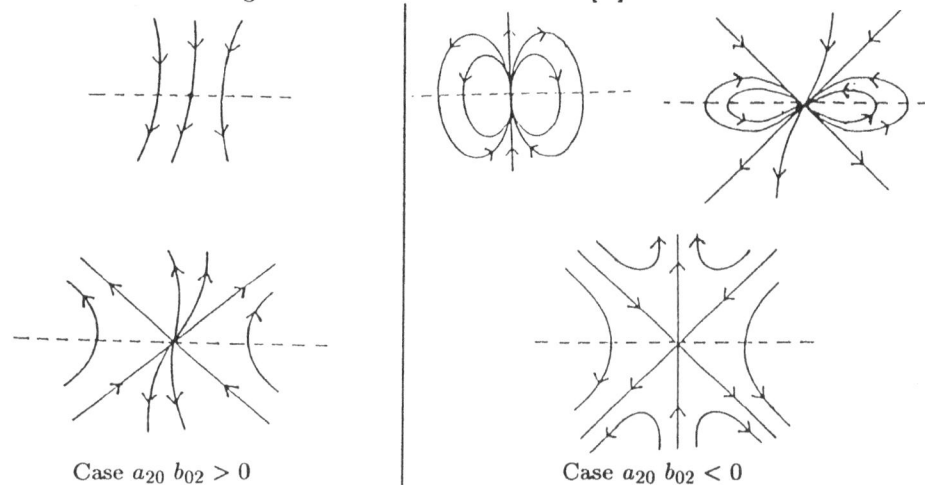

Case $a_{20}\,b_{02} > 0$ Case $a_{20}\,b_{02} < 0$

Figure 2. The G_2−singularity

4. Bifurcation diagram: Regular Case

In this section we discuss the bifurcation diagram (of fixed points) of a generic arc f_α in Ω passing through an element $f = f_0$ in G_1. As usual we do this by analysing the associated flow map X_α of f_α, where $f_\alpha = f + H_\alpha$ and H_α is given in Lemma 2.2.

4.1. Remark. We use the following notations for a critical point of a vector field or fixed point of a diffeomorphism:

A- attractor focus or node, R- reppeler focus or node, F- focus, N- node,
S- saddle point, E- elliptic point, BT- Bogdanov-Takens singularity,
SN- saddle-node singularity.

Theorem 1. *Let f_α be a one parameter family of diffeomorphisms in Ω, as above. Then f_α has the following bifurcation diagrams (Figures 3,4,5 below) in sufficiently small neighborhood of the origin in the phase space and for sufficiently small values of the param-*

318

eter α.

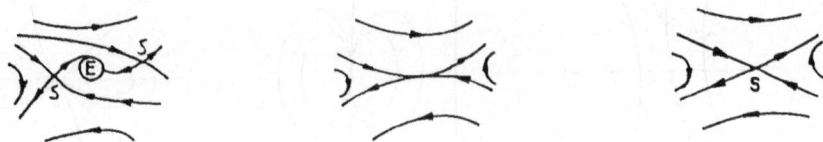

Figure 3. Bifurcation diagram of a G_{11} (or G_{12})-singularity (saddle)

Figure 4. Bifurcation diagram of a G_{11} (or G_{12})-singularity (elliptic)

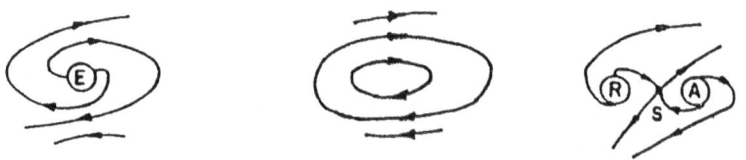

Figure 5. Bifurcation diagram of a G_{11} (or G_{12})-singularity (focus)

The proof of Theorem 1 comes from the next propositions:

4.2. Proposition. Let f_α be in Ω such that $f = f_0$ is in G_{11}. Then Case

i) $a_{11}b_{02} < 0$ (elliptical or focal cases):

if $\alpha < 0$ then f_α has a unique fixed point $P_0 = (0,0)$ which is of elliptical type ; if $\alpha > 0$

then f_α has tree fixed points $P_0 = (0,0)$ (saddle type) , $P \approx (-\alpha/(a_{11}^2), \pm(\alpha/a_{11}b_{02})^{1/2})$ (reppeler and attractor nodes) (the bifurcation diagram is illustrated in Figure 3).

Case ii) $a_{11}b_{02} < 0$ (saddle case):

if $\alpha > 0$ then f_α has a unique fixed point $P_0 = (0,0)$ which is of saddle type ; if $\alpha > 0$ then f_α has tree fixed points $P_0 = (0,0)$ (elliptical type) , $P \approx (-\alpha/(a_{11}^2), \pm(\alpha/a_{11}b_{02}^{1/2})$ (saddles)   (the bifurcation diagram is illustrated in Figure 8).

Proof.

The flow map associated to f_α is of the form:

$\dot{x} = \alpha x + (2\alpha + \alpha^2)y + a_{20}x^2 + a_{11}xy +$ h.o.t
$\dot{y} = x + \alpha y + b_{20}x^2 + b_{11}xy + b_{02}y^2 +$ h.o.t

The critical points of the system are given by:

$P_0 = (0,0)$ for $(\alpha/(a_{11}b_{02})) < 0.$
and
$P_1 \approx (-\alpha/a_{11}, \pm(\alpha/a_{11}b_{02})^{1/2})$ for $(()) ¿ 0.$

Moreover:

Det $X'(P_0) = -2\alpha,$ $Det X'(P_1) = \alpha + o|\alpha|^2$ and
$\sigma(X(P_1) = \pm(a_{11} + 2b_{02})(\alpha/a_{11}b_{02})^{1/2}.$

Now the conclusion of the proposition is direct.

The bifurcation diagrams of a G_{12}–singularity are of those ones of a G_{11}–singularity.

The proof of the next proposition is straigthforward.

4.3. **Proposition.** Let f_α be in Ω such that $f = f_0$ is in G_{12}.

Case i) Assume $a_{03} > 0$:

if $\alpha > 0$ then f_α has a unique fixed point $P_0 = (0,0)$ which is of saddle type ; if $\alpha < 0$ then f_α has tree fixed points $P_0 = (0,0)$ (elliptical type) , $P \approx ((2b_{02}\alpha)/a_{03}), \pm(-2\alpha/a_{03})^{1/2})$ (saddles).

Case ii) Assume $a_{03} < 0$:

if $\alpha < 0$ then f_α has a unique fixed point $P_0 = (0,0)$ which is of elliptical type ; if $\alpha > 0$ then f_α has tree fixed points $P_0 = (0,0)$ (saddle type) , $P \approx (-\alpha/(a_{11}^2), \pm(\alpha/a_{11}b_{02})^{1/2})$ (reppeler and attractor nodes).

5. Bifurcation Diagram: Singular Case

In this section we derive the fixed points bifurcation diagram of generic two-parameter families $f_{\alpha\beta}$ of diffeomorphisms such that:

i) $f_{00} = f$ is in G_2 and   ii) $f_{\alpha\beta} = f + (H_{\alpha\beta} -$ Id$)$ with $H_{\alpha\beta}$ given in Section 2.

320

Theorem 2. *Let $f_{\alpha\beta}$ be a two parameter family of diffeomorphisms in Ω, as above. Then $f_{\alpha\beta}$ has the following bifurcation diagrams (Figure 6, 7, 8 and 9 below) in sufficiently small neighborhood of the origin in the phase space and for sufficiently small values of the parameters α and β.*

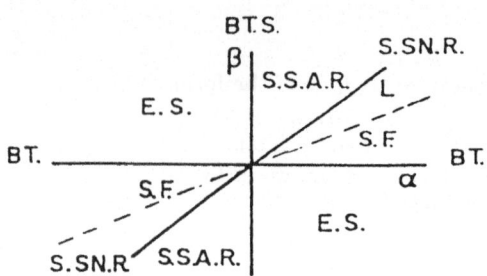

Figure 6. Bifurcation diagram of a G_2–singularity: $b_{20} > 0, b_{02} > 0$

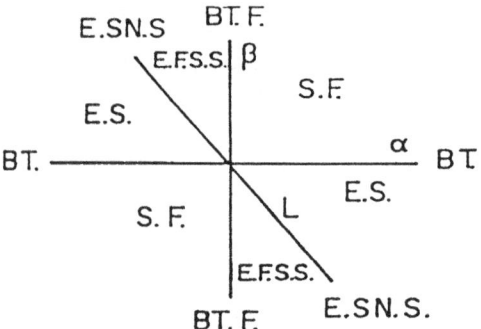

Figure 7. Bifurcation diagram of a G_2–singularity: $b_{20} < 0, b_{02} < 0$.

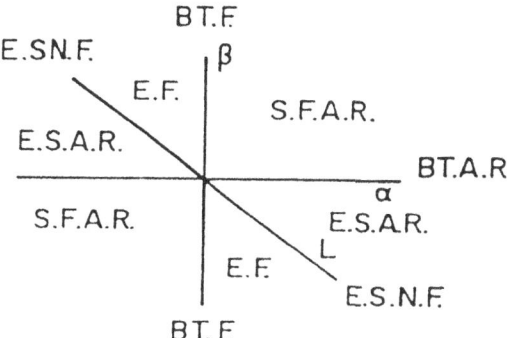

Figure 8. Bifurcation diagram of a G_2-singularity: $b_{20} < 0, b_{02} > 0$

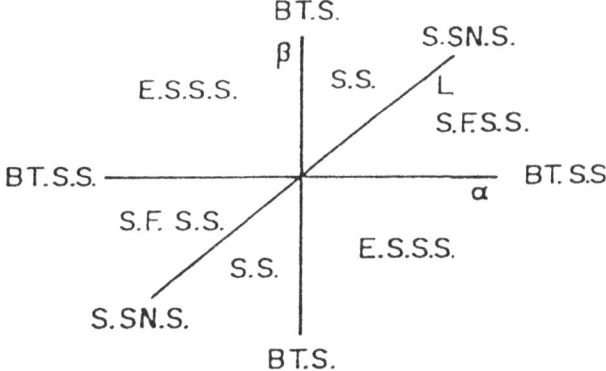

Figure 9. Bifurcation diagram of a G_2-singularity: $b_{20} > 0, b_{02} < 0$.

As usual, we analyse the flow map $X_{\alpha\beta}$ given by:

$$\dot{x} = \alpha\beta x + (2\alpha + \alpha^2\beta)y + a_{11}xy + \text{h.o.t.}$$

(5.1)

$$\dot{y} = \beta x + \alpha\beta y + b_{20}x^2 + b_{02}y^2 + \text{h.o.t.}$$

By means of a following change of coordinates of the form
$X = a_{11}x + 2\alpha + \circ|(\alpha, \beta, x, y)|^2$ and $Y = y$,   (5.1) is written as

$$\dot{x} = a_{11}xy + \circ|(\alpha, \beta, x, y)|^3$$

(5.2)

$$\dot{y} = 2a_{11}^2\alpha m - a_{11}^2 nx + lx^2 + \circ|(\alpha,\beta,x,y)|^3 + b_{02}y^2$$

where $m = (2b_{20}a^\alpha - a_{11}\beta), n = (4b_{20}\alpha - a_{11}\beta)$ and $l = b_{20}/a_{11}^2$.

We may and do assume, throughout the paper, that $a_{11} = 1$.

We get that the critical points of $X_{\alpha\beta}$ are given by:

0) $P_0 \approx (2\alpha, 0)$ which is a saddle (resp. elliptic) point of $X_{\alpha\beta}$ for $\alpha\beta > 0$ (resp. $\alpha\beta < 0$). We look for the others critical points of the vector field.

1) $P_1 = (m/b_{20}, 0) + \circ|(\alpha,\beta)|^2$,

2) $P_2 \approx (0, \pm y_2)$ where $y_2 = (-2\alpha m/b_{02})^{1/2}$.

Denote by L the curve given by $\{\alpha\beta^2 - 4b_{20}m \approx 0\}$ which characterises the existence of the critical points of $X_{\alpha\beta}$ of the form $sta \approx P_2$ (see below).

It should be mention that the expression 5.2 contains terms in "$\alpha\beta x$" and "$\alpha\beta y$".

We have that :

1) $X_{\alpha\beta}$ has 4 critical points (P_0, P_1, P_2^+ and P_2^-) provided that $\beta \neq 0$ and $(\alpha m) < 0$.

2) $X_{\alpha\beta}$ has 2 critical points (P_0 and P_1) provided that $\beta \neq 0$ and $(\alpha m) > 0$.

The bifurcation set of the family is concentrated on the curves "$\{\alpha \approx 0\}$", "$\{\beta \approx 0\}$" and "$\{\alpha\beta^2 - 4b_{20}m \approx 0\}$" .

Moreover:

i) On the curve $L_1 = \{\alpha \approx 0\}$ there are 2 critical points: $\tilde{P}_0 \approx (0,0)$ and $\tilde{P}_1 \approx (-\beta/b_{20}, 0)$;

ii) On the curve $L_2 = \{\beta \approx 0\}$ there are either one critical point ($\overline{P}_0 \approx (2\alpha, 0)$) or 3 critical points ($\overline{P}_0$ and $\overline{P}_2^{\pm} \approx (0, \pm 2(-b_{20}/b_{02})^{1/2}\alpha)$;

iii) On L there are 3 critical points.

The determinant and trace of $DX_{\alpha\beta}$ have the following form:

$\text{Det}(X_{\alpha\beta})\ (x,y) = 2b_{02}y^2 - 2\alpha\beta - (\beta + 4b_{20}\alpha)x - 2b_{20}x^2 + \text{h.o.t.}$

$\sigma(X_{\alpha\beta})\ (x,y) = (1 + 2b_{02})y + \text{h.o.t.}$

5.1. Proof of Theorem 2

First of all observe that $P_0 \approx (2\alpha, 0)$ is a saddle (resp. elliptic) point provided that $\{\alpha\beta > 0\}$ (resp. $\{\alpha\beta < 0\}$) and $P_1 \approx (m/b_{20}, 0)$ is a saddle (resp. focus) point provided that $\{\beta m/b_{20} < 0\}$ (resp. $\beta m/b_{20} > 0\}$) in $\{\alpha\beta \neq 0\}$.

The critical points $P_2 \approx (0, \pm(-2\alpha m/b_{02})^{1/2})$ of $X_{\alpha\beta}$ exist provided that $(\alpha m/b_{02}) < 0$ in such a way that they are saddle points for $b_{02} > 0$ and node critical points (one attractor and one repeller) for $b_{02} < 0$.

On L_1 we have that $\tilde{P}_0$ is a Bogdanov-Takens singularity and $\tilde{P}_1$ is generically either a saddle point (for $b_{20} > 0$) or a focus (for $b_{20} < 0$)

On L_2 we have to separate the cases:

i) if $(b_{20}/b_{02}) > 0$ then one has just a critical point $\overline{P}_0 \approx (2\alpha, 0)$ which is a Bogdanov-Takens singularity;

ii) if $(b_{20}/b_{02}) < 0$ then there exist 3 critical points for $X_{\alpha\beta} : \overline{P}_0$ as above and $\overline{P}_2 \approx (0, \pm 2(-b_{20}/b_{02})^{1/2}\alpha)$ in such a way that they are of nodal types (one attractor and one repeller) provided that $b_{02} > 0$ and of saddle type provided that $b_{02} < 0$.

On $L, X_{\alpha\beta}$ has 3 critical points:

i) $\overset{\approx\pm}{P}_1 \approx (-\alpha\beta, (1 \pm (1 - h_{\alpha\beta})^{1/2}/2b_{02})$ where $h_{\alpha\beta} = 4b_{20}\alpha(2m - (b_{02} - 1)\alpha\beta^2)$. They can be of elliptical, nodal, focal or saddle types (see the nature of them in the bifurcation diagrams).

ii) $\overset{\approx}{P}_2 \approx (0, -\alpha\beta/2b_{02})$ which is generically a saddle-node bifurcation provided that $b_{02} \neq 1/2$.

The above assertions can be easily checked by the expressions of $\Delta = \text{Det} X_{\alpha\beta}$ calculated at each critical point. So:

(1) $\Delta(P_0) \approx -\alpha\beta,$ (2) $\Delta(P_1) \approx \beta m/b_{20},$ (3) $\Delta(P_2) \approx -4am$

(4) $\Delta(\tilde{P}_1) \approx -\beta^2/b_{20}$ (5) $\Delta(\tilde{P}_2) \approx -2b_{02}y_0^2$, where $y_0^2 = -4b_{20}\alpha^2/b_{02}.$

(6) $\Delta(\overset{\approx}{P}_1) \approx 2b_{02}[2 \pm 2(1 - h_{\alpha\beta})^{1/2} - h_{\alpha\beta}]$ ($h_{\alpha\beta}$ is given above).

(7) $\Delta(\tilde{P}_0) \approx \begin{vmatrix} 0 & 0 \\ \beta & 0 \end{vmatrix},$ (8) $\Delta(\overline{P}_0) \approx \begin{vmatrix} 0 & 2\alpha \\ 0 & 0 \end{vmatrix},$ (9) $\Delta(\overset{\approx}{P}_0) \approx \begin{vmatrix} \alpha\beta(1 - 1/2b) & 0 \\ 0 & 0 \end{vmatrix}.$

This finishes the proof.

6. Appendix 1

−AXIOM program finding 2-reversibility conditions
−composition

```
p:=(1+a)*x+(a**2+2*a)*y+a[2,0]*x**2+a[1,1]*x*y+a[0,2]*y**2;
q:=x+(1+a)*y+b[2,0]*x**2+b[1,1]*x*y+b[0,2]*y**2;
pp:=eval(p,[y=-y]);
qq:=eval(q,[y=-y];
p:=eval(p,[x=pp,y=-qq])
Q:=eval(q,[x=pp,y=-qq])
```

```
- constructing list of definin equations
E:=nil;
for i in 0.. 2 repeat
    for j in 0.. 2 — (j + i) = 2 repeat
        E:=cons(coefficient(P, [x,y], [i, j])=0, E)
        E:=cons(coefficient(Q, [x,y], [i, j])=0, E)
```

```
- constructing list of variables Va: =reduce(concat, [[a[i,j] for i in 0.. 2 — (i+j)=2]
for j in 0.. 2])
    Vb: =reduce(concat, [[b[i,j] for i in 0.. 2 — (i+j)=2] for j in 0.. 2])
```

V:= concat(Va, Vb)

– reduction-resolution using AXIOM solver involving Groebner basis solve(E, V)

7. REFERENCES

[A1] V. Arnold, Reversible systems, in Nonlinear and Turbulent Process in Physics, 3, ed. R. Sadgeev, 1984, 1161-1174.

[A2] V. Arnold, Chapitres suplementaires de la theorie des equations differentielles ordinaires, Ed. Mir, Moscow, 1980.

[AS] V. Arnold, M. Sevryuk, Oscillations and bifurcations in reversible systems , in Nonlinear Phenomena in Plasma Physics and Hydrodynamics, ed. R. Sadggev, 1986, 31-64.

[B] R.I. Bogdanov, Versal deformation of a singularity of vector fields on the plane with nilpotent linear part in the case of zero eigenvalues, Selecta Mathem. Sovietica, V. 1, 4, 1981, 389-421.

[DRR] F. Dumortier, P. Rodrigues, R. Roussarie, Germs of diffeomorphisms in the plane, Lecture Notes in Math., 902, Springer, 1981.

[DRS] F. Dumortier, R. Roussarie, J. Sotomayor, Generic 3-parameter families of vector fields on the plane with nilpotent linear part. The cusp case. Ergodic theory and dynam. syst.,7, 1987, 375-413.

[DRSZ] Dumortier F, R. Roussarie, J. Sotomayor and H. Zoladeck, Bifurcations of planar vector fields, Lecture Notes in Math., 1480, Springer,1991.

[K] F. Khechichine, Familles generiques à quatre parametres de champs de vecteurs quadratiques dans le plan: singularité à partie lineaire nulle, These de Doctorat, Université de Bourgogne, Dijon, France.

[PCQW] T. Post, H. Capel, G. Quispel, J. van der Weele, Bifurcations in 2-dimensional reversible maps, Phisica A 164, 1990, 625-662.

[QC] G. Quispel, H. Capel, Local reversiblity in dynamical systems, Phys. lett. A 142,1989, 112-116

[RQ1] J. Roberts, G. Quispel, Reversible mappings in the plane, Phys. Lett. A, 132, 1988, 161-163.

[RQ2] J. Roberts, G. Quispel ,Chaos and time-reversal symmetry. Order and chaos in reversible dynamical systems, Physics Reports, 216, 2, 1992, 63-177.

[Ta] F. Takens, Forced oscillations and bifurcations, in : Applications of Global Analysis ,
      Comm. of Math. Utrecht University, 3 , 1974.

[T1] M.A. Teixeira, Local and simultaneous structural stability of certain diffeomorphisms,
      V. 898, eds.D. Rand and L.S. Young, Springer, 1981, 382-390.

[T2] M.A. Teixeira, Stability conditions for discontinuous vector fields, J. of Diff. Eq. 88,
      1980, 15-29.

ON A NEW PHENOMENON IN BIFURCATIONS OF PERIODIC ORBITS

Dr. Hans True
Institute of Mathematical Modelling, Building 321
The Technical University of Denmark, DK–2800 Lyngby, Denmark
and
ES–Consult, Staktoften 20, DK–2950 Vedbaek, Denmark

ABSTRACT.

An example is presented of a nonlinear dynamic problem, in which the slope of
a bifurcating periodic orbit is finite in the bifurcation point.

INTRODUCTION.

In her master's thesis Eva Slivsgaard examined the dynamics of a single–axle
bogie under a railway car running with constant speed along a straight and
horizontal railway track. She continued the work by Knudsen et al. [1], where
the dynamical system describing the motion can be found.

The model of the loaded rolling single–axle–bogie has two degrees of freedom -
lateral and yaw motion – and a set of constraints expressing that the wheelset
and the rails remain in contact. The equations for the vertical displacement
and roll of the bogie frame are assumed to uncouple from the equations for the
horizontal motion, so they need not to be considered. This assumption is tacitly
implied in the configuration of the support in the model.

The resulting fourth order system has two nonlinearities. The first results
from the creepage–creep force relation in the wheel–rail contact points. The
other nonlinearity stems from the modelling of the wheel flange restoring force
as a very stiff, linear spring with a dead band.

The force relation has a discontinuity in its second derivative at the origin,
which represents the steady undisturbed rolling along – and centred in – the
track. The trivial solution is a solution of the problem for all values of the
speed, V, of the vehicle. V is the control parameter in the problem, and it is
obviously real.

The problem has a non–trivial bifurcation in $V=0$, which is a physical reality.
There exist ranges of lateral displacements and yaw respectively in which the
wheel is in static equilibrium, held by the adhesion forces in the rail–wheel
contact points. For any $|V| \neq 0$ and sufficiently small, however, the trivial

327

P. Chossat (ed.), Dynamics, Bifurcation and Symmetry, 327–331.

solution is locally asymptotically stable. The global stability is under investigation, but it is assumed to apply – an assumption which is supported by experience.

When we consider the trivial solution and let V grow (V can be assumed positive without loss of generality), a bifurcation point V_0 is reached where bifurcation of a periodic solution takes place. The bifurcation is supercritical, and the bifurcating stable solution represents a time–periodic oscillation of the moving wheelset around the track center line.

Eva Slivsgaard discovered that the bifurcation is unusual, since the slope of the bifurcating solution at the origin seems to be <u>finite</u>. She determined the bifurcation numerically, blew the graph of the amplitude of the state variable x_1 versus V up and changed the scales of x_1 and V so the finite slope easily could be observed. Figure 1 is taken from Eva Slivsgaard's thesis.

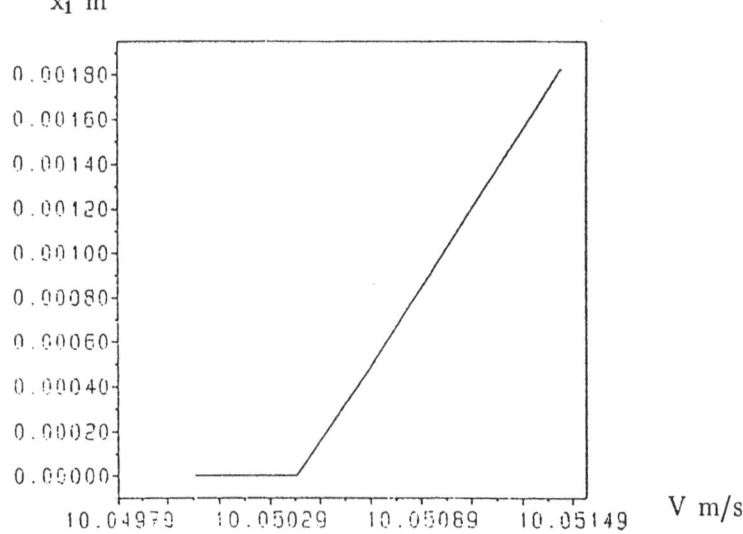

Figure 1   The bifurcation of a periodic solution from the trivial solution at $V = V_0$. Only stable solutions are indicated.

The question arises, whether this bifurcation behaviour is a numerical fluke or generic in the sense that it is connected with a singularity in the system ?

Next a simple two–dimensional model with the same singularity at the origin will be presented. It is simple to analyze analytically, and we shall find, that Eva Slivsgaard's bifurcation actually does exist.

THE BIFURCATION PROBLEM.

Let us consider the following model of a mechanical oscillator with nonlinear damping. It has one degree of freedom.

$$\ddot{x} + (k \sqrt{x^2 + \dot{x}^2} - \mu) \, \dot{x} + x = 0 \qquad\qquad \text{or}$$

(1)
$$\begin{cases} \dot{x} = y \\ \dot{y} = - (k \sqrt{x^2 + y^2} - \mu) y - x \end{cases}.$$

$x(t)$ is real, $\quad t > 0, \quad \dot{x} := \dfrac{dx}{dt}.$

$k \in \mathbb{R}/0$ and $\mu \in \mathbb{R}$ are constants. μ is the bifurcation parameter.

It is easily seen that $x = y \equiv 0$ is a solution of (1) for all μ, and that the associated linear system

(2) $\qquad \ddot{X} - \mu\dot{X} + X = 0 , \qquad X = X(t) , \qquad t > 0 ,$

has only bounded and exponentially damped solutions when $\mu < 0$, and when $\mu > 0$ only the trivial solution is bounded, but it is unstable.

We define

$$f(x,y) = \left[\begin{array}{c} y \\ -y(k\sqrt{x^2 + y^2} - \mu) - x \end{array} \right] .$$

The Jacobian is found to be

(3) $\qquad \dfrac{\partial f}{\partial(x,y)} = \left[\begin{array}{cc} 0 & 1 \\ -1 - \dfrac{kxy}{\sqrt{x^2 + y^2}} & \mu - k\dfrac{x^2 + 2y^2}{\sqrt{x^2 + y^2}} \end{array} \right] .$

Since we are interested in bifurcations from the trivial solution we evaluate the Jacobian at $(0,0)$:

$$\dfrac{\partial f}{\partial(x,y)} \bigg|_{(0,0)} = \left[\begin{array}{cc} 0 & 1 \\ -1 & \mu \end{array} \right]$$

with the eigenvalues $\lambda_{1,2} = \frac{1}{2}(\mu \pm \sqrt{\mu^2 - 4}).$

$\mu = 0$ is a bifurcation point, where $\lambda_{1,2} (0) = \pm i$, i.e. a periodic solution bifurcates at $\mu = 0$.

It is easily seen, that the trivial solution is asymptotically stable for $\mu < 0$ and unstable for $\mu > 0$.

When we change to polar coordinates

$$x = r \cos\Theta , \quad y = r \sin\Theta , \quad \text{with}$$

$$\dot{r} = \frac{1}{r} (x\dot{x} + y\dot{y}) \quad \text{and} \quad \dot{\Theta} = \frac{1}{r^2} (x\dot{y} - \dot{x}y)$$

we obtain

(4)
$$\begin{cases} \dot{r} = (\mu - kr) \, r \, \sin^2\Theta \\ \dot{\Theta} = (\mu - kr) \cos\Theta \, \sin\Theta - 1 \end{cases} .$$

System (4) has a solution $\mu = kr$, $(r \geq 0)$,

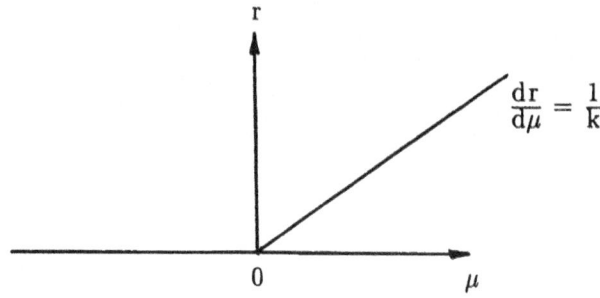

Figure 2   The bifurcation at $\mu = 0$.

and under the assumptions made, the bifurcating solution will have a finite slope at $\mu = 0$.

We find that the bifurcation is supercritical when $k > 0$, and then the bifurcating periodic solution is stable, and the bifurcation is subcritical when $k < 0$ and then the bifurcating periodic solution is unstable.

The resulting finite slope of the bifurcating solution in the bifurcation point is clearly an effect of the singularity of the damping function at the origin.

The singularity in the single–axle–bogie problem is of the same form as the one in this example, but the single–axle–bogie problem is four dimensional and the bifurcation parameter V enters in a more complicated way. The damping in the single–axle–bogie problem, however, also depends on the bifurcation parameter in such a way that the amplitude of the time dependent damping decreases with growing parameter values – i.e. with the speed in the bogie problem.

REFERENCE.

[1]   Knudsen, C., Feldberg, R. and True, H., Bifurcation in a model of a rolling wheelset. Phil. Trans. R. Soc. Lond. A **338**, 1992, 455–469.

ACKNOWLEDGEMENT.

The work was done during a visit to Dipartimento di Matematica, Università di Napoli in 1993. The author wishes to thank Professor Salvatore Rionero and Università di Napoli for their hospitality and support and CNR (The Italian National Research Council) for its financial support.

AN INHOMOGENEOUS PICARD-FUCHS EQUATION

S A. VAN GILS
Faculty of Applied Mathematics
University of Twente
The Netherlands

1   Introduction

In the study of singularities of vector fields on the plane the analysis of perturbations of Hamiltonian systems is crucial. The number of isolated limit cycles in the perturbed system is related to the number of zeros of periods (Abelian integrals). If the Hamiltonian function is algebraic, then the there are finitely many independent periods. They satisfy a matrix linear homogeneous differential equation, the Picard-Fuchs equation. See for instance [BK81]. The average of the perturbed vector field over the periodic orbits of the unperturbed problem can be expressed in those periods.

A systematic way to find the Picard-Fuchs equation in the elliptic case has been worked out in [SC].

A disadvantage is that dimension of the Picard-Fuchs equation may be high, while the averaging procedure may involves only two periods. See for instance [CCH85, CSG87, vGH86, Z84]. In some applications the equation for those two relevant periods takes the form

$$\begin{cases} q(h)\dot{I}_1 = a\,I_0 + b\,I_2 + e\,R(h) \\ q(h)\dot{I}_2 = c\,I_0 + d\,I_2 + f\,R(h), \end{cases} \tag{1}$$

where $a - f$ are real constants and q, R are smooth functions of h. This type of equations was used for the first time in a paper of Carr, Chow and Hale [CCH85]. Another example can be found in [CSG87]. For the method that we apply we do not only need as a starting point an equation of the form (1), but also the Hamiltonian function should have nice properties. We will indicate in the proof what is needed.

P. Chossat (ed.), Dynamics, Bifurcation and Symmetry, 333–341.

In this contribution we analyse the following system:

$$\begin{cases} \dot{x} = y - \frac{x^2}{2} + \epsilon_2 x^3 \\ \dot{y} = -x + \epsilon_1 \end{cases} \qquad (2)$$

which, at $\epsilon = 0$, has a conserved quantity

$$H = H(x,y) = -e^{-y}\left(y - \frac{x^2}{2} + 1\right) = h. \qquad (3)$$

This problem occurs in a study of canard phenomena by Dumortier and Roussarie

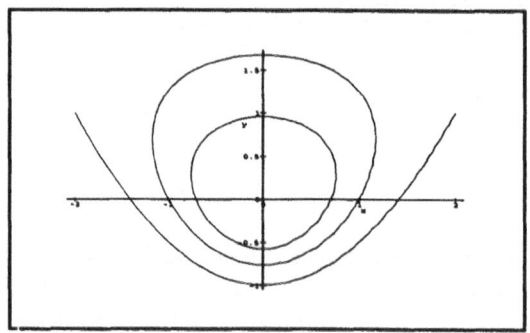

Figure 1: Some level sets of H

[DR]. In their analysis they need the following monotonicity result. Consider the periods

$$I_1 = \oint_{\gamma(h)} e^{-y}\,dx, \qquad I_2 = \oint_{\gamma(h)} e^{-y} x^3\,dy. \qquad (4)$$

Is it true that for $h \in (-1,0)$:

$$\frac{\partial}{\partial h}\left(\frac{I_2(h)}{I_1(h)}\right) > 0. \qquad (5)$$

We will prove that this is indeed the case (see Theorem 1).

A consequence of this result is that for ϵ_1, ϵ_2 small enough this system (2) has a unique limit cycle in a open region in the (ϵ_1, ϵ_2)-plane with part of the boundary tangent to the cone $0 < \frac{\epsilon_1}{\epsilon_2} < 3$. Hopf bifurcation occurs at a line tangent to the line

$\epsilon_1 = 3\epsilon_2$ and the line of homoclinic connections is tangent to the ϵ_2-axis.

This consequence can be proved more easily using a theorem of Zhang Zhi-fen [Zha86], as was pointed out to the author by H. Zoladek. However, the uniqueness of limit cycles is *not* equivalent to the monotonicity of the quotient of periods.

2 The monotonicity result

Most of the proofs in this section are given in the next one.

We will denote the closed level sets of H by $\gamma(h)$ and we denote the periods as follows:

$$\delta_m^n = \delta_m^n(h) = \oint_{\gamma(h)} e^{-y} y^n x^m dy$$

$$I_1 = \delta_1^0, \quad I_2 = \delta_3^0, \quad R = \delta_{-1}^0 \tag{6}$$

$$Q = Q(h) = I_2/I_1.$$

Note that the definition of I_1 in (6) is in agreement with (4), as is seen after integration by parts. The periods δ_m^n satisfy a recursion relation which is given in the next lemma.

Lemma 1 $\delta_m^{n+1} = \frac{n}{m+2}\delta_{m+2}^{n-1} + \frac{m}{2(m+2)}\delta_{m+2}^n.$ *(RE)*

The next lemma states that I_1 and I_2 satisfy a linear inhomogeneous differential equation of the type (1).

Lemma 2

$$\begin{cases} h\frac{d}{dh}\delta_1^0 = \delta_1^0 - \delta_{-1}^0 \\ \\ h\frac{d}{dh}\delta_3^0 = -3\delta_1^0 + \delta_3^0. \end{cases} \tag{7}$$

In terms of I_1, I_2, R this reads

$$\begin{cases} h\dot{I}_1 = I_1 - R \qquad\qquad (\dot{} = \frac{d}{dh}) \\ \\ h\dot{I}_2 = -3I_1 + I_2. \end{cases} \tag{8}$$

We work towards the result that Q is a monotone function of the level sets. We begin to give the value at the end points of the interval $[-1, 0]$. On the interval $[-1, 0)$, H has closed bounded level sets.

Lemma 3 $Q(-1) = 0$, $Q(0) = 3.$

To prove the monotonicity of Q we will show that the second derivative is of one sign at points where the first derivative vanishes, thus showing that this set is empty. First we give the expression for this second derivative at a point where the first derivative vanishes.

Lemma 4 *If $\dot{Q}(\bar{h}) = 0$ then at $h = \bar{h}$:*

$$\left(\frac{-\bar{h}I_1}{Q}\right)\ddot{Q} = \bar{h}\ddot{I}_1 + \left(1 - \bar{h}\frac{\dot{I}_1}{I_1}\right)\dot{I}_1. \tag{9}$$

In the next lemma an apriory bound for the quotient is given.

Lemma 5 *For $-1 < h < 0 : 0 < Q(h) < 3$.*

The problem with the right hand side of (9) is that there is another quotient of periods and the second derivative of I_1 appear. We cannot perform two differentiations under the integral, using $\frac{\partial x}{\partial h} = \frac{exp(y)}{x}$, without getting divergent integrals. As

$$e^{-y}\frac{x^2}{2} = h + e^{-y}(y + 1),$$

and $e^{-y}(y + 1)$ is maximal at $y = 0$, it follows that

$$\max_{\gamma(h)} e^{-y}\frac{x^2}{2} = h + 1.$$

Hence

$$e^{-y}(y + 1) - \max_{\gamma(h)} e^{-y}(y + 1) = e^{-y} - y^{-y} - 1,$$

which is independent of h. This fact is crucial for the proof of the next lemma. In fact we use that we can write the first integral in the form

$$x^2 f(y) - g(y) - k(h) = 0,$$

where the maximum of g is obtained at the value of y of the center. This motivates the introduction of the function J from $[-1, 0]$ into $\mathbb{R}$ defined by

$$J(h) = \oint_{\gamma(h)} \left\{\frac{x^2}{2}e^{-y} - (h + 1)\right\} xe^{-y}dy. \tag{10}$$

We collect the useful properties of this function in the next lemma.

Lemma 6

$$(i) \quad J(h) \le 0$$

$$(ii) \quad \dot{J}(h) \le 0$$

$$(iii) \quad \dot{J}(h) = \tfrac{1}{2}I_1 - (h+1)\dot{I}_1$$

$$(iv) \quad \ddot{J}(h) = -\tfrac{1}{2}\dot{I}_1 - (h+1)\ddot{I}_1$$

$$(v) \quad \ddot{J} = 3 \oint \frac{g(y)e^y}{x} dy$$

with

$$g(y) = -\frac{1}{6}\left(\frac{e^{-y}y^2 + e^{-y}y + y + 2e^{-y} - 2}{y^2}\right)$$

Proof. Note that

$$\frac{x^2}{2}e^{-3} - (h+1) = e^{-y}y + e^{-y} - 1 \le 0 \quad \text{for} \quad -1 \le y < \infty.$$

Furthermore, xdy is positive along the level set. This proves (i). As $\frac{x^2}{2}e^{-y} - (h+1)$ does *not* depend on h it follows that

$$\dot{J}(h) = \oint_{\gamma(h)} \{\frac{x^2}{2}e^{-y} - (h+1)\}\frac{dy}{x}$$

$$= \frac{1}{2}I_1 - (h+1)\dot{I}_1 \le 0.$$

This proves (ii)& (iii). Differentiation yields (iv). Finally, the last statement is obtained by partial integration as follows:

$$J(h) = \oint \left\{\frac{x^2}{2}e^{-y} - (h+1)\right\} xe^{-y}dy$$

$$= \oint (e^{-y}y + e^{-y} - 1)xe^{-y}\frac{\left(\frac{x^2}{2} - x\frac{\partial x}{\partial y}\right)}{y}dy$$

$$= \oint \left(\frac{e^{-y}y + e^{-y} - 1}{y}\right) x^3 e^{-y}dy$$

$$-\frac{1}{3}\oint \left(\frac{e^{-y}y + e^{-y} - 1}{y}\right) e^{-y}dx^3$$

$$\overset{\text{p.i.}}{=} \oint g(y)x^3 e^{-y}dy.$$

As x is raised to the power 3 we can differentiate this integral representation for $J(h)$ twice! This gives the expression in (v) ¶

Lemma 7

$$h\ddot{I}_1 + \left(1 - \frac{h\dot{I}_1}{I_1}\right) \dot{I}_1 \geq \frac{1}{h+1} \oint \frac{e^y}{x} P(h,y) dy,$$

where

$$P(h,y) = \frac{1}{y^2} \left\{ \frac{1}{2} hy^2 e^{-y} + \frac{1}{2} hye^{-y} + \frac{1}{2} hy + he^{-y} - h + y^2 e^{-y} \right\}.$$

Proof. From Lemma 6 (ii)& (iii) we infer that

$$-h\frac{\dot{I}_1}{I_1} \geq \frac{-h}{2(h+1)},$$

$$\ddot{I}_1 = -\frac{1}{2(h+1)} \dot{I}_1 - \frac{1}{(h+1)} \ddot{J}.$$

Using this we get

$$h\ddot{I}_1 + (1 - \frac{h\dot{I}_1}{I_1})\dot{I}_1 \geq \frac{1}{h+1} \oint \frac{e^y}{x} \left(-3hg(y) + e^{-y}\right) dy$$

$$= \frac{1}{h+1} \oint \frac{e^y}{x} P(h,y) dy,$$

where

$$P(h,y) = \frac{1}{y^2} \left(\frac{1}{2} hy^2 e^{-y} + \frac{1}{2}^{-y} + \frac{1}{2} hy + he^{-y} - h + y^2 e^{-y} \right).$$

¶

P is not positive on all of $\mathbb{R}^2$, but we only need to consider P on

$$\Omega = \{ (h,y) : -1 \leq y < \infty, \ -e^{-y}(y+1) \leq h \leq 0 \}.$$

Lemma 8 $P > 0$ *on the domain* Ω.

Proof First we observe that the derivative of P with respect to h is sign definite:

$$\frac{\partial P}{\partial h} = \frac{1}{2y^2} \left\{ e^{-y}y^2 + e^{-y}y + 9 + 2e^{-y} - 2 \right\} \geq 0.$$

We compute the value of P on the lower boundary of Ω: $\{(h,y) : h = -e^{-y}(y + 1), y \in [1, \infty)\}$

$$P(-e^{-y}(y+1), y) = \frac{1}{2y^2}(y^2 + y + 2)(e^{-y} - ye^{-2y} - e^{-2y}) \geq 0.$$

¶

Corollary 1 If $\dot{Q}(h) = 0$ then $\ddot{Q}(h) > 0$.

Theorem 1 $\forall h \in (-1, 0) : \quad \dot{Q}(h) > 0$.

Proof. If $\dot{Q} = 0$ then $\ddot{Q} > 0$. Combined with the a priori estimate in Lemma 5 the result follows. ¶

3 Proofs of Lemma 1-5

Lemma 1. Multiply the identity

$$e^{-y}x\,dx = e^{-y}(\frac{x^2}{2} - y)dy$$

with $x^m y^n$ and integrate by parts over $\gamma(h)$.

Lemma 2.
$$\frac{d}{dh}\delta_1^0 = \frac{d}{dh}\oint xe^{-y}dy = \oint \frac{dy}{x}.$$

This implies that

$$h\frac{d}{dh}\delta_1^0 \quad = \quad \oint -e^{-y}(y - \frac{x^2}{2} + 1)\frac{1}{x}dy$$

$$= \quad -\delta_{-1}^1 + \frac{1}{2}\delta_1^0 - \delta_{-1}^0$$

$$\overset{(RE)}{\underset{n=0,m=-1}{=\!=\!=}} \quad \delta_1^0 - \delta_{-1}^0.$$

Similarly,

$$h\frac{d}{dh}\delta_3^0 \quad = \quad -3\delta_1^1 + \frac{3}{2}\delta_3^0 - 3\delta_1^0$$

$$\overset{(RE)}{\underset{n=0,m=1}{=\!=\!=}} \quad \delta_3^0 - 3\delta_1^0.$$

Lemma 3.

$$Q(0) = 2\int_{-1}^{\infty} x^3 e^{-y} dy \left(2\int_{-1}^{\infty} x e^{-y} dy\right)^{-1}$$

$$= \int_{-1}^{\infty}(2(y+1))^{3/2} e^{-y} dy \left(\int_{-1}^{\infty}(2(y+1))^{1/2} e^{-y} dy\right)^{-1}$$

$$= \frac{2^{3/2}}{2^{1/2}}\frac{\Gamma(5/2)}{\Gamma(3/2)} = 3.$$

$$Q(-1) = \lim_{h\downarrow-1} \frac{\int e^{-y} x^3 dy}{\int e^{-y} x dy}$$

$$= \lim_{h\downarrow-1} \frac{\int\int 3x^2 e^{-y} dx dy}{\int\int e^{-y} dx dy} = 0.$$

Lemma 4.

$$h\dot{Q} = h\left(\frac{\dot{I_1}}{I_1} - \frac{I_2 \dot{I_1}}{I_1^2}\right)$$

$$\stackrel{(8)}{=} \frac{-3I_1 + I_2}{I_1} - \frac{I_2}{I_1}\left(\frac{I_1 - R}{I_1}\right).$$

Differentiation yields:

$$\dot{Q} + h\ddot{Q} = \dot{Q}\frac{R}{I_1} + Q\frac{\dot{R}}{I_1} - \frac{QR\dot{I_1}}{I_1^2}.$$

Therefore, if $\dot{Q} = 0$ at $h = \bar{h}$ then

$$\bar{h}\ddot{Q} = Q\frac{\dot{R}}{I_1} - \frac{QR\dot{I_1}}{I_1^2}$$

$$= \frac{Q}{I_1}\left(\dot{R} - R\frac{\dot{I_1}}{I_1}\right).$$

We eliminate $\dot{R}$ using the identity $\bar{h}\dot{I_1} = I_1 - R$. Then the result follows.

Lemma 5. If $\dot{Q} = 0$ then

$$h\frac{\dot{I_1}}{I_1} Q = (Q - 3).$$

As $h\frac{\dot{I_1}}{I_1}$ is negative on the interval $(-1, 0)$ it follows that $\frac{Q-3}{Q} < 0$ at a point where $\dot{Q}$ vanishes. As Q is continuous on this interval, the result follows.

References

[BK81] E. Brieskorn and H. Knörrer. *ebene algebraische kurven*. Birkhäuser, Basel, 1981.

[CCH85] J. Carr, S.N. Chow, and J.K. Hale. Abelian integrals and bifurcation theory. *J. Differ. Eq.*, 59:413–436, 1985.

[CSG87] J. Carr, J.A. Sanders, and S.A. van Gils. Nonresonant bifurcations with symmetry. *SIAM. J. Math. Anal.*, 18:579–591, 1987.

[DR] F. Dumortier and R. Roussarie Canard cycles and center manifolds preprint Laboratoire de Topologie, Université de Bourgogne, 1994.

[SC] J.A. Sanders and R. Cushman. Abelian integrals and global hopf bifurcation. In Broer Braaksma and Takens, editors, *LNiM 1125*, pages 87–90, .

[vGH86] S.A. van Gils and E. Horozov. Uniqueness of limit cycles in planar vector fields which leave the axes invariant. In *Contemporary Mathematics 56*, pages 117–129, 1986.

[Ż84] H. Żołądek. On the versality of a family of symmetric vector fields in the plane. *Math. USSR Sb.*, 48:463–492, 1984.

[Zha86] Z.F. Zhang. On the uniqueness theorem of limit cycles of generalized lienard equations. *Applicable Analysis*, 23:63–76, 1986.

HOPF BIFURCATION IN SYMMETRICALLY COUPLED LASERS

MICHAEL WEGELIN
Institut für Informationsverarbeitung
Köstlinstr. 6
D-72074 Tübingen
Federal Republic of Germany

ABSTRACT. The theory of equivariant Hopf bifurcation with $\mathbf{D}_m \times \mathbf{O}(2)$-symmetry is used to determine the possible periodic patterns of polarized light produced in a ring of lasers with saturable absorber. For three coupled lasers the coupled Maxwell-Bloch equations are integrated numerically, displaying many of the theoretically classified periodic solutions with maximal and submaximal symmetry.

1. Introduction

In this contribution we study a symmetrically coupled ring of three lasers with saturable absorber. The laser array is shown schematically in Figure 1. We will exploit the symmetries of the Maxwell-Bloch equations describing the dynamics of this laser array and employ the methods of equivariant bifurcation theory [10] to determine and to classify the solutions at onset of an oscillatory instability (Hopf bifurcation). The theoretical analysis reveals eight different possible periodic patterns of polarized light the array is able to produce. These patterns are distinguished by their symmetry properties. Many of the analytically classified solutions are found in numerical integrations of the Maxwell-Bloch equations near the instability.

Lasers with saturable absorber contain two different media in the resonator cavity. One, called the active or lasing medium, is pumped so that its atoms have a positive population inversion. The other, called passive medium or absorber, has a negative population inversion. Due to the competing effects of destabilizing pumping and stabilizing absorption, already the primary laser instability may lead to temporal oscillations in the electrical field amplitude of the emitted light. Besides this oscillatory mode of operation, lasers with saturable absorber may also produce short pulses of high intensity or show bistability of two different states of continuous lasing [7, 13].

In a single wave plane wave approximation, the electrical light field within a single laser does not possess any transversal structure. The vector of the electrical field always is perpendicular to the wave vector, but is allowed to take any orientation

343

P. Chossat (ed.), Dynamics, Bifurcation and Symmetry, 343–354.

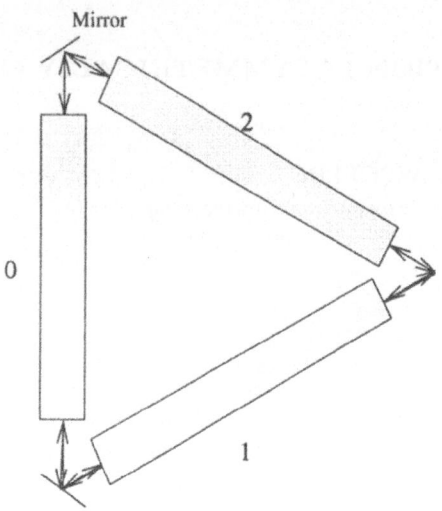

Figure 1: Ring of three symmetrically coupled lasers with saturable absorber.

within this plane, i.e. it may take any linearly polarized state. This is reflected by an O(2)-symmetry of the Maxwell-Bloch equations describing the dynamical behavior of lasers with saturable absorber at exact resonance. The oscillatory instability of the trivial steady state via Hopf bifurcation then leads either to standing waves, in which the field vector pertains its direction, but oscillates slowly in magnitude (amplitude modulated, linearly polarized light), or to standing waves, in which the magnitude of the field vector remains the same, but the orientation of the vector slowly rotates (circularly polarized light) [3].

Thus, coupled lasers are an example of symmetrically coupled cells, where the individual cells have an internal symmetry $\mathcal{L}$. Depending on whether the coupling of the identical individual cells is equivariant or invariant under the action of this internal symmetry, the symmetry of the whole system is a direct product $\mathcal{G} \times \mathcal{L}$ or a wreath product $\mathcal{G} \wr \mathcal{L}$ of the global and internal symmetries $\mathcal{G}$ and $\mathcal{L}$ [5, 8].

Here we focus on the direct product case and choose the dihedral group $\mathbf{D}_m$ as the global symmetry $\mathcal{G}$. Similar studies have been carried out, where the internal symmetry $\mathcal{L}$ is also a dihedral group $\mathbf{D}_n$, and equivariant Hopf bifurcation theory with symmetry $\mathbf{D}_m \times \mathbf{D}_n$ was used to study examples of hierarchically coupled van der Pol- or neural oscillators [1, 2, 15, 21]. Other examples and a general group theoretical framework for bifurcations of coupled cells with internal symmetries are presented in [5, 8].

2. Hopf bifurcation in the coupled Maxwell-Bloch equations

¿From the Maxwell equations for the electrical field $\vec{E}$ and the dielectric displacement $\vec{D}$ and from the Schrödinger equation for the wave functions of the two media semi-classical equations for the laser with saturable absorber may be derived. The coupling between the Maxwell equations and the Schrödinger equation is on the one hand via the dipole interaction in the Schrödinger equation, coupling the field to the media, on the other hand via the expectation value of the dipole moment, which is proportional to the macroscopic polarization $\vec{P}$, coupling the media via $\vec{D} = \vec{E} + 4\pi\vec{P}$ to the field again.

Assuming exact resonance between the transition frequencies in active and passive medium and the cavity, the Maxwell-Bloch equations in a single mode plane wave approximation are [3, 22]

$$
\begin{aligned}
\dot{a}_i &= \rho(-a_i + Ap_i + r_1(1 - C)q_i) + G(a_{i-1}, a_i, a_{i+1}), \\
\dot{p}_i &= a_i(1 - d_i) - p_i, \\
\dot{q}_i &= ka_i(1 - e_i) - r_1 q_i, \\
\dot{d}_i &= \sigma[-d_i + \frac{1}{2}(a_i\bar{p}_i + \bar{a}_i p_i)], \\
\dot{e}_i &= \sigma[-r_2 e_i + \frac{1}{2}(a_i\bar{q}_i + \bar{a}_i q_i)],
\end{aligned}
\tag{1}
$$

$i = 0, \ldots, m - 1$. Here the indices i are understood to be taken modulo m and number the individual lasers in the array.

The variables in Eq. (1) have been scaled such that the system is dimensionless. The amplitude of the electrical field in laser i is proportional to a_i, p_i and q_i are proportional to the atomic polarization in the active medium and in the absorber, d_i and e_i are proportional to the population inversion in the active and in the passive medium of each laser. The parameters ρ, r_1, σ and σr_2 are ratios of phenomenologically introduced relaxation constants, k is the ratio of the two matter-field coupling constants of absorber and lasing medium. A is proportional to the number of excited atoms in the active medium. In the passive medium, this number is denoted with $(1 - C)$. Since A can easily be varied by the rate of pumping in a wide range, it is chosen as a bifurcation parameter. For gas lasers, the absorption rate C can also be varied in a wide range by varying the partial pressure of the absorber. All parameters in (1) are positive.

The variables a_i, p_i and q_i are complex with an over-bar denoting complex conjugation. Therefore, Eq. (1) can also catch polarization effects [3]. The argument of a_i is the direction of polarization of the electrical field of laser i in a plane perpendicular to the wave vector. Choosing as reference direction ($0°$-direction) a horizontal direction, the real parts of a_i are proportional to the horizontal components of the electrical field amplitude, the imaginary parts to the vertical components. The equations allow any linearly polarized state, because they are equivariant with respect to an action of O(2).

The rotation $\Theta \in \mathbf{O}(2)$ acts as a rotation of the direction of polarization, the flip κ as inversion of the vertical component of the amplitude-vector. Hence, $\mathbf{O}(2)$ acts as

$$\Theta\,(a_i, p_i, q_i, d_i, e_i) = (e^{\imath\Theta}a_i, e^{\imath\Theta}p_i, e^{\imath\Theta}q_i, d_i, e_i), \tag{2}$$

$$\kappa_2\,(a_i, p_i, q_i, d_i, e_i) = (\bar{a}_i, \bar{p}_i, \bar{q}_i, d_i, e_i). \tag{3}$$

The coupling is via the electrical fields only. The coupling function G is chosen in a way that the complete system is $\mathbf{D}_m \times \mathbf{O}(2)$-equivariant. The dihedral group $\mathbf{D}_m$ is generated by a rotation Θ_m of the array by one laser and a reflection κ_1 across a diameter of the laser ring. On (1) $\mathbf{D}_m$ acts by permuting the indices according to

$$\Theta_m(a_i, p_i, q_i, d_i, e_i) = (a_{i+1}, p_{i+1}, q_{i+1}, d_{i+1}, e_{i+1}) \tag{4}$$

$$\kappa_1(a_i, p_i, q_i, d_i, e_i) = (a_{-i}, p_{-i}, q_{-i}, d_{-i}, e_{-i}) \tag{5}$$

As a model for the coupling we will choose

$$G(a_{m-1}, a_0, a_1) = \alpha(a_{m-1} + a_1) + \beta(a_{m-1}\bar{a}_1 + \bar{a}_{m-1}a_1)a_0 \tag{6}$$

with real α and β. With this coupling, equation (1) possesses $\mathbf{D}_m \times \mathbf{O}(2)$-symmetry. The linear part $\alpha(a_{m-1}+a_1)$ is used in the literature to model the coupling to externally injected light [22, 23]. Our investigations showed that degeneracies occur, if only this linear part is taken into account. We therefore also include a cubic part $\beta(a_{m-1}\bar{a}_1 + \bar{a}_{m-1}a_1)a_0$ into the coupling functions.

For $m = 3$ and small negative linear coupling coefficient α, the trivial steady state of (1) looses stability in a Hopf bifurcation at

$$A_c = \frac{(\rho + r_1 - \alpha_1)\,[k\rho r_1(C-1) + (r_1+1)(\rho+1-\alpha_1)]}{\rho(1 + \rho - \alpha_1)}, \tag{7}$$

with $\alpha_1 = 2\alpha\cos(2\pi/3)$. If A is increased above this value, the trivial steady state becomes unstable against oscillations. The critical eigenvalues $\pm\imath\omega$ have multiplicity four, and the eight dimensional center manifold is attracting.

3. Hopf bifurcation with $\mathbf{D}_m \times \mathbf{O}(2)$ symmetry

In this section we want to review some results for the Hopf bifurcation with $\mathbf{D}_m \times \mathbf{O}(2)$-symmetry [5, 8, 21]. Here, we focus on the action of $\mathbf{D}_m \times \mathbf{O}(2)$ on the eight dimensional center eigenspace and the isotropy subgroups of this action.

Performing a center manifold reduction and a transformation to normal form [6] of (1) at the Hopf bifurcation point yields a system of ordinary differential equations for $z = (z_1, z_2, z_3, z_4) \in \mathbf{C}^4$. Near the bifurcation point, the dynamics of the normal form is qualitatively the same as that of the full system (1). In addition to $\mathbf{D}_m \times \mathbf{O}(2)$ the normal form is also equivariant with respect to $\mathbf{S}^1$. This circle group action reflects the temporal phase shift invariance of periodic solutions. The normal form is equivariant with respect to the following representation of $\mathbf{D}_m \times \mathbf{O}(2) \times \mathbf{S}^1$ on $\mathbf{C}^4$:

$$(\Theta_m, \Theta) \begin{pmatrix} z_1 \\ z_2 \\ z_3 \\ z_4 \end{pmatrix} = \begin{pmatrix} z_1 e^{\imath(j\Theta_m + \Theta)} \\ z_2 e^{\imath(-j\Theta_m + \Theta)} \\ z_3 e^{\imath(-j\Theta_m - \Theta)} \\ z_4 e^{\imath(j\Theta_m - \Theta)} \end{pmatrix}, \tag{8}$$

$$\kappa_1 \begin{pmatrix} z_1 \\ z_2 \\ z_3 \\ z_4 \end{pmatrix} = \begin{pmatrix} z_2 \\ z_1 \\ z_4 \\ z_3 \end{pmatrix}, \quad \kappa_2 \begin{pmatrix} z_1 \\ z_2 \\ z_3 \\ z_4 \end{pmatrix} = \begin{pmatrix} z_4 \\ z_3 \\ z_2 \\ z_1 \end{pmatrix}, \tag{9}$$

$$\Psi z = z e^{\imath \Psi} \quad \Psi \in \mathbf{S}^1. \tag{10}$$

In the following we look only at the case $j = 1$, since the results for the isotropy subgroups and the stability of solutions do not depend on j [21].

The kernel Γ_c of this action consists of all elements acting trivially on each $z \in \mathbf{C}^4$. For odd m it is given by

$$\Gamma_c = \{(0,0,0),(0,\pi,\pi)\}, \tag{11}$$

for even m by

$$\Gamma_c = \{(0,0,0),(0,\pi,\pi),(\pi,0,\pi),(\pi,\pi,0)\}. \tag{12}$$

The equivariant Hopf Bifurcation theorem [9] guarantees the existence of a unique branch of periodic solutions with symmetry Σ bifurcating from a trivial steady state if the dimension of the fixed point subspace of Σ is equal to two. The isotropy subgroups of the action of $\mathbf{D}_m \times \mathbf{O}(2) \times \mathbf{S}^1$ are given in Table 1. This Table shows a representative for each conjugacy class of isotropy subgroups together with its fixed point subspace. Each group in Table 1 has to be supplemented with the elements from the kernel Γ_c of the representation, which acts trivially on $\mathbf{C}^4$. Some special group elements have been given an own symbol:

$$\tilde{\Theta}_m = (\Theta_m, 0, -\Theta_m), \quad \tilde{\Theta}_2 = (0, \Theta, -\Theta), \quad \Theta_m^c = (\Theta_m, -\Theta_m, 0),$$
$$\tilde{\kappa}_1 = (\kappa_1, 0, \pi), \quad \tilde{\kappa}_2 = (0, \kappa_2, \pi).$$

The patterns belonging to isotropy subgroup 1 are rotating waves. Each laser produces circularly polarized light. At a given time the directions of polarization are shifted by $2\pi/m$ between adjacent lasers. Observing only one component of the electrical field vector (e.g. in the vertical polarization direction) yields a discrete travelling wave through the arrangement of lasers. This solution is therefore denoted as a TTW (travelling-travelling wave). The first 'T' denotes a discrete travelling wave through the laser array, the second 'T' refers to the continuous travelling wave in the direction of polarization. The conjugacy class of this isotropy subgroup consists of four different groups: the light is either left or right circularly polarized, and the discrete wave is traveling clockwise or counterclockwise through the array of lasers.

Nr.	Name	Fix(Σ)	Σ	Generators	dim Fix(Σ)
0		$(0,0,0,0)$	$D_m \times O(2) \times S^1$	$\kappa_1, \kappa_2, \Theta_m, \Theta, \Psi$	0
1	TTW	$(z_1,0,0,0)$	$\tilde{Z}_m \times \widetilde{SO}(2)$	$\tilde{\Theta}_m, \tilde{\Theta}_1$	2
2	S_0TW	$(z_1,z_1,0,0)$	$Z_2 \times \widetilde{SO}(2)$	$\kappa_1, \tilde{\Theta}$	2
3	S_πTW	$(z_1,-z_1,0,0)$	$\tilde{Z}_2 \times \widetilde{SO}(2)$	$\tilde{\kappa}_1, \tilde{\Theta}$	2
4	TS_0W	$(z_1,0,0,z_1)$	$\tilde{Z}_m \times Z_2$	$\tilde{\Theta}_m, \kappa_2$	2
5	SW	$(z_1,0,z_1,0)$	D_m	$\kappa_1\kappa_2, \Theta_m^c$	2
6	S_0SW	(z_1,z_1,z_1,z_1)	$Z_2 \times Z_2$	κ_1, κ_2	2
7	S_πSW	$(z_1,-z_1,-z_1,z_1)$	$\tilde{Z}_2 \times Z_2$	$\tilde{\kappa}_1, \kappa_2$	2
8	AW	(z_1,z_2,z_1,z_2)	Z_2	$\kappa_1\kappa_2$	4
9		$(z_1,z_2,0,0)$	$1 \times \widetilde{SO}(2)$	$\tilde{\Theta}$	4
10		$(z_1,0,0,z_2)$	$\tilde{Z}_m \times 1$	$\tilde{\Theta}_m$	4
11		$(z_1,0,z_2,0)$	Z_m	Θ_m^c	4
12		(z_1,z_1,z_2,z_2)	$Z_2 \times 1$	κ_1	4
13		$(z_1,-z_1,z_2,-z_2)$	$\tilde{Z}_2 \times 1$	$\tilde{\kappa}_1$	4
14		(z_1,z_2,z_2,z_1)	$1 \times Z_2$	κ_2	4
15		(z_1,z_2,z_3,z_4)	Γ_c		8

Table 1: Isotropy subgroups of $D_m \times O(2) \times S^1$ and their fixed point subspaces.

Solutions of isotropy type 2 are also rotating waves in the polarization, but standing waves in the array of lasers. Two lasers facing each other across a reflection axis along a diameter of the ring of lasers oscillate in phase, with the same wave form and amplitude. The conjugacy class of this isotropy subgroup therefore has $2m$ members because, again, the light is either eft or right circularly polarized, and there are m axis of reflection symmetry. Solutions of this types are called S_0TW (standing in phase-travelling waves).

Solutions of isotropy type 3 look similar, but now two lasers oscillate identically, but out of phase by π and are therefore called S_πTW. The remaining laser for odd m, lying on the reflection axis, has to be shifted in phase by π against itself. Therefore it oscillates with zero amplitude.

In states of type 4 each laser produces linearly polarized light with the same direction of polarization. Because of the $\tilde{Z}_m$ part of the isotropy subgroup the times in which each laser reaches the maximum of its electrical field envelope are separated by T/m, where T is the period of the oscillating field envelope. Therefore a discrete travelling wave of linearly polarized light is running through the laser array, which is denoted as TSW. The conjugacy class of this isotropy subgroup consists of two complete circles.

In solutions of type 5 each laser exhibits linearly polarized light, but the directions of polarization are shifted by $2\pi/m$ against each other, and the maxima are reached at the same time. This wave form, coming also in two complete circles, is denoted as SW. Note that this isotropy is not a direct product of the form $\Sigma_1 \times \Sigma_2$, where Σ_1 and Σ_2 are isotropy subgroups of $\mathbf{D}_m \times \mathbf{S}^1$ and $\mathbf{O}(2) \times \mathbf{S}^1$, respectively.

Waves with isotropy 6 also show linearly polarized light, and again two lasers oscillate in phase. There are m complete circles of this type denoted by $\mathrm{S_0SW}$.

Solutions of type 7, finally, show the same structure as those of type 6, but now with two lasers by π out of phase. Again, for odd m one laser oscillates shifted by π against itself, hence it has zero amplitude. This solution is named $\mathrm{S_\pi SW}$.

The stability of the periodic solutions guaranteed by the equivariant Hopf theorem can be calculated from the coefficients for the Birkhoff-Poincaré normal form. This has been done elsewhere for $m = 3$ [20, 21]. One can also show that submaximally symmetric periodic solutions with isotropy 8 may exist [20, 21]. These are called alternating waves (AW).

4. Three coupled lasers

In this section we present some results of numerical integrations of (1) near the Hopf bifurcation point at different parameters. First we fixed the parameter values to

$$\rho = 0.13, \quad r_1 = 0.84, \quad r_2 = 1.23, \quad \sigma = 3.9,$$

the coupling constants to

$$\alpha = -0.2 \quad \text{and} \quad \beta = 0.1$$

and varied k and C in a range of $0.4 \leq k < 1.6$ and $50 \leq C < 250$. In this range we performed a reduction of (1) to normal form and calculated the stabilities of the bifurcating periodic solutions near the Hopf bifurcation point [20, 21]. The subsequent numerical integrations of (1) at various points of parameter space showed the validity of these calculations near the bifurcation.

Figure 2 shows the result of a numerical integration at $k = 1.0$ and $C = 90$. The Hopf bifurcation point is $A_c = 72.7965$, for the integration we choose $A = 75$. The upper part of Figure 2 shows projections of the trajectories onto the complex a_i-planes $(i = 0, \ldots, 2)$ of the three lasers. The lower part shows the imaginary parts (i.e. the vertical components) of the electrical field amplitudes within the lasers versus time. All three lasers show circularly polarized light. Observing only the vertical components of the electrical field amplitudes shows a discrete travelling wave running through the laser array. Hence this solution has isotropy subgroup 1.

Figure 3 shows the result of a numerical integration at the same parameter values as in Figure 2 but with different initial conditions. Again the system relaxes to a periodic state but with submaximal isotropy. Lasers 0 and 1 emit elliptically polarized light with the two major axis perpendicular to each other, laser 2 emits circularly polarized light. In the picture of the temporal evolution we plotted the vertical components of the electrical field amplitudes of lasers 0 and 2, but the *horizontal* component of laser 1 versus time. Lasers 0 and 1 have nearly same amplitudes and are shifted in phase by

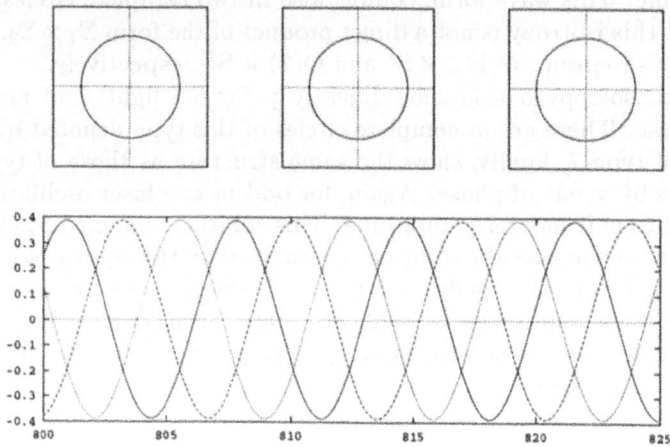

Figure 2: Solution with isotropy subgroup 1.

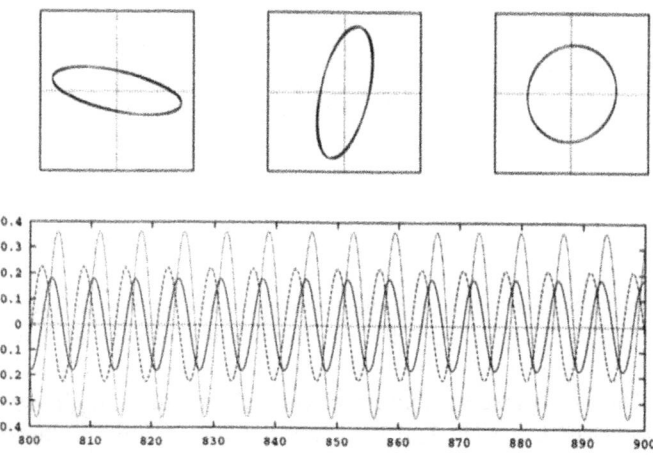

Figure 3: Solution with submaximal isotropy subgroup 8.

$\pi/2$ against each other. This indicates that we now have a periodic orbit (z_1, z_2, z_1, z_2) with $z_1 \sim e^{\iota\omega t}$ and $z_2 \approx \iota z_1$ with submaximal isotropy subgroup 8.

Figure 4 shows a simulation at $k = 0.54$ and $C = 230$. At this point solutions with isotropy 4, 6 and 7 bifurcate subcritically, hence they are unstable near the bifurcation. The other periodic solutions of maximal and submaximal isotropy bifurcate

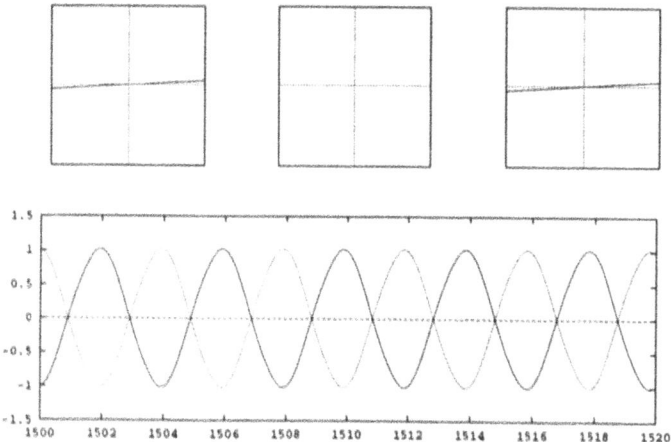

Figure 4: S_πSW-solution with isotropy subgroup 7.

supercritically, but are all unstable. The Hopf bifurcation point is $A_c = 96.902$. Figure 4 shows the result of a simulation at $A = 96.8$ directly below the bifurcation point. The numerically obtained solution has isotropy subgroup $\tilde{\mathbf{Z}}_2 \times \mathbf{Z}_2$ (isotropy subgroup 7). Although the simulation was done very close to the bifurcation point, the solution shows a relatively large amplitude and remains stable beyond the bifurcation point. The S_πSW-solution with isotropy subgroup 7 which bifurcates supercritically at A_c therefore seems to undergo a saddle-node bifurcation to the left of the bifurcation point. Between this limit point and the Hopf bifurcation point a stable and an unstable periodic solution with isotropy subgroup 7 exist, the latter with smaller amplitude. Beyond the Hopf bifurcation point only the stable branch continues to exist.

Finally, Figure 5 shows the result of a numerical simulation of (1) at

$$\rho = 0.13, \quad r_1 = 0.84, \quad r_2 = 1.23, \quad \sigma = 3.9$$

and coupling parameters

$$\alpha = -0.2 \quad \text{and} \quad \beta = 0.25,$$

i.e. with larger β than in the previous simulations. The Hopf bifurcation is at $A_c = 66.6067$, Figure 5 was obtained with $A = 68$. Figure 5 again shows in the upper part the projections of the trajectories onto the three complex planes a_0, a_1, and a_2, in the lower part the vertical components of the a_i are plotted versus time. The numerically obtained periodic solution has isotropy $\mathbf{D}_m$ (isotropy subgroup 5).

Hence in direct numerical integrations near the Hopf bifurcation point we find many of the theoretically predicted and classified solutions. Moreover, the results of the normal form reduction and the stability calculations from the normal form calculations are in excellent agreement with the numerical observations [20, 21].

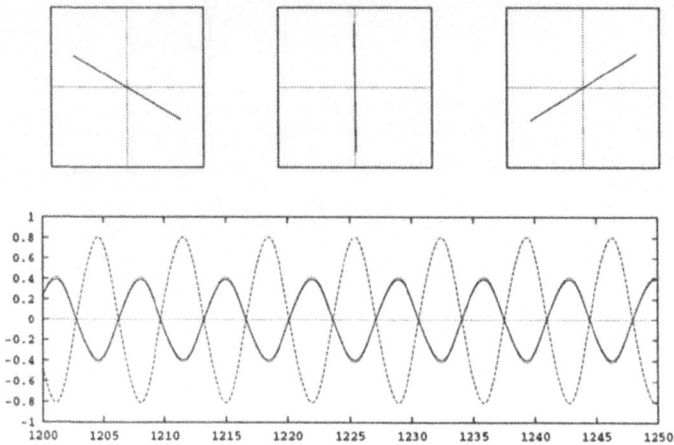

Figure 5: SW-solution with isotropy subgroup 5.

5. Discussion and conclusions

In this paper we identified and classified seven maximally symmetric and one sub-maximally symmetric periodic solutions to the semi classical Maxwell-Bloch equations (1) of m $\mathbf{D}_m$-symmetrically coupled lasers with saturable absorber. This part of the analysis relied solely on the representation of the group $\mathbf{D}_m \times \mathbf{O}(2) \times \mathbf{S}^1$ on the eight dimensional center manifold.

For $m = 3$ we integrated the full system (1) numerically. In the investigated range of parameters we found many of the maximally symmetric periodic solutions and the submaximally periodic solution with isotropy subgroup 8. Near the Hopf bifurcation, the results of the normal form analysis and the numerical solutions were in excellent agreement.

The $\mathbf{O}(2)$-symmetry of a single laser has been exploited in theoretical investigations only recently [3, 4, 19]. Our methods and results are most similar to those of [3] in which a codimension two Takens-Bogdanov bifurcation for a single laser is investigated.

In recent investigations of coupled lasers symmetry arguments have not been used. In the mainly numerical work on ring arrays [11, 12, 14, 16] or globally coupled arrays [17, 18] of semiconductor or gas lasers without absorber oscillatory bifurcations occur only as secondary instabilities. In contrast we explicitly exploited the symmetry of the equations enabling us to classify the primary oscillating patterns according to their residual symmetry. This classification proved to be extremely helpful in the choice of parameter values for the simulations.

Further away from the bifurcation the simulations however showed that coupled lasers are able to produce much more complicated patterns than we have captured in our analysis. Our work is therefore only a starting point into the investigation of the rich dynamical behavior of symmetrically coupled laser arrays.

Acknowledgments

I thank Martin Golubitsky and Ian Stewart for stimulating discussions on the structure of symmetrically coupled cell arrays with internal symmetries. I am especially grateful to Gerhard Dangelmayr for his help and advice throughout. This work was supported by a fellowship of the Studienstiftung des deutschen Volkes.

References

[1] G. Dangelmayr, W. Güttinger, J. Oppenländer, J. Tomes, and M. Wegelin. Coupled neural oscillators with $D_3 \times D_3$-symmetry. Preprint, 1992.

[2] G. Dangelmayr, W. Güttinger, and M. Wegelin. Hopf bifurcation with $D_3 \times D_3$-symmetry. *Journal of Applied Mathematics and Physics (ZAMP)*, 44:595–638, 1993.

[3] G. Dangelmayr and M. Neveling. Codimension–two bifurcations and interactions between differently polarised fields for laser with saturable absorber. *Journal of Physics A: Mathematical and General*, 22:1291–1301, 1989.

[4] E. J. D'Angelo, E. Izaguirre, G. B. Mindlin, G. Huyet, L. Gil, and J. R. Tredicce. Spatiotemporal dynamics of lasers in the presence of an imperfect $O(2)$ symmetry. *Physical Review Letters*, 68(25):3702–2705, 1992.

[5] B. Dionne, M. Golubitsky, and I. Stewart. Arrays of oscillators with internal and global symmetries. In preparation.

[6] C. Elphick, E. Tirapegui, M. E. Brachet, P. Coullet, and G. Iooss. A simple global characterization for normal forms of singular vector fields. *Physica D*, 29:95–127, 1987.

[7] T. Erneux and P. Mandel. Stationary, harmonic, and pulsed operations of an optically bistable laser with saturable absorber. II. *Physical Review A*, 30(4):1901–1909, 1984.

[8] M. Golubitsky, B. Dionne, and I. Stewart. Coupled cells: Wreath products and direct products. These proceedings.

[9] M. Golubitsky and I. Stewart. Hopf bifurcation in the presence of symmetry. *Archive of Rational Mechanics and Analysis*, 87:107–165, 1985.

[10] M. Golubitsky, I. Stewart, and D. G. Schaeffer. *Singularities and Groups in Bifurcation Theory. Volume II*, volume 69 of *Applied Mathematical Sciences*. Springer, 1988.

[11] R.-D. Li and T. Erneux. Preferential instability in arrays of coupled lasers. *Physical Review A*, 46(7):4252–4260, 1992.

[12] R.-D. Li, P. Mandel, and T. Erneux. Periodic and quasiperiodic regimes in self-coupled lasers. *Physical Review A*, 41(9):5117–5126, 1990.

[13] P. Mandel and T. Erneux. Stationary, harmonic, and pulsed operations of an optically bistable laser with saturable absorber. I. *Physical Review A*, 30(4):1893–1901, 1984.

[14] P. Mandel, R.-D. Li, and T. Erneux. Pulsating self-coupled lasers. *Physical Review A*, 39(5):2502–2508, 1989.

[15] J. Oppenländer. Zur Dynamik hierarchischer Oszillatorennetze. Diplomarbeit, Universität Tübingen, Fakultät für Physik, Institut für Informationsverarbeitung, 1992.

[16] K. Otsuka. Self-induced turbulence and chaotic itinerancy in coupled laser systems. *Physical Review Letters*, 65(3):329–332, 1990.

[17] K. Otsuka and J.-L. Chern. Synchronization, attractor fission, and attractor fusion in a globally coupled laser system. *Physical Review A*, 45(7):5052–5055, 1992.

[18] K. Otsuka and J.-L. Chern. Dynamical spatial-pattern memory in globally coupled lasers. *Physical Review A*, 45(11):8288–8291, 1992.

[19] R. L. Ruiz, G. B. Mindlin, and C. P. Garcia. Mode-mode interaction for a CO_2 laser with imperfect $O(2)$ symmetry. *Physical Review A*, 47(1):500–509, 1993.

[20] M. Wegelin. Patterns of polarized light in symmetrically coupled lasers. In preparation.

[21] M. Wegelin. *Nichtlineare Dynamik raumzeitlicher Muster in hierarchischen Systemen*. Dissertation, Universität Tübingen, Fakultät für Physik, Institut für Informationsverarbeitung, 1993.

[22] C. O. Weiss and R. Vilaseca. *Dynamics of Lasers*. VCH Verlagsgesellschaft, 1991.

[23] H. Zeghlache and V. Zehnlé. Theoretical study of a laser with injected signal. I. Analytical results on the dynamics. *Physical Review A*, 46(9):6015–6027, 1992.

The manufacturer's authorised representative in the EU is Springer
Nature Customer Service Centre GmbH, Europaplatz 3, 69115 Heidelberg,
Germany. If you have any concerns regarding our products, please
contact ProductSafety@springernature.com

Printed and bound by CPI Group (UK) Ltd, Croydon, CR0 4YY
23/04/2026
02095624-0006